EXPLORING CONTEMPORARY MIGRATION

Exploring Contemporary Migration

PAUL BOYLE, KEITH HALFACREE AND
VAUGHAN ROBINSON

LONGMAN

Addison Wesley Longman Limited
Edinburgh Gate
Harlow
Essex CM20 2JE
United Kingdom
and Associated Companies throughout the world

Published in the United States of America
by Addison Wesley Longman, New York

First published 1998

ISBN 0 582 25161 3

British Library Cataloguing in Publication Data
A catalogue record for this book is available from the British Library.

Library of Congress Cataloguing-in-Publication Data
A catalog record for this book is available from the Library of Congress.

Set by 3 in 9/11pt Times Roman
Produced by Longman Singapore Publishers (Pte) Ltd.
Printed in Singapore

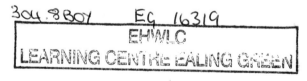

For

Gillian and Francis

Hazel and Barry

Jayne and Bryn

Contents

Preface

Many authors agree that population migration is becoming an increasingly important aspect of everyday life for more people and more places. The academic interest in this phenomenon is evident from the wide range of journals devoted to the topic and the increasing numbers of articles and books being written on the subject, and this interest spans numerous disciplines. In geography at least, migration used to be taught simply as a subset of the broader field of population geography but, increasingly, the topic crops up in other courses. Indeed, in some departments human migration is a course topic in its own right. It is therefore something of a surprise that there is no contemporary textbook on the subject and this is the gap we hope to fill here.

This text, therefore, presents a theoretically informed account of some of the major migration patterns and processes that are currently of interest and importance. Examples are drawn from both developed and developing world contexts from a wide range of academic sources. The book also integrates the analysis of international and internal migration. Often these types of movement are treated independently, but there are strong arguments why they need to be considered together. First, for many migrants an international move, between neighbouring countries in Europe for example, may involve less social disruption and may occur over a shorter distance than long-distance migration within a single nation, such as China. To some extent, the international/internal distinction can thus be seen as relatively arbitrary. Second, there is an increasing realisation that international and internal migration may be strongly linked and are not the discrete phenomena that they may at first appear. These points are taken up in later chapters.

The book is written by geographers and the geographical realisation of such a spatial process as migration is undoubtedly a major focus throughout. However, the material included in the book bridges the various disciplinary divides and there has been a conscious attempt to include the work of sociologists, economists and others, such that the final product is as comprehensive as possible, given the limited space available. It seems to us rather sad in these days of inter-disciplinary research that more collaborative endeavour has not been undertaken in the field of human migration and it is hoped that this book will suggest some of the benefits that can come from addressing the topic from different but complementary perspectives.

Rather than simply providing a descriptive account of the varied work that exists we have aimed in the book to present a case for particular viewpoints and methods that we think deserve more attention. It is not that we necessarily reject other perspectives. In fact, an important message that we hope this book provides is that the implementation of a wide variety of approaches is to be applauded in migration research, with the final choice of approach for a specific piece of work only being decided upon sensibly when the exact research question that is to be addressed has been stated. For example, the long tradition of migration modelling, much of which is founded implicitly or explicitly upon a neoclassical economic framework, appears to have been the dominant approach for a long time. However, this approach can be regarded as being unduly restrictive if applied to all areas of migration work and, consequently, we advocate more acceptance and identification of the cultural implications of the migration process. Such a perspective would adopt much more qualitative methods and would take a very different philosophical tack. Even so, it is not our aim to deride the modelling work that has been or is being done, as this would be to throw out the baby with the bathwater! Quite simply, we wish to make the case for some of the research gaps that we think exist, while accepting the worth of work that is perhaps more traditional.

The book is very detailed in places and includes a wide range of examples and specific findings relating to a broad spectrum of migration research. Consequently, for students, the book may most usefully be used as a textbook that is dipped into whenever a particular topic is addressed in the relevant class, rather than one that needs to be read cover to cover. Indeed, it may be employed most effectively as a means of bringing together a migration course that is near its conclusion. It certainly should prove useful for lecturers selecting and focusing their migration teaching and set readings. However, it is also hoped that the book will be of interest to researchers who may be developing their ideas about migration research, or who have focused for some time on a particular research direction. We hope that, besides providing a useful summary perspective on contemporary migration patterns and processes, some of the suggestions we make about the apparent reluctance among many researchers to adopt mixed approaches, which draw upon various ideas and methods, may be of particular interest to this group.

In Chapter 1, we introduce the concept of migration and demonstrate its importance by presenting some initial examples of its impact across the world at a general level. This provides an initial 'taste' before migration is explored more critically in subsequent chapters. Chapter 2 begins by discussing the definition of 'migration', the relevant migration data sources and appropriate qualitative and quantitative methods of analysis. The sheer diversity of data sources and analytical techniques is stressed and the chapter provides an introduction for those who wish to undertake migration research. This chapter is complemented by Chapter 3, which sets the philosophical context for such work, reviewing general theories of migration from the nineteenth century onwards. While population geography as a sub-discipline has been criticised as being theoretically outdated, this chapter demonstrates that migration research is at the forefront of theoretical developments in geography as a whole. The chapter charts the development of a more complex conceptualisation of migration, from a focus on income maximising to an appreciation of the more cultural aspects of migration. Hence, it looks in detail at economic, behavioural and biographical approaches, concluding that we need to accept a more pluralistic understanding of the theoretical migration process. Consequently, the thematic chapters should be viewed as complementary perspectives rather than referring to clearly distinct types of migration.

To date, the majority of research on migration has been concerned with its links with employment. Reflecting this focus, Chapter 4 is concerned with labour migration, discussing the neoclassical interpretation and its associated modelling literature before considering the various criticisms of this approach. In particular, structural and behavioural interpretations of labour migration are examined. The form of labour migration is shown to vary considerably between spatial scales, illustrating clearly its complexity.

Besides economic considerations, a key determinant of migration is the life-cycle or, less deterministically, the life-course. As individuals and households move through different stages of their lives, significant events, such as the birth of a child or the breakdown of a marriage, often result in residential migration. In Chapter 5, the life-cycle model, which attempts to identify strictly bounded stages in people's lives, is analysed critically and it is argued that the life-course approach, which acknowledges the diversity of life stories, is more realistic. The important connections between people's changing lives and migration are identified and particular attention is paid to the way that changing trends in modern lifestyles have altered migration outcomes.

Chapter 6 follows neatly from the previous chapter by summarising the literature on migration and quality of life. People usually move to improve their overall living circumstances, and this chapter analyses different spatial patterns of migration from such a perspective, concentrating on 'environmental' reasons for moving. It discusses the 'lure of the city', in both the developed and developing worlds, as a place of economic, social and cultural betterment and freedom from constraints. Examples are given of the 'non-economic' determinants of urbanisation and of recent forms of urbanisation, such as gentrification. The chapter also details and compares the corresponding 'lure of the countryside', as represented by counterurbanisation in the developed world. Attention is also paid to quality of life migration linked less clearly to images of either the rural or the urban environment.

The strong political dimension of migration is demonstrated in Chapter 7, which deals with the role of nation-states in engineering migration flows. At both the international and national level, official and popular concerns about issues such as ethnicity, race, religion, economic buoyancy and strategic political interest are reflected in the range of migration policies that exist. Rather than interfere with fertility and mortality, migration is usually seen as the tool with which the various elements of the national character can be manipulated. Such interference, however, is often highly contentious and politically contested.

A key issue of global political concern is that of refugees and refugee movements. This topic is covered in Chapter 8, which explores forced migrations. Besides detailing the scale, location, character and composition of what has been seen as a global refugee crisis, the policies adopted towards refugees by host nations are detailed. The chapter also investigates the varied fates of political asylum seekers worldwide, as well as forced migrations caused by economic development, industrial and ecological disasters, and state slavery. The significance and contested issue of defining groups such as refugees are emphasised.

The final chapter, Chapter 9, explores the extent to which migration is both infused with cultural values and infuses such values. It also discusses the cultural impacts caused by migration flows. This chapter provides something of a conclusion to the book, integrating many of the findings from the other thematic chapters and tying in with the theoretical discussion. It shows how the selectivity of migration by cause and by age, class, and other factors generates distinct 'cultures of migration' and how such cultures feed back into the selectivity of migration. The chapter argues that, through such cultures, selective migration experiences are naturalised and normalised. The chapter contrasts typically migratory lifestyles with more sedentary lifestyles, interpreting their cultural causes and outcomes. For certain groups, migration is shown to be a distinct feature of their cultural identities, while for other groups the act of migrating is culturally less important than it is to those living in the receiving areas.

Finally, it hardly seems worthwhile to acknowledge the shortcomings of this book as there are many. The very volatility of migration and the daily alteration in the patterns of this movement mean that much will have changed since this book was sent to print. Similarly, the wealth of research material on the subject and its various themes in both the developed and the developing world means that it is impossible to do justice to it all; we simply hope to have identified some of the more important contributions.

Acknowledgements

In putting together this book we have been assisted by a wide range of individuals too numerous to acknowledge individually. However, we wish to extend our particular gratitude to Sally Wilkinson for encouraging us to embark on the project in the first place and for eternal patience as we struggled to complete by our deadlines. Special thanks must also go to our cartographic support in both Swansea and Leeds – Nicola Jones, Guy Lewis, Anna Ratcliffe and Alistair French – to the editorial staff at Addison Wesley Longman – Tina Cadle, Shuet-Kei Cheung and Matthew Smith – and to our friends and colleagues in the Department of Geography at the University of Wales Swansea and the School of Geography at the University of Leeds. Finally, special thanks are extended to Professor Tony Champion, Professor Allan Findlay and Dr Robin Flowerdew for their sound academic advice at numerous stages of the project.

We are grateful to the following for permission to reproduce copyright material;

International Music Publications Ltd for excerpts from the lyrics to the songs *Thousands are Sailing* by Philip Chevron, *Sally Maclennane* and *Fairytale of New York* by Shane MacGowan, all © Perfect Songs, London W11 1DG; International Thomson Publishing Services for an adapted extract from *Migrants, emigrants and immigrants: a social history of migration* by K. Bartholemew (published by Routledge, 1991); MCA Music Ltd for an excerpt from the lyrics to the song *Song to Woody* by Bob Dylan. Reproduced by kind permission; Routledge Ltd. and the editors for an extract from *Class and space* by R. Forrest & A. Murie – edited by Nigel Thrift & Dr Peter Williams (1987); Croom Helm for figure 1.2 and table 7.4; Macmillan Press Ltd. for figures 1.11, 1.12, 1.13 and 1.14; HMSO for figure 2.2 © Crown copyright; Pion Ltd for figure 2.4; Blackwell Publishers for figures 3.3b, 3.3c, 4.4, the table in box 6.1 and the figure in box 2.5; Arnold for figure 3.4 and the table in box 3.2; the Ohio State University Press for figure 3.5 and the figure in box 5.3; Paul Chapman Publishing Ltd. for figure 3.1 and table 6.3; the Association of American Geographers for figure 4.4; Elsevier Science Ltd. for figures 4.3, 4.2, 4.6 and the figures in boxes 1.2 and 6.7; INSEE for the figure b in box 1.2; Harcourt, Brace & Company for figure 6.4; Kluwer Academic Publishers for figures 5.2b, 7.1 and 9.3; John Wiley & Sons Ltd for figure

1.9; the Royal Geographical Society for figures 4.1, 5.5, 5.6, 7.2 and the table in box 6.3; The Gerontological Society of America for figure 5.4; Human Sciences Press for the figure in box 5.1; Carfax Publishing Company for figures 4.5 and 5.8; David Fulton Publishers Ltd. for figure 5.6 and the figure in box 5.8; Oxford University Press for tables 5.1 and 5.2; Dr Barrett A. Lee for figures 3.3d and 6.2; The Royal Dutch Geographical Society for figure 6.3; Routledge for tables 6.1 and 6.2; the European Society for Rural Sociology for tables 6.6 and 6.8; V. H. Winston & Son Ltd for box figure 6.6; Cambridge University Press for tables 7.1, 7.2 and 8.2; Liverpool University Press for table 7.6; the Canadian Association of Geographers for table 8.3; the Guilford Press for figure 9.2 and the Gomer Press for figure 9.5.

Whilst every effort has been made to trace owners of copyright material, in a few cases this has proved impossible and we would like to take this opportunity to apologise to any copyright holders whose rights we may have unwittingly infringed.

CHAPTER 1

Introduction: the spatial impact of migration

Migration as a feature of daily life

At first glance, writing an introductory text on migration seems to be a relatively straightforward task. Migration is a simple concept. People move between places and we are interested, as geographers, in describing and understanding these patterns. However, when you think about and study the topic in detail, it becomes clear that migration is, in reality, extremely complex and multi-faceted. This explains why there is such a large literature on migration in geography and other disciplines, and perhaps the hardest task is to order systematically such a diverse set of material in a logical and coherent manner. Migration includes international flows of large numbers of refugees stimulated by wars, famine or political unrest; young adults moving between regions in search of employment; middle-aged professionals moving back to the land in their search for a rural retreat; families moving down the road to satisfy changing housing requirements; and gypsies and other nomadic peoples for whom mobility is a way of life. Consequently, it is difficult to identify an optimum structure for a text such as this. Do we distinguish between different types of migration based on the distance moved? Do we stress the differences between broad regions or groups of countries, such as between the developed and developing world? Or do we identify the migration themes that underlie the different migration processes?

We have adopted the latter structure. A thematic approach captures the migration experience because it relates directly to the varied underlying causes of migration. However, because particular migration events and individuals' migration experiences do not fall into neat categories, inevitably there is considerable overlap between any themes that we focus our attention on. As we demonstrate in this book, the act of moving rarely involves one factor, even if the move is motivated primarily by one overriding issue. Rather migration is firmly embedded within the complexity of people's everyday lives and experiences. Indeed, migration decisions are often influenced by factors that are much less obvious than might appear at first sight. Such an appreciation of the embeddedness of migration within daily life was captured by Chambers (1994: 2) when he observed that 'migration ... is ... deeply inscribed in the itineraries of much contemporary reasoning'. It is with the processes bringing about this inscription that this book is primarily concerned.

This chapter sets the scene for the remainder of this book. It provides a brief overview of the significance of migration in shaping people's lives and experiences in the contemporary world. This is achieved in a number of ways. First, attention is given to the migration histories of the three authors, which shows very directly that migration has been a central feature of our lives. Second, a more abstract account of the importance of migration is sketched. Third, the bulk of the chapter is taken up with describing some of the principal spatial patterns of migration within both the developed and developing worlds and at the international scale. These flows, which are set in their historical context, provide an empirical starting point for the thematic chapters' attempts to highlight in more detail the explanations for and experiences of migration in the contemporary world. Finally, although we have yet to define what we mean by migration more precisely – this issue is considered in some detail in the next chapter – a 'common-sense' understanding of the term allows this chapter to present an immediate impression of the phenomenon that forms the subject matter of this book.

Three migration biographies: an illustration

The complexity of the migration process can be gleaned immediately from the migration biographies of the authors of this book (Figure 1.1). These demonstrate the variety of factors that have influenced the migration experience of three individuals, despite the facts that they currently work in similar jobs and their migration paths converged when they worked in the Geography Department at the University of Wales in Swansea during the early 1990s.

Paul Joseph Boyle was born in Felixstowe, England in 1964. For his formative years he lived in a small rented flat until his English mother and Irish father bought a property on the other side of town, where they remain to this day. Their decision to move was prompted by the imminent arrival of their second son. In 1983 Paul went to university in Lancaster, expecting to spend three years in the area. However, he moved to Boulder, Colorado, for the second year of his degree on a student exchange scheme and travelled widely in the United States and Mexico. He spent his third year back in Lancaster. On finishing his degree he returned to Felixstowe to earn enough money to allow him to travel, and he worked in Israel and Malta for periods of a few months and cycled across Europe to Istanbul at the end of this 'year off'. During this time he was officially resident with his parents in Felixstowe but he spent little time there during the year, making his migrant status more complicated. He returned to Lancaster to begin a three-year Ph.D. For the first time he lived in Lancaster city in a large rented house rather than on campus. The travelling bug returned during this time and he spent three months working in Kenya with a charity organisation at the end of 1988, only to be persuaded to return to Kenya for a second spell of three months early in 1989. After completing his Ph.D. in 1991, he took up a lectureship in the Department of Geography in Swansea, spending four months renting a room in a nearby house, until he became an owner-occupier, purchasing a property 300 metres away. In 1993 he met his partner, Rhona, on a train. She was studying medicine in Cardiff and then took on a series of short posts in Shrewsbury, Hereford, Abergavenny and, finally, Swansea as part of her training. The decision to take this job finally allowed them to live in the same place. In 1995 Paul moved jobs to the University of Leeds, sold his house and rented a flat with his partner in an old seventeenth-century building. Rhona managed to find work in Leeds quite easily, and has worked in Barnsley, Harrogate, Bradford, Hull and Leeds itself. In February 1997 they moved to Christchurch, New Zealand, where Paul took up a one-year visiting lectureship at the University of Canterbury, while Rhona found work in an out-of-hours clinic. Accommodation was provided in a small flat above the clinic, and after this contract ended they moved into a rented flat close to the city centre. At the end of the year they plan to return to Leeds.

Born in Brent, north London, in 1965, the young Keith Harold Halfacree moved with his parents to the small commuter town of Sandy in the county of Bedfordshire two years later. This move resulted from a quality of life decision made by his parents, although the act of migrating was stimulated by his birth and his increasing demand for space. At four, he moved to an isolated house, tied with a gamekeeping job in the Devon countryside when his mother remarried. During the long years at secondary school, Keith spent most of his time lodging in the nearby town of Tiverton with various relatives and their neighbours because of the difficulty of travelling from home daily. As such, he became quite used to a 'nomadic' lifestyle. He finally moved away from home in 1984, when he went off to the University of Bristol to study geography. He was there for three years and lived at three different addresses. His educational career continued in 1987, with a move to Lancaster University, where he studied for a Ph.D. He lived at four different addresses during the four years he was there. It was here that he met Paul Boyle for the first time! The last geographically significant move brought him to Swansea, to take up a job in the Geography Department, since when he has moved from a privately rented house (where he lived with Paul Boyle and various postgraduate students for a short time) to his own property, 200 metres around the corner.

Vaughan Robinson is the elder statesman among the trio, having started life in a village in the Trough of Bowland in Lancashire in 1957. When, eight years later, his father was promoted to the headship of a large school in Darwen, he had his first experience of long-distance commuting, travelling daily with his father to the new school for a term, while his mother sought a headship in the same area. Once she had achieved this, the family moved to a commuter

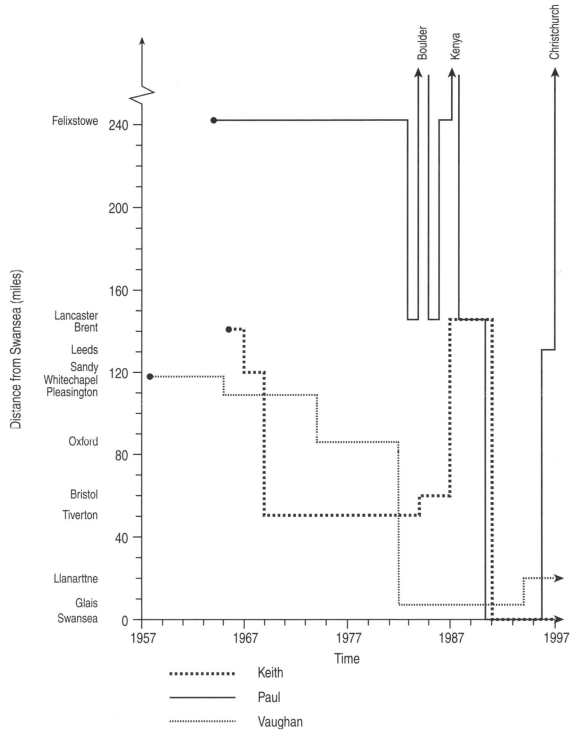

Figure 1.1 Simplified migration 'map' for the three authors. Note: only residential periods of six months or more shown.

village near Blackburn, where it remained until the retirement of his father. Vaughan, meanwhile, had departed to Oxford University in 1977, where he lived both in halls of residence and the usual unpleasant rented lodgings. He then embarked upon a D.Phil. at Oxford, and moved from one college (St Catherine's) to another (Nuffield). While at Nuffield he lived, with his partner, in a rented ground floor flat, horrifying the affluent neighbours by replacing the front lawn with an organic vegetable plot. His first job (as a research fellow) was also at Nuffield College and, having gained some financial security, he married and the Robinsons bought their first house. However, the cost of housing in Oxford necessitated a move out of the city to Bicester and a commute of two and a half hours per day. In 1982 the couple moved to Swansea when Vaughan took up a lectureship in geography. There, they bought a small cottage in a semi-rural area half an hour outside the city. Being a tied migrant moving to an area with few company head offices, Jayne Robinson was forced to accept demotion and, in her subsequent search for continued promotion, has commuted daily from Swansea to jobs in Bridgend, Cardiff, Caerphilly, Brecon and Hereford. When she gained promotion in the early 1990s, the couple began to look for a bigger house in a rural area; quality of life had become more important to them than accessibility to Swansea. Their search ended when Vaughan's mother, who had retired to west Wales prior to the death of his father in 1990, decided that the detached home that they had bought as a couple was now too large. Vaughan and Jayne bought a nineteenth-century villa near Llandeilo, 50 kilometres west of Swansea, and hope to raise a family there alongside their sheepdog, Bryn. Both of them travel with their jobs, Jayne having been seconded to a company in Oxford and also undertaking regular daily travel to Edinburgh, Alsager and London, and Vaughan working in the Sudan, Russia, Australia, India and Pakistan, as well as taking sabbaticals in Oxford. Indeed, travelling is one of their shared passions and they have visited China, Hong Kong, Indonesia, Mexico, Guatemala and Thailand.

From the brief evidence presented for just these three individuals, it is clear immediately that migration and travel have played an important part in their biographies – in making them who they are today. Critical decisions often involved moving, either temporarily or permanently, and the very fact that this book came to fruition results from their co-location in Swansea for a relatively long period.

Migration is a key element in most people's lives and consequently is an important phenomenon for academic study.

The scale and importance of migration

The academic significance of migration is demonstrated further by the wide interest in the topic among people from various disciplines outside geography, including demographers, economists, sociologists, anthropologists, historians, political scientists, psychiatrists and psychologists (Richardson 1967). As geographers we aim to consolidate these perspectives, since migration is not just an inherently spatial phenomenon but is now more than ever an important element of international, national and local affairs; in short, migration is the very stuff of geography.

Numerically, the size of migration flows is becoming increasingly impressive. The precise number of international migrants is unknown but the International Organisation for Migration (1990) estimated that there were over 80 million persons (1.7 percent of the world's population) resident outside their country of birth; by 1992 estimates had risen to over 100 million, although these differences may be attributable to source materials, rather than real trends (Burgers and Engbersen 1996). About one-fifth of these migrants were refugees or asylum seekers, a third were labour migrants and the remainder include officially economically inactive family members (Castles and Miller 1993). Flows within broad regions of the world are also both considerable and significant. In the preface to a major study of migration within the European Union (EU) (Rees *et al.* 1996), Harry Cruijsen of Eurostat (the EU's statistical division) drew attention to the sheer scale of migration within the EU alone. In the early 1990s, at least 25 million EU residents changed their place of main residence each year, with internal migration being the key demographic phenomenon in numerical terms. In comparison with this number of internal migrants, 1994 saw 4.0 million births, 3.7 million deaths, 2.0 million immigrations and 1.0 million emigrations. Within Europe, the 1960s saw a surge of migration within Spain, with 4.5 million people moving between municipalities, and high levels of migration have continued since that time (González and Puebla 1996). Likewise, in Britain it is generally accepted that around one person in ten changes their regular place of residence in any one year. Further evidence to support the rise of both internal and international migration is readily to hand. For example, in Asia, Hugo (1996: 96) observed:

The last two decades have seen a growth of international population movements in the region which has gathered increasing momentum in the last decade and will continue to do so into the new millennium.

All of these moves have a number of overlapping dimensions that merit study and understanding:

- The need to gain an understanding of contemporary migration patterns and processes represented through the figures is reflected in migration's strongly *cultural* character and in the cultural impacts that migration can have, both for the migrants themselves and for the places they migrate from and move into. The experience of migration is written into almost all contemporary cultures and is expressed through art, the media and, perhaps most ubiquitously, through the common-sense practices of daily life.
- A cultural perspective blends into a *psychological* appreciation of migration. For example, for the migrant, the act of moving can promise much and/or can be a highly stressful and disruptive experience. For international migrants especially, migration is frequently associated with leaving a familiar home environment and the often disquieting experience of settling into a culturally very different place. New horizons can be reached as former lives are left behind, with all the existential rewards and traumas that such a change can bring.
- The *economic* impacts of migration are also immense, especially as economic considerations underpin such a high percentage of migration flows. Although it is impossible to provide precise estimates, international labour migrants are believed to remit over $67 billion annually to their homelands, making this second only to oil in world trade figures (Martin 1992). Certain national economies rely heavily on these remittances and actively encourage short-term emigration. In contrast, in other nations immigrants are frequently unfairly blamed for high unemployment levels.
- Migration also has considerable *political* significance, both on the world stage and domestically. The reunification of Germany, the Zionist-inspired repopulation of Israel, the arrival of thousands of Vietnamese boat people in Hong Kong, and the illegal migration of Mexican

workers into the United States across the Rio Grande all involve migration events that have had considerable political impacts. Migration issues form a central aspect of the contemporary political scene, both at the domestic and at the international scale.
- Finally, we can return to focus on migration as a *spatial* event. As geographers we are obviously interested in the impacts of migration across space and it is important that the range, scale and complexity of migration patterns are demonstrated. Consequently, the majority of the remainder of this chapter is concerned with describing some of the key spatial patterns of migration, in order to set the scene for the thematic chapters, which elaborate upon the reasons underlying these empirical results. This discussion of spatial patterns provides a historical grounding within which the contemporary themes explored in subsequent chapters can be better understood.

Migration as a spatial event

Internal migration in the developed world

Urbanisation

The 'first' urban revolution occurred in the fifth millennium BC, following the agricultural revolution in the Near and Middle East, and identifiable towns and cities emerged (Figure 1.2) (overleaf). Usually, however, discussions of the beginnings of the urbanisation process focus on the period after the Industrial Revolution in late eighteenth-century Britain, which resulted in the growth of large metropolises and the emergence of a dynamic urban-industrial society (D. Clark 1982). Between 1751 and 1801 the population rose rapidly, increasing by approximately 45 percent as a result of interrelated demographic factors, agrarian change and economic development (Lawless and Brown 1986).

While the absolute population continued to grow during the nineteenth century, the percentage increases were not spectacular. However, the increasingly urban character of this population was spectacular and this was caused by the rural–urban migration that dominated this period. Complex patterns of migration resulted in regional population shifts towards the growing urban centres, and the basic population map of England was quickly re-

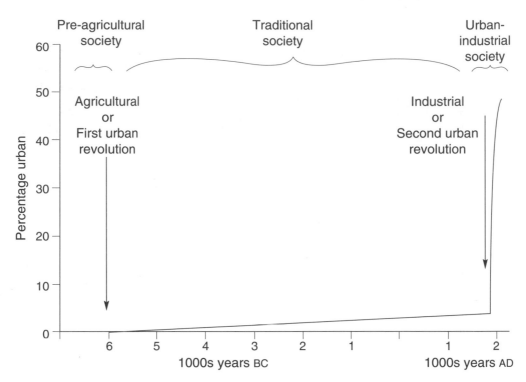

Figure 1.2 The growth of world urban population (source: Clark 1982, reproduced by permission of Croom Helm).

versed. The out-migration of young adults from the rural 'reservoirs' provided the impetus for rapid urbanisation, despite the high mortality rates in these centres. From the 1780s the pull towards London was increasing, but the population of new economic centres was also beginning to rise rapidly. Thus, between 1781 and 1830, the most marked migration gains were seen in the coalfield areas of South Wales, Lancashire and Durham and the growing industrial areas of the North West and Midlands (Figure 1.3).

During the nineteenth century, the centre of gravity of the British population continued to be pulled northwards, where industry was concentrating, but more detailed data show how population redistribution through migration was spatially concentrated into relatively few areas between 1841 and 1911, not all of these being in the north. Primarily, it was London, the larger towns, the residential towns and the colliery districts that gained substantially through migration, while the industrial towns began to witness net out-migration during this period. The rural exodus that helped to fuel the urban population increases was striking: while rural areas lost 4.5 million people through net out-migration, the

towns and colliery districts gained 3.3 million (Table 1.1) (page 9).

Moch (1989) warns that we need to delve deeper than simple urban and rural net migration figures if we are to understand fully the complexity of the urbanisation process. In France, the 'rather faceless population concentration and urbanisation' (p. 97) actually involved many types of migration other than simple rural–urban flows. These included temporary moves; exchanges between towns and cities within regions; long-distance moves of professionals and bureaucrats; and intense movement between cities and their immediate hinterlands, or *basins démographiques* (Figure 1.4) (page 10). A second migration process, which has received less attention, is the movement into and out of small towns. While cities grew, rural areas declined and larger towns stagnated:

small towns were deceptively active; they received rural people and sent their own citizens to regional centres and other cities ... they were not backwaters, but rather the receiving areas for rustics and the cradle for urbanites.

(*ibid.*: 99).

It is also misleading to interpret the net urban

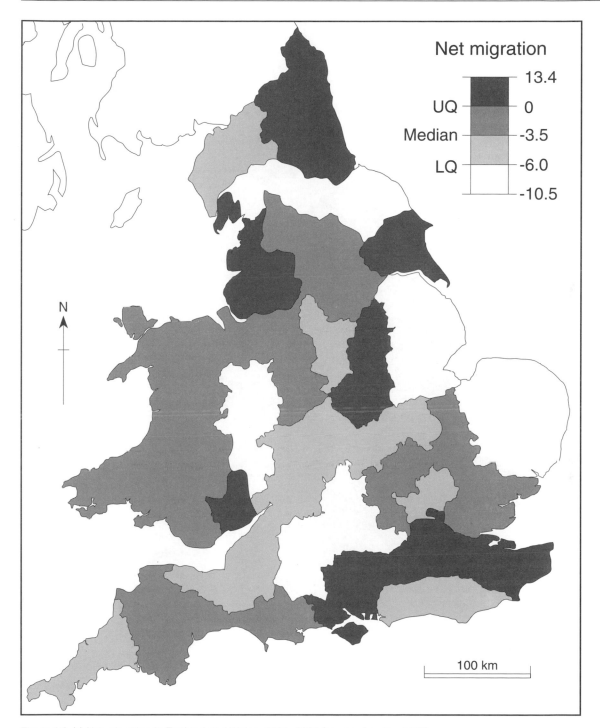

Figure 1.3 (a) Net migration in England and Wales, 1781–1800 (source: Lawton 1978).

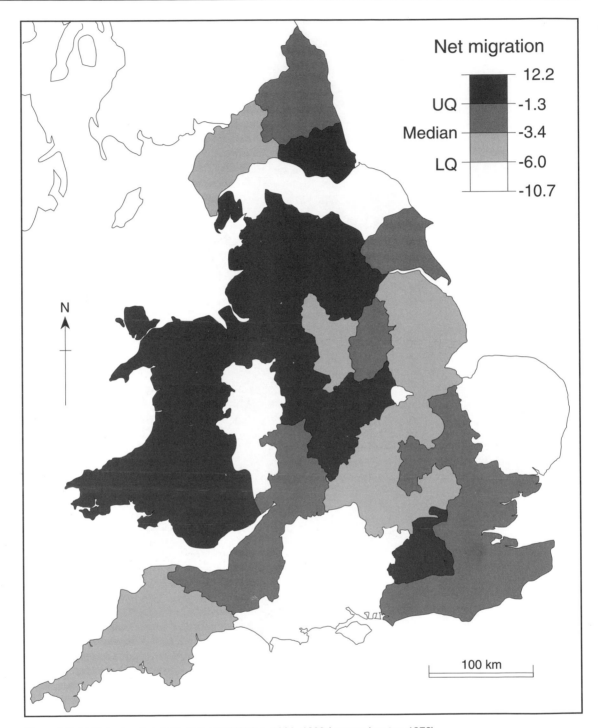

Figure 1.3 (b) Net migration in England and Wales, 1801–1830 (source: Lawton 1978).

Table 1.1 Urban and rural population growth in England and Wales, 1841–1911.

Location	Population 1841	Population 1911	Net migration 1841–1911	Net migration as a ratio of natural increase
Large towns				
London	2,261,525	7,314,738	+1,250,511	+32.9
8 Northern	1,551,126	5,191,769	+893,337	+32.5
Textile towns				
22 Northern	1,386,670	3,182,382	+89,933	+5.3
Industrial towns				
14 Northern	603,214	1,812,219	−152,994	−11.2
11 Southern	196,009	708,693	−15,679	−3.7
Old towns				
7 Northern	289,819	648,769	+15,944	+4.6
13 Southern	664,782	1,375,651	−22,004	−3.0
Residential towns				
9 Northern	206,897	559,022	+140,230	+66.2
26 Southern	692,185	1,770,030	+327,362	+43.6
Military towns				
16 Southern	470,821	1,212,413	+124,948	+20.3
Colliery districts				
9 Northern	1,320,342	5,334,002	+650,548	+19.3
Rural remainder				
12 Northern	2,425,614	2,875,113	−1,643,770	−78.6
12 Southern	3,740,228	4,085,691	−2,863,266	−89.2

Source: Lawton 1978.

gains and rural losses that were witnessed in much of Europe during this period as evidence of permanent migration. Langton and Hoppe (1990) argue that the picture of rapid demographic urbanisation that results from cross-sectional data, where everyone is classified as living in *either* urban *or* rural areas, hides the fact that:

Very few individuals are either 'rural' or 'urban', peasant or wage-earning, 'traditional' or 'modern' throughout their lives, or exclusively one or the other for significant lengths of time.

(Langton and Hoppe 1990: 140).

Net migration figures disguise the movement of short-stay migrants and there is evidence that circulation, rather than rural–urban migration, was characteristic of many parts of late nineteenth-century Europe (Anderson 1982; Hochstadt 1981; Johnson 1979). In Vadstena, Sweden, longitudinal household registers (Chapter 2) allow the migration histories of nineteenth-century inhabitants to be detailed. It appeared:

normal for people to change their place of residence at all stages of the life-cycle. Although there was a tendency for

this mobility to be at its most frenetic when people were in their twenties, everyone had moved, some two or three times, by their early teens, and everyone continued to move about until their sixties and seventies.

(*ibid.*: 147).

The urbanisation processes described above involved a complexity of migration forms that are not identified from the more traditional cross-sectional sources of information. Overall, however, migration was redistributing population towards urban areas and away from the rural periphery, such that urbanisation was the dominant migration process of this period, just as it is in much of the developing world today.

Suburbanisation

As the number of urban in-migrants swelled, city growth could not continue unabated. Suburbanisation, or the gradual decentralisation of population at the city edge, is a process that has been occurring since the earliest cities began:

Our property seems to me the most beautiful in the world. It is so close to Babylon that we enjoy all the advantages of

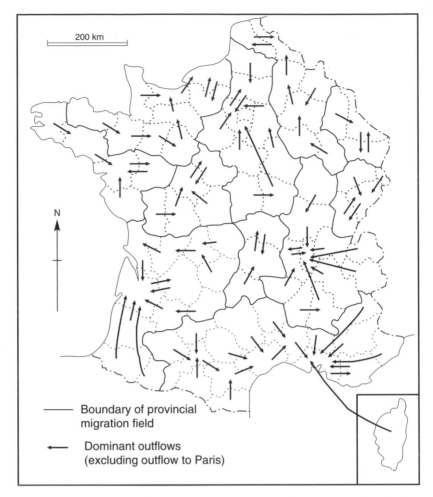

Figure 1.4 Lifetime migration in France, 1911. Note that the map depicts dominant out-migration from each *département*, excluding movement to Paris (source: White 1989).

the city, and yet when we come home we are away from all the noise and dust.

(Anonymous letter to the King of Persia in 539 BC on a clay tablet, quoted in Jackson 1985: 12).

Indeed, suburban villa developments with spacious gardens existed in Ur in southern Mesopotamia from about 2300 BC.

More recently, however, suburbanisation has been most prolific and culturally distinctive in the United States, where it is the most outstanding residential characteristic of everyday life in the twentieth century. Sustained growth at the edges of cities had occurred since the early nineteenth century and, in 1880, the United States Bureau of the Census was the first to employ the term 'suburb' in its analysis of Greater New York. Central to an understanding of

this process is a set of beliefs and values that were, and are, ingrained in American society (Chapter 6), but more practical transportation considerations are also important.

Pre-industrial cities, or 'walking cities', were characteristically disaggregated between the opulent centre and the inferior periphery. The most fashionable addresses were close to the city centre, as the lack of transportation meant that the optimum residential sites were close to the central facilities. Thus, in Philadelphia in 1790, the wealthy lived in the centre, while the city's first suburb contained a variety of blue collar artisans (carpenters, shoemakers and tailors) or those whose work was associated with the sea (Table 1.2). Indeed, as Foster noted in 1849, the most remote suburbs often housed the least privileged:

Table 1.2 The occupations of residents in Southwark, Philadelphia, United States, 1790.

White collar (17%)		Blue collar (45%)			
Sea captains	37	Labourers	128	Cabinet makers	4
Merchants	26	Ships' carpenters	56	Plasterers	4
Innkeepers	22	Mariners	45	Painters	4
Grocers	20	Shoemakers	39	Porters	4
Shopkeepers	18	House carpenters	32	Ships' joiners	4
Schoolteachers	15	Tailors	30	Caterers	3
Pilots	14	Blacksmiths	29	Caulkers	3
Lodgehousekeepers	11	Coopers	26	Mantua makers	3
Gentlemen	10	Weavers	17	Brewers	3
Gentlewomen	7	Bakers	15	Wheelwrights	3
Clerks	5	Rope makers	15	Silversmiths	3
Doctors	4	Mates	12	Sailmakers	3
Justices of the Peace	4	Joiners and cabinet makers	11	Sailors	2
Ministers	4	Bricklayers	7	Potters	2
Tobacconists	3	Ships' caulkers	7	Tinmen	2
Attorneys	2	Butchers	6	Printers	2
Constables	2	Mast makers	5	Barbers	2
Others	9	Seamstresses	5	Shallop men	2
		Boat builders	4	Others	25
		Bees housekeepers	4		

Source: United States Bureau of the Census 1908, adapted from Jackson 1985.

Nine-tenths of those whose rascalities have made Philadelphia so unjustly notorious live in the dens and shanties of the suburbs ... the core of the rottenest and most villainous neighbourhood ever peopled by human beings.

(quoted in Taylor 1969: 34).

The development of public transport within urban centres heralded the beginnings of population migration towards suburban communities. Between 1815 and 1875, American cities were turned 'inside out' as population growth occurred at the city edge and people's average journey to work increased. Brooklyn has been identified as the first 'commuter suburb', transformed by the regular steam ferry service that ran to New York City and offering a highly desirable residential environment. The daily commuting experience was captured by the celebrated writer Walt Whitman:

In the morning there is one incessant stream of people – employed in New York on business – tending toward the ferry. .... It is highly edifying to see the phrenzy exhibited by certain portions of the younger gentlemen ... they rush forward as if for dear life, and woe to the ... unwieldy person of any kind, who stands in their way.

(quoted in Jackson 1985: 28).

By 1860, the various urban ferries were carrying just under 33 million passengers a year rising to 50 million by 1870 (Jackson 1985). This reflected the suburb's faster growth rate than the city, doubling its population every decade from 1800 until the Civil War.

Demand for non-water-based transportation systems grew as congestion increased; prior to 1825 no city possessed a mass transit system. The first horse-drawn omnibus was developed by a French army officer called Baudry in Nantes (western France), and this innovation was copied in American cities during the 1820s and 1830s. By the middle of the century these were a standard feature of urban life, and a further key turning point in the history of the American suburb came with the invention of the electric street-car in 1884. The result was a series of cities developing along linear routes out of the city centre. These new suburban territories revolutionised the social geography of metropolitan areas, with middle-class higher-income workers migrating further from the city centre.

In 1905, Henry Ford introduced the mass-produced Model T, which substantially reduced the cost of car ownership. Two million cars were registered by 1916, rising to 33 million by 1940 (Muller 1981). Concurrent with this increase in private vehicles, concrete and asphalt roads were developed rapidly, along with tunnels and bridges spanning urban waterways. By the 1930s, suburban population growth overtook that of the cities (Figure 1.5) (over-

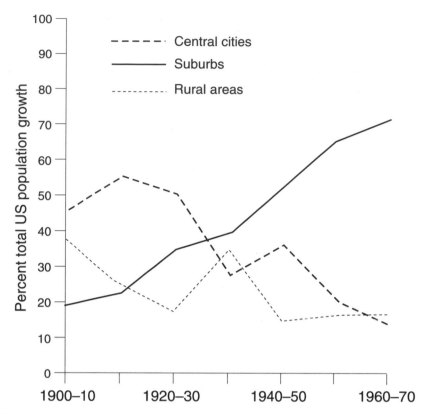

Figure 1.5 Percentage share of United States population change, 1900–1975 (source: Muller 1981).

leaf) and after the Second World War the growth of suburbs was monumental; the suburban population increased by 3 percent between the wars but by 13 percent between 1945 and 1960 (Muller 1981). By 1980, only one of the fifteen largest metropolitan areas in the United States (Houston) had a majority of its residents in the central city, although defining city and suburban boundaries makes comparisons problematic (Table 1.3).

Whether suburbanisation will continue so strongly into the future is debatable, although current trends show that suburbs remain the focus for middle-class growth (Frey and Fielding 1995). Indeed, many commentators have argued that suburbanisation is an inherent characteristic of American society that is unlikely to change in the future. For example, the sociologist Herbert Gans wrote in the *New York Times Magazine* in 1968 that 'Nothing can be predicted quite so easily as the continued proliferation of suburbia' (Stave 1989).

Table 1.3 Suburban population in the fifteen largest metropolitan areas in the United States, 1980.

Metropolitan area	Metropolitan population	Suburban population	Suburban percentage
Boston	3,448,122	2,885,128	83.7
Pittsburg	2,263,894	1,839,956	81.3
St Louis	2,355,276	1,902,191	80.8
Washington	3,060,240	2,422,589	79.2
Atlanta	2,029,618	1,604,596	79.1
Detroit	4,618,161	3,414,822	73.9
Cleveland/Akron	2,834,062	2,023,063	71.4
Philadelphia	5,547,902	3,859,682	69.6
San Francisco Bay	5,179,784	3,524,972	68.1
Los Angeles/ Anaheim	11,497,568	7,620,560	66.3
Baltimore	2,174,023	1,387,248	63.8
Chicago	7,869,542	4,864,470	61.8
Dallas/Fort Worth	2,974,878	1,685,659	56.7
New York/ New Jersey	16,121,297	8,721,019	54.1
Houston	2,905,350	1,311,264	45.1

Source: Jackson 1985.

Box 1.1 The Levittown developments

Advances in transportation technology allowed people to live further from their place of work but it has been argued that this was an enabling factor rather than the cause of suburbanisation. In the post-1945 United States, there was a rapid growth of large residential builders:

There was nothing new about suburban development in America. What was new in this period was the developed capacity of large builders to take raw suburban land, divide it into parcels and streets, install needed services, apply mass production methods to residential construction, and sell the finished product to unprecedented numbers of consumers.

(Checkoway 1984: 153).

The impact of such developments is clear from the following table of new housing units in the United States between 1938 and 1959:

Year	New housing units started	Houses built by large builders (%)
1938	406,000	5
1949	1,466,000	24
1959	1,554,000	64

Levitt and Sons exemplified the importance of large builders. Founded in 1929 by Abraham Levitt, the company was forced to build low-cost housing in government defence areas during the war years and this provided experience in building prefabricated, mass-produced housing. In 1947, the company bought around 560 hectares of farmland on Long Island, 50 kilometres from New York City, and by 1948 it was completing more than 35 houses per day. The house-building industry had been revolutionised and more than 17,000 identical homes for 70,000 people were constructed in uniform rows in Levittown. Selling at $7990, they were widely regarded as the best value for money in the United States (Larrabee 1948). A second Levittown was developed outside Philadelphia, in Lower Bucks County, and when the first houses were opened for inspection in 1951 more than 50,000 people came to view during the first weekend. On the first two days, more than $2,000,000 worth of houses were sold!

These impressive developments were achieved by transferring Fordist mass-production techniques from the factories to the construction industry. Each site was an assembly line with people and machines moving through in teams to perform one of 26 operations, using parts that had been pre-assembled or pre-cut in factories beforehand. Levitt also gained considerable financial backing from the government and invested more money in consumer research than any other builder. The homes were well advertised and the salespeople were remarkably efficient; the entire financing and title transaction took one hour.

Mass suburban development in post-war America was therefore, at least to some extent, structurally determined. While the role of individual preferences and the search for socially homogenous enclaves were important determinants of the drive towards suburbanisation, the role of large developers and the federal government support provided to them were crucial factors. Thus, we must:

question those studies which fail to explain the impossibility of inferring the spatial dynamics and decision behaviour of large operators and government partners from the residential aspirations and satisfaction of the eventual suburban consumers, or which fail to specify the narrow range of alternatives actually available, or which fail to emphasize the fact that consumers were important but not decisive actors in the decisions which produced the choices they made. Consumers made a logical choice among alternatives developed elsewhere.

(Checkoway 1984: 167–168).

Source: Checkoway 1984.

Counterurbanisation

The decentralisation of population from urban centres into surrounding suburbs dominated migration patterns in much of the developed world after 1945 and urban growth continued at the expense of the rural periphery. More recently, this pattern of migration has been superseded by a radical reversal of the trend towards urbanisation. First witnessed in the United States (Beale 1975), *counterurbanisation* was the term used to describe both the population growth that was occurring in non-metropolitan American counties and the population decline in the metropolitan counties during the 1960s and 1970s. Indeed, it was shown that the 'rural renaissance' (Morrison and Wheeler 1976) was gaining pace faster in remote rural areas than in rural areas neighbouring metropolitan counties; it was not simply an extension of metropolitan expansion (McCarthy and Morrison 1978; Sternlieb and Hughes 1977).

Several multinational studies have found similar patterns of counterurbanisation in much of north west Europe, Japan, Australia, New Zealand and parts of Scandinavia (Berry 1976a; Champion 1989;

Table 1.4 Population change in English East Midlands parishes of under 1,000 residents, 1971–1981.

County	Total parishes	Population loss	Population gain	Parishes depopulating (%)
Derbyshire	169	83	86	49
Leicestershire	206	93	113	45
Lincolnshire	452	217	235	48
Northamptonshire	203	87	116	43
Nottinghamshire	169	72	97	43
East Midlands total	1,199	552	647	46

Source: Weekley 1988.

Vining and Kontuly 1978; Vining and Pallone 1982). Various reasons for this radical reversal of population concentration have been suggested (Champion 1992). Perhaps the most persuasive are those related to quality of life decisions (Chapter 6) and economic restructuring and industrial (re)location (Chapter 4).

Other studies have been more sceptical of the extent of counterurbanisation. Weekley (1988) used parish data in the East Midlands of England to show that, while at larger scales it appears that rural areas are growing rapidly, there was considerable variation between local areas, with many continuing to experience depopulation (Table 1.4). Elsewhere, Flowerdew and Boyle (1992) used local-level 1981 census data in Hereford and Worcester county in England to show that the vast majority of in-migration from the nearby West Midlands metropolitan conurbation was over very short distances and could more convincingly be described as an extended suburbanisation than typical counterurbanisation. Furthermore, Boyle (1995a) statistically modelled (Chapter 2) the flows into and within rural districts in England over the 1980–1981 period and showed that flows from the most urban areas into the most remote rural areas were not unusually large. Thus, metropolitan population losses and rural gains were not necessarily directly linked. Even with such qualifications, however, it is generally agreed that the 1960s and 1970s were periods of rural population growth at the expense of urban areas in many developed nations.

The urban revival

The experience of the 1980s has suggested yet another unexpected turnaround in population redistribution. It was in the United States that counterurbanisation occurred first and appeared to be most strongly developed but, from the beginning of the

1980s, metropolitan growth began to rise once more (Long and DeAre 1988). As demonstrated in Table 1.5, the metropolitan revival between 1980 and 1990 was even greater in the large metropolitan areas than in the smaller ones. Contrary to the beliefs of many researchers writing in the early 1980s, it is thus possible that counterurbanisation may have been a relatively short-term blip in the general trend towards metropolitan growth.

Various theories have been suggested to explain this turnaround. Berry (1988) compared the pattern of net migration change in the urban centres of the United States with economic cycles (Figure 1.6) (page 17). The relatively consistent waves of gross national product (GNP) are negatively related to the familiar *Kondratiev waves* of prices, which rise during lean spells but, according to Berry, there is also a strong positive relationship between economic growth cycles and urban in-migration in the United States.

Geyer and Kontuly (1993) developed a broader *differential urbanisation* model, which distinguishes between large, intermediate and small cities, which appear to go through successive phases of fast and slow growth (Figure 1.7) (page 18). Initially, population concentration, through net migration, dominates as primate cities grow rapidly. Subsequently, new urban centres begin to expand and economic

Table 1.5 Population change in the United States, 1960–1990 (percentage).

Period	Large metropolitan areas	Other metropolitan areas	Non-metropolitan areas
1960–1970	18.5	14.6	2.2
1970–1980	8.1	15.5	14.3
1980–1990	12.1	10.8	3.9

Source: Frey 1992.

Box 1.2 Counterurbanisation in France

While work has shown a widespread pattern of urbanisation and counterurbanisation in many developed societies, France may represent an exception. Here, the urbanisation process was unusual because of the slow rate of population growth during the nineteenth century, which restricted the development of industrial agglomerations and left Paris as a dominant primate city (van de Walle 1979). Urbanisation and counterurbanisation were telescoped in time, as evidence for a rural revival was identified in the 1975 census results (Courgeau 1978). France has since displayed some of the classic features of counterurbanisation, with rural *communes* experiencing population gains after a century of loss. The diagram below shows that, between 1975 and 1982, with the exception of the smallest, all the *communes* under 20,000 population experienced net in-migration, while the larger *communes* witnessed net declines. Even so, some authors are less than convinced by the counterurbanisation thesis in France, arguing that these trends are more apparent than real because of problems with data sources and with the definition of 'urban' and 'rural' (Noin and Chauviré 1987).

Winchester and Ogden showed that, in 1982, it was the rural *communes* within the functional urban regions (*zones de peuplement industriel et urbain*) that experienced population gains, while depopulation continued in the remoter rural areas. Decentralisation appeared to be more important than deconcentration.

Source: Winchester and Ogden 1989.

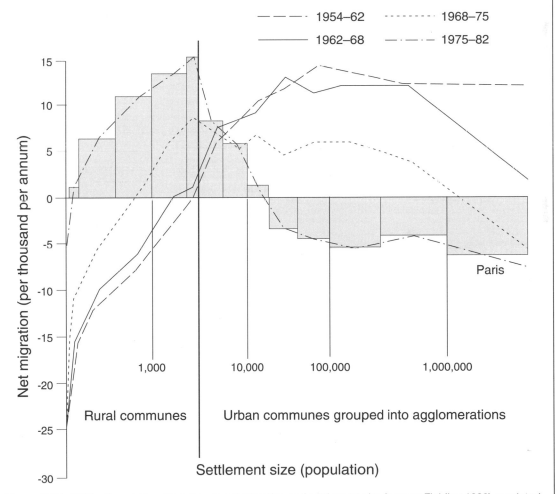

France 1954–1982 – the relationship between net migration and settlement size (source: Fielding 1982), reprinted from *Progress in Planning*, **17**: 1–52, 'Counterurbanization in Western Europe', © 1982, with kind permission from Elsevier Science Ltd, The Boulevard, Longford Lane, Kidlington, OX 51 GB, UK.

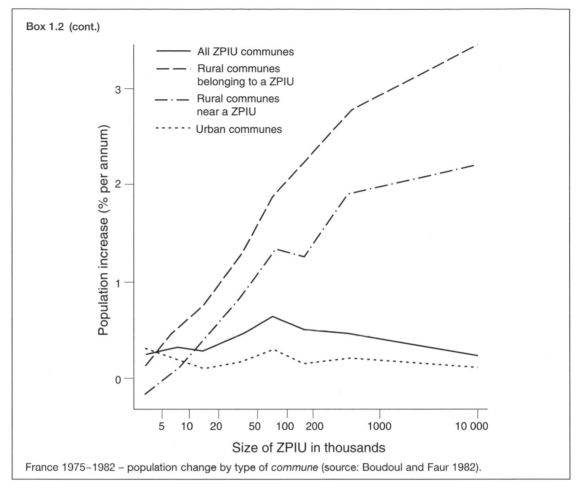

Box 1.2 (cont.)

— All ZPIU communes

— — Rural communes belonging to a ZPIU

—·— Rural communes near a ZPIU

····· Urban communes

France 1975–1982 – population change by type of *commune* (source: Boudoul and Faur 1982).

development disperses, but the primate cities continue to grow rapidly. Third, there is a period of suburban growth, which is followed by a 'small city', or 'counterurbanisation', stage, where small cities and more rural areas grow faster than the large centres. However, this stage does not last indefinitely, according to this model, and it is followed by a new concentration phase in which large metropolitan areas once again attract net in-migrants. Applying this model to France, the Republic of Korea and India they suggest that it can be used to characterise the degree of development in and within various countries spanning the developed and developing worlds.

Frey (1993) also examined the urban revival in the United States in depth using 1990 census data. Two suggested causes were, first, the shifts in industrial structure and the growth of service sector employment (Chapter 4) and, second, the expanded growth of minority populations. Continued immigration from Latin America and Asia, combined with high levels of natural increase among native-born minorities, has increased the relative size of the minority population, and much of this growth is geographically concentrated in metropolitan areas (Figure 1.8) (page 19). The precise patterns vary for the three main minority groups, with Asian gains in large metropolitan areas throughout America, large Hispanic gains in the largest southern metropolitan areas, and black gains in southern metropolitan areas at the expense of northern metropolitan areas and in communities of all sizes in western America.

This second migration turnaround is not as uniform within the developed world as the move towards counterurbanisation, however. While there was evidence of core growth in Finland, Italy, Japan, Norway, Spain and Sweden, in north west Europe the patterns were not so obvious. There are also variations within as well as between nations. For

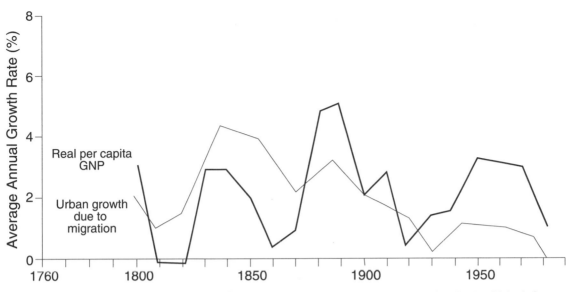

Figure 1.6 The cyclical relationship between economic growth and urbanward migration in the United States, 1790–1980 (source: Berry 1988).

example, Stillwell, Rees and Boden (1992a) showed that rural areas in the south of Britain performed well during the 1980s, in terms of net migration, but that rural areas in the north were less attractive migrant destinations.

Regional trends

While net flows to and from urban centres are important, broad regional patterns of population redistribution add another migration dimension in many nations. Counterurbanising flows dominated migration trends during the 1960s and 1970s in Britain, but concomitantly there has been a persistent drift of population from the north to the south of the country. Similarly, in the United States net migration from the snowbelt to the sunbelt has been a prominent feature of internal population migration (Cadwallader 1991), although some snowbelt regions, such as parts of New England, are experiencing recent growth from in-migration (Lichter and De Jong 1990). These regional patterns generally involve long-distance migration, and underlying structural reasons often influence them.

In Italy, there has been a long-term net flow of population from the south to the north (Table 1.6). During the 1950s, the industrialisation of the north west 'industrial triangle' was a potent trigger for large-scale migration from the depressed south and north east. In the 1960s, socio-economic developments in the north east improved the region's migration balance, while the situation in the south worsened. However, during the 1970s, inter-regional migration patterns changed and the net migration balance of the south improved significantly. This is not necessarily a reflection of improving economic conditions, however, as the southern economy remains largely dependent on the rest of the country.

Table 1.6 Internal migration in Italy, 1956–1987 (000s).

Years	Total	Intra-regional	Inter-regional	South to north	North to south
1956–1957	1,349	907	442	116	39
1966–1967	1,487	978	509	179	82
1976–1977	1,211	799	412	142	92
1986–1987	1,141	830	311	104	72

Source: Bonaguidi and Terra Abrami 1996.

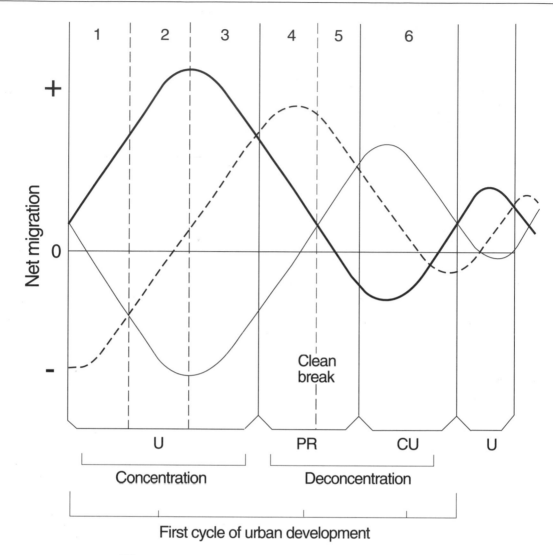

Figure 1.7 The differential urbanisation model (source: Geyer and Kontuly 1993).

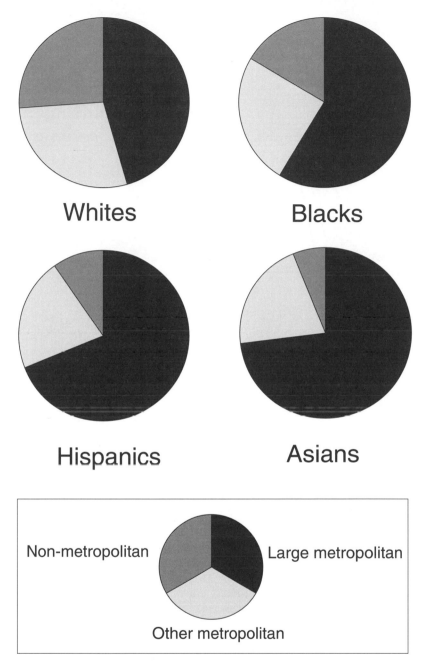

Figure 1.8 Geographical distribution of whites, blacks, Hispanics and Asians in the United States, 1990 (source: Frey 1993).

In fact, the age-standardised curves for flows between the north west and the south show that the south continues to lose young adults but gains families with children and the elderly, who are motivated more by environmental considerations (Figure 1.9) (overleaf). The improved migration balance in southern Italy also resulted from a rapid decline in out-migration rather than increased in-migration. Overall, the inter-regional redistribution of population was declining at the expense of intra-regional

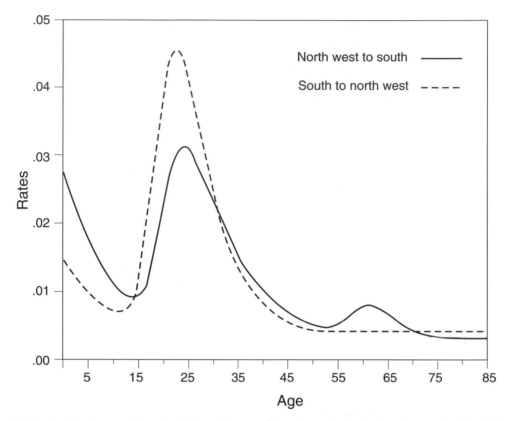

Figure 1.9 Standardised age profiles of south to north west and north west to south migration rates in Italy, 1980–1982 (source: Bonaguidi and Terra Abrami 1996, © John Wiley & Sons Ltd. Reprinted by permission).

moves (Bonaguidi 1990). In Italy, therefore, inter-regional migration trends have become less important than intra-regional (mainly urban–rural) flows.

Internal migration in the developing world

Urbanisation

According to the World Bank (1992), the urban population in 'low- and middle-income countries' rose from 557 million in 1965 to 1,820 million in 1990, while the annual rate of growth rose from 3.7 percent between 1965 and 1980 to 6.6 percent during the 1980s. It is generally agreed that the pace of urban growth within the developing world is without precedent and that most of this growth stems from natural change (Gugler 1988). Even so, about two-fifths is fuelled by rural–urban migration.

In many parts of the developing world this urbanisation process is biased towards primate cities. For example, in Sudan, 8 percent of Sudanese lived in urban areas according to the 1955 census. By 1983, this had risen to 20 percent. Despite this, Sudan was relatively non-urbanised, even by African standards, and urbanisation was strikingly uni-directional, with the capital Khartoum attracting large numbers of rural–urban migrants (Figure 1.10). Between 1983 and 1987, migration towards the Khartoum conurbation increased considerably due to drought, famine and military activity in the south, such that the population rose by 123 percent from 1.34 million to 3.0 million. Typical of many developing nations, economic development has also been focused in the primate city and some of the other large metropolitan areas, while the smaller towns have attracted little investment. Thus, in 1983, the Khartoum conurbation housed only 6 percent of the population but 75 percent of the industrial establishments, 65 percent of the banking and commercial companies, 57 percent of the physicians and 90 percent of enrolments in higher education (El-Bushra 1989).

Overall, net rural–urban migration does not ap-

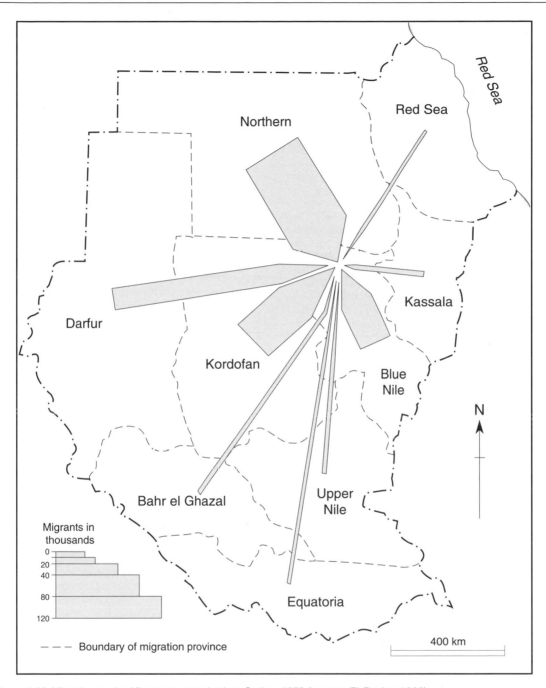

Figure 1.10 Migration to the Khartoum conurbation, Sudan, 1973 (source: El-Bushra 1989).

pear to be accelerating in developing countries, although there are substantial variations in urban growth rates between countries. Gilbert (1993) contrasted the continually rapid growth in sub-Saharan Africa, where urbanisation has been a relatively recent phenomenon, with the slowing growth in North Africa, the Middle East and, especially, Latin America, concluding that the least urbanised countries continue to endure a flood of people towards urban centres, while the more urbanised countries

Box 1.3 Urbanisation in China

The scale of rural–urban migration in China has been described as an 'enigma' because of the various definitions of 'urban' and 'rural' that have been used in recent years (Chan 1994). Unlike most nations, where a person is defined as 'urban' if they live within a pre-defined urban boundary, in China *individuals* are categorised as 'agricultural' or 'non-agricultural'; the decision is individually rather than geographically based.

Until recently, rural people worked within various 'People's communes', through which many economic activities were organised. In general, however, they were not encouraged to participate in non-agricultural production, which was assigned to the urban 'non-agricultural' population, nor were they able to move freely into urban areas as a strict residential registration system existed and there were tight controls on employment opportunities and the distribution of grain (Kirkby 1985). It has now been recognised that there were surplus agricultural labourers, who needed to be transferred to other sectors of the economy and, after the reform period that began in 1978, manufacturing was encouraged in rural areas and the controls on migration into urban areas were relaxed. Consequently, the rural–urban population shift was composed of a number of elements:

- Rural–urban migrants recruited directly by the formal urban sector and whose registration status changed from 'non-agricultural' to 'agricultural';
- Rural–urban migrants who moved to cities and towns to engage in self-employed business and whose registration status has remained 'agricultural', except in a few places, such as Zhejiang Province, where their status may be altered if they pay a sum of money to the urban authority;

- 'Agricultural' people living in urban areas whose registration status is changed to 'non-agricultural' because of employment change.

This issue is complex as 'official' definitions of the urban population differ over time. The pre-1982 definition excluded all the 'agricultural' population, regardless of where they lived. On the other hand, the 1982 census definition included all those living in urban areas, even if they were registered as 'agricultural'. The 1990 census adopted a mid-way definition treating the 'agricultural' population resident in city districts (not towns or other parts of cities) as urban.

The result is a complex set of definitions that make it difficult to agree on the actual urban population in China. What is agreed is that the official number of cities increased from 191 in 1978 to 464 in 1990 and the number of towns increased from 2,819 in 1982 to 11,392 by 1990. Thus, many rural residents have changed status simply through the redefinition of urban areas. Some idea of the rural population of China over the 1978–1990 period is given in the following table:

	Rural population (%)			
Year	Pre-1982 definition	1982 census definition	1990 census definition	'Agricultural' population (%)
1978	87.1	82.1	n.a.	84.2
1982	85.5	79.2	n.a.	82.4
1987	81.9	53.4	n.a.	79.9
1990	81.5	46.8	73.6	78.4

Source: Shen 1995

are seeing a decline in rural–urban migration. The gulf between primate and secondary cities is narrowing in these countries and fears of increasing polarisation between the two kinds of cities seem to be waning. However, in some instances, the slowing of primate city growth may simply be a reflection of the urban area spreading into neighbouring municipal areas and of a subsequent failure to update city boundaries (Castles 1991). The suburbanisation of industrial and residential activities is certainly becoming more widespread throughout the developing world, as urban diseconomies in the major cities increase to a critical level and infrastructure improvements gradually enable such developments (Gilbert 1993).

Unlike work on migration to urban areas in the developed world, the gender dimension of urbanisation in the developing world is well-researched, mainly because 'migration is virtually always gender-selective to some degree in all developing regions' (Chant and Radcliffe 1992: 2). Women dominate urbanward migration flows in much of Latin America and parts of East and South East Asia, which are the most urbanised areas of the developing world (Table 1.7). The high level of female migration into urban areas is not necessarily *associational* migration, linked to the former movement of a male member of the family. Rather, much of this movement is *autonomous*, as many women move independently (Dickenson 1983). While men are

Table 1.7 Urban sex ratios in selected developing
countries.

Region Country, year	Men per 100 women	Percentage of men in urban areas	Percentage of women in urban areas
Africa			
Botswana 1986	106	23.2	20.2
Egypt 1986	106	44.1	43.7
Equatorial Guinea 1983	97	28.2	27.0
Gambia 1980	104	18.3	18.1
Ghana 1984	95	31.6	32.4
Malawi 1987	108	14.2	12.5
Tunisia 1984	103	52.7	53.0
Zaire 1984	100	28.2	28.1
Zimbabwe 1982	114	25.6	21.6
Latin America and Caribbean			
Argentina 1985	95	81.5	84.4
Bolivia 1985	95	47.2	48.3
Brazil 1980	95	65.1	67.7
Chile 1988	94	80.4	84.0
Columbia 1985	91	64.8	69.6
Costa Rica 1985	92	42.6	46.4
Cuba 1986	97	70.1	73.1
Ecuador 1985	97	50.9	53.2
Haiti 1985	80	23.4	27.5
Honduras 1985	94	38.3	41.1
Mexico 1980	95	65.3	67.2
Panama 1988	87	50.1	54.4
Peru 1980	100	68.3	69.3
Puerto Rico 1980	92	65.7	67.8
Virgin Islands 1980	88	38.3	39.8
West Asia			
Iran 1986	106	54.4	53.9
Syria 1988	107	50.6	49.4
South Asia			
Bangladesh 1981	126	16.4	13.9
India 1987	112	26.5	25.1
Nepal 1981	115	6.7	6.1
Pakistan 1981	115	28.9	27.7
Sri Lanka 1981	110	22.1	20.9
East/South East Asia			
China 1982	110	20.8	20.3
Indonesia 1985	99	26.3	26.2
Korea 1985	99	65.0	65.7
Malaysia 1980	100	37.4	37.0
Philippines 1980	95	36.3	38.3
Thailand 1980	96	16.8	17.3

Source: Chant and Radcliffe 1992.

usually more mobile than women, dominating flows between and into rural areas, women outnumber men in flows into urban areas. In contrast, males dominate the flows into urban areas in South Asia and most of Africa.

Chant and Radcliffe (1992) suggested some gen-

eral factors that account for these broad differences. First, female involvement in the agricultural sector explains the contrasts between Africa, where women have a strong role in agricultural production, and Latin America, where women traditionally make up under 20 percent of the agricultural workforce. Women are therefore more enmeshed in rural life in Africa, and in some parts of the continent they make up more than 50 percent of the agricultural labour force. Second, the relative demand for female labour in cities is especially high in South East Asia, where young women have been drawn into the workforces of multinational companies, and in Latin American cities, where service industries have employed large numbers of women (De Oliveira 1991). These female employment opportunities are rarer in South Asian and African cities (Townsend and Momsen 1987). Third, female migration is constrained by 'social and cultural constructions of gender' (Chant and Radcliffe 1992: 7) and much of the apparently autonomous migration of women towards cities is often the result of decisions made within households (Chapter 5).

Circulation

For some time it was generally assumed that the urbanisation process in the developed world involved relatively permanent resettlement of rural dwellers in urban areas. It is only more recently that the importance of circulatory movement has been accepted. In the developing world, however, the importance of *circulation* has long been recognised and may take many forms (Chapter 2).

Unfortunately, the cross-sectional nature of standard census data sources fails to illuminate circulatory patterns. Consequently, we need to obtain:

interlocking sets of data derived from field censuses, ongoing mobility registers, family genealogies, retrospective movement histories, and oral historiographic reconstructions of residential transfers.

(Chapman and Prothero 1985).

Fortunately, some studies have utilised such data sources. Nair (1985) provided an example from Fiji, where the circulatory practices of 199 Melanesian Fijian households and 201 Indo-Fijian (Asian immigrants) migrant households were contrasted using questionnaire and migrant history data (Chapter 2). Considerable differences were identified between these groups. Indo-Fijians who had moved to the capital, Suva, were much more likely to wish to remain there and had fewer ties to their home settle-

---

Box 1.4 Rural–urban migration and household reproduction in Guanacaste, Costa Rica

Neoclassical accounts of urbanisation identify the expansion of economic activities and the subsequent employment and income opportunities as key factors in attracting rural in-migrants. This interpretation has been questioned using data collected from 350 low-income households who moved from rural areas to three towns (Liberia, Cañas and Santa Cruz) in Guanacaste province in north west Costa Rica. The province is one of the poorest in Costa Rica, with traditionally low levels of urban growth, and it is somewhat surprising that rates of in-migration to urban areas increased significantly during the 1980s. Urban employment opportunities do not appear to have increased and most of the males in the families interviewed were forced to commute away from the urban centres for employment in rural areas. Women do have better work opportunities in the towns, but these are usually poorly paid and could not be relied upon to support the family:

If 'productive' factors were to constitute the primary motivation, then households would probably have moved wholesale to areas where opportunities for regular work, particularly for men, were much more favourable. Instead, small towns in Guanacaste with a largely permanent female population appear to be in the process of becoming sites for the reproduction of labour power that in turn depends on temporary male out-migration to external labour markets.
(Chant 1991a: 246).

Four factors were identified that, in combination, explain these surprising patterns of rural-urban migration:

- Financial constraints prevent households moving to other parts of the country where employment opportunities are better;
- The spatial dispersal of household earners, between towns and rural areas, minimises economic risk;
- Access to housing, school and medical services is better in the towns;
- Familiarity with the region and access to kin who provide useful support networks discourage migrants from moving further away.

Urbanisation appears to relate primarily to reproductive imperatives of household survival, such as housing and welfare provision, and the spatial divisions of labour are closely intermeshed with gender divisions of labour. However, female employment is poorly rewarded, women can rarely survive without male income, and:

one major corollary of the spatial separation of men's and women's activities is that this basic pattern not only *derives* from sexual divisions of labour, but may also *reinforce* them.
(*ibid.*: 247).

Source: Chant 1991a.

---

ments than Melanesians. This stemmed from cultural traditions, as the Melanesian Fijians had a powerful sense of commitment to the home *koro* (village), where a strong form of communalism existed. These Fijians regularly returned home for ceremonies and they were taught both formally and informally that their traditions and lifestyles should be retained at all costs. It is not surprising, therefore, that circulation was especially important for these individuals.

Other patterns

Finally, it is important to note that many developing countries also show some of the other spatial patterns of migration represented in the developed world. Sometimes these trends appear to mirror previous patterns of migration within the developed world, giving rise to generalisations about the evolution of migration patterns through time – such as Zelinsky's mobility transition model (Chapter 3) – and sometimes they do not. Thus, although less commonplace than in many developed countries, there

are examples of developing world counterurbanisation (Chapter 6). More notably, regional migration trends are commonplace, including large redistributions of the population engineered by the state (Chapter 7).

International migration

The European legacy

International migration is by no means a new phenomenon, although volumes were low and patterns relatively simple until recently. Of course, we should not disregard the international and intercontinental movement of free and enslaved people within the expanding Roman Empire, nor can British readers sideline the maritime migrations of the Vikings, which have been so influential in shaping the contemporary culture and administrative systems of Britain. Nevertheless, large-scale international migration blossomed only after the development of new transport technologies and new opportunities in the nineteenth century. Even then, patterns were geographically straightforward, with H. Jones (1990: 230) arguing that:

by about the seventeenth century long-distance migrations throughout the world began to comprise a single network organized in the interests of the politically and economically dominant states of north-western Europe.

As European empires spread out into North and South America, Africa and Asia, so the volumes of migration increased. European administrators, traders and military personnel were despatched by the colonial cores to extract maximum value from the new acquisitions. New societies and social structures were the result, with social status and 'race' often becoming synonymous (see Lowenthal 1972, for example, on the Caribbean). In addition, where indigenous labour was insufficient in the expanding agricultural economies, slave labour was introduced. Between 11 million and 12 million Africans were 'exported' from Africa between the beginning of the seventeenth century and the abolition of slavery in the mid-nineteenth century (Curtin 1969). Even after that time, involuntary slavery was often replaced by indenture, where Indian or Chinese peasants signed up for a period of temporary voluntary slavery (Tinker 1977), a system abolished as recently as 1914. However, in some cases the movement of labour from one colonial territory to another was totally voluntary, as with the Indian labourers who built the East Africa Railway (Twaddle 1995).

European colonial expansion was thus a driving force behind the growth of international migration, but it was not until Europeans themselves began to get involved in substantial numbers in the migration (as opposed to organising the migration of non-Europeans) that migration began to increase sharply. The combination of cheap fertile land in countries such as the United States, Canada, Australia, South Africa and New Zealand, and poverty, famine and high population growth rates at home, produced what has been described as 'arguably the most important migratory movement in human history' (H. Jones 1990: 231). Between 55 million and 65 million settlers are thought to have left Europe between 1820 and 1930, with over one million leaving Ireland alone in the five years of the potato famine (Cousens 1960). The United States was the key destination for many of these international migrants and, when immigration reached its peak in 1900–1909, over 8.2 million immigrants were recorded as entering the country.

By the inter-war years the European empires were in decline, if not yet dissolution, and the main 'countries of immigration' were beginning to revise their immigration policies. As southern and eastern Europeans came to dominate immigration, and even non-Europeans began to seek entry to countries such as the United States, so regulation became more commonplace. The 1924 Quota Act in the United States ensured that annual national quotas were set for immigration, with the size of quota for each country directly linked to the size of the existing settler population of that nationality within the United States. The composition of migrant flows was thus artificially ossified, with certain European nationality groups, such as the British, being given preferential treatment. The Australian Immigration Restriction Act of 1901 had a similar deliberate effect (Chapter 7). The story of pre-war international migration was dominated by European power, European policies and European people.

Post-colonialism and the rise of international labour migration

In the immediate aftermath of the Second World War, circumstances might have seemed little different. Europe was preoccupied with repatriating, settling or exporting the human detritus resulting from six years of 'total war' (Marrus 1985), and former European colonies were anxious to assist, by encouraging or accepting migrant settlers (Collins 1991). However, change was already apparent. As decolonisation and independence gathered momentum, so too did international migration in the developing world. Newly independent countries struggled to assert new identities and disentangle previously mixed populations. For example, the partition of the Indian subcontinent led to a massive redistribution of population according to religion. In other cases, ethnic minorities that had been implanted by colonial rule were rejected, as was the case with the Ugandan Asians (Robinson 1995). In a third type of post-colonial migration, European settlers returned from former colonies to their metropolitan 'motherlands'. Zlotnik (1996) estimates that a total of 500,000–800,000 Portuguese citizens returned to Portugal from Angola, Cape Verde, Guinea-Bissau and Mozambique after these countries gained their independence.

Despite these stirrings, Zlotnik argues that Europeans still dominated major intercontinental migration flows until the early 1960s. Around that time, a significant change occurred, in that Europe

Box 1.5 Migration and the partition of India, 1947

The partition of the Indian subcontinent into what were originally two new countries in 1947 created a predominantly Hindu India and an overwhelmingly Muslim Pakistan. Along the new international border in the north west between the Indian Punjab and the Pakistani Punjab (announced in August 1947), many families found themselves on the wrong side of what they thought was a religious divide. Over 8.5 million chose to rectify this by crossing the international border into a new country during the period September–December 1947, with 4.1 million Muslims moving into Pakistan and 4.4 million Hindus and Sikhs transferring to India. Contemporary press reports described the transfers as 'the greatest migration of modern times' and recorded how, at its peak, 60,000 per day were crossing the new international border. After an initial period of uncoordinated and individual migration, from September 1947 both governments intervened and cooperated to plan and control the population transfers. On 11 September alone, 400,000 Hindus and Sikhs were marched on foot out of West Punjab (Pakistan). The British Overseas Airways Corporation was dragooned into helping and eventually carried 35,000 passengers in 40 days, while the railway system carried 2.3 million people in three months. This was in itself a dramatic migration, but the short- and long-term consequences were no less significant. The city of Lahore, which had been home to 300,000 Hindus and Sikhs, contained less than 10,000 after the transfers had taken place. Longer term, scholars have also argued that the density of settlement forced on the new Indian Punjab by the larger number of people moving into this area than had left it for Pakistan was a major contributory factor to the later international migration of Punjabis to countries such as the United Kingdom from the mid-1950s onwards (Robinson 1986a).

Source: various editions of *The Times*.

became a net importer of labour. The diffusion of post-war economic growth, the rising aspirations of educated European workers and greater social mobility, which allowed more Europeans to access white-collar work, all combined to create a vacuum at the base of the occupational hierarchy that could not be filled by local labour. In response, many European countries adopted positive or permissive policies towards the recruitment of foreign labour.

These policies took different forms. In Germany, the guest worker system was instigated (Chapters 7 and 9), in which young workers from Turkey, Yugoslavia or North Africa were permitted to enter Germany and work for a limited period of time (Castles and Kosack 1973). They were granted few rights and were denied access to welfare support, and they were expected to repatriate when labour demand fell away. By 1973, there were over 2.6 million guest workers in Germany, with concentrations in particular industries, cities and residential quarters (O'Loughlin 1980; O'Loughlin and Glebe 1984). In the United Kingdom, different policies were adopted. The labour shortfall was initially made up from displaced refugees from other parts of Europe but, as this supply dried up, British industry (rather than government) looked to alternative sources. Migrants were recruited from traditional sources, including marginal and less developed parts of Europe such as Ireland (Peach *et al.* 1988) and Italy (Colpi 1991). Employers such as London Transport and the British Hoteliers Association also looked further afield to Commonwealth countries in the Caribbean for new sources of labour. The arrival of the liner *Empire Windrush* in 1948 thus marked the beginning of substantial migration of non-white labour to the United Kingdom, not only from the Caribbean but later also from the Indian subcontinent. As Commonwealth citizens, these workers initially had the right to permanent settlement, political participation and welfare benefits. They also had the right to bring dependants with them. Consequently, by 1971, the ethnic minority population of the United Kingdom had grown to 1.2 million (Robinson 1993a). In the United States, too, economic growth and post-war recovery also demanded extra labour but, in this case, workers were drawn from adjacent labour markets such as Mexico (Cornelius 1991), either as legal migrants or as tolerated illegals (Chapter 7). Legislation was also introduced to abandon the old system of quota immigration and allow labour to enter the United States from a much wider range of sources. Table 1.8 shows how the developed world was increasingly turning to the developing world for labour in the period prior to 1974.

In the wake of economic recession

The 1970s – in particular, the oil 'crisis' of 1973, when the price of oil increased three-fold as a result of concerted action by the oil-producing countries – marked yet another turning point in international

Table 1.8 Average annual size of gross migration to Canada, Australia, the United States, Israel, New Zealand, Belgium, Germany, the Netherlands, Sweden and the United Kingdom, 1960–1974, (000s).

Origin	1960–1964	1965–1969	1970–1974
North Africa and West Asia	97	168	329
South Asia	11	53	62
East and South East Asia	18	65	129
China	6	18	22
Central America and Caribbean	81	135	194
South America	26	31	42
Sub-Saharan Africa	18	45	66

Source: Zlotnik 1996.

Table 1.9 Growth of the ethnic Indian population of England and Wales.

1951	30,800
1961	81,400
1966	223,600
1971	375,000
1976	550,000
1981	676,000
1983–1985	763,000*
1986–1988	786,500*
1991	840,255

Source: Robinson 1996b.
Note: * average values for these years.

migration patterns. The shock wave of oil price rises and underlying economic restructuring in much of the developed world combined to reduce sharply the need for labour, especially in the manufacturing sector. Economic arguments against further labour migration also became entwined with other attitudinal shifts, which reflected growing unease about the presence and growth of communities of 'visible' minorities. Unemployment of indigenous workers became an issue around which racists and bigots coalesced and, in country after country, barriers were erected to further mass labour migration, and existing minority groups were put under pressure to leave. Recently in France, for example, not only were policies introduced to encourage repatriation, but an extreme right-wing political party, LePen's Front Nationale, acquired considerable political support for its simple message that '2 million unemployed = 2 million immigrants too many' (Hainsworth 1992a). In Germany there was a complete halt on further guest worker recruitment (Castles, Booth and Wallace 1984). Despite this, in most European countries, the size of 'foreigner populations' continued to expand, as successful labour migrants were joined by their families – the process of family reunion. Table 1.9 therefore shows how, despite increasing regulation of primary migration from India, the size of the ethnic Indian population in England and Wales has continued to grow through family reunion and associated fertility.

In complete contrast, as the economic fortunes of the oil consumers deteriorated, so those of oil producers improved. This had the effect of spreading economic wealth more widely and ensuring that de-veloping countries also became destinations for new flows of labour migration. As discussed in Chapter 7, Venezuela, for example, began to recruit labour from Spain, Portugal and the neighbouring countries of the Andean Pact. The states of the Persian Gulf became major destinations for workers from other Arab countries and also later from India, Pakistan and Bangladesh. Capital-rich but labour-poor countries, such as Saudi Arabia, Kuwait and the United Arab Emirates, had by 1980 attracted so many immigrants that half their active labour forces were foreigners (Birks, Seccombe and Sinclair 1986), with 2.8 million foreign workers in the ten leading Arabian oil-producing countries. More recently, though, declining oil revenues and a shift to contract labour led to a reduction in migration to the Gulf and a re-orientation towards East Asian sending countries such as Taiwan and Korea.

The 1980s saw both an intensification of trends apparent in the previous decade and the eruption of new patterns and trends associated with the end of the Cold War. Gould and Findlay (1994) and Gould (1994) argue that three macro-level processes have reshaped international migration in this period: first, the collapse of one of the world's two superpowers, the former Soviet Union; second, the globalisation of business activity and production systems; third, an increase in economic inequalities within and between regions, nations and international blocs. In response, developed countries tightened their migration controls and sought to attract only skilled labour. The new international division of labour (Chapter 4) saw manufacturing employment relocating from high to low labour cost countries (Frobel, Heinrichs and Kreye 1980), with many of the 'developing' countries themselves becoming centres of economic growth and immigration.

Channels of migration in the contemporary world

Given the complexity of the global migration system and the speed with which it – and its component parts – change, it is very difficult to summarise all the pertinent patterns and trends in a short account. Nevertheless, Figures 1.11–1.14 show the main channels of migration in the contemporary world. Figure 1.11 shows the major migratory movements worldwide since 1973, while Figures 1.12–1.14 disaggregate these into a series of *regional migration networks* (Salt 1989) or regional theatres. The figures demonstrate that the reach of migratory attraction is now huge, with migrants travelling thousands of kilometres between continents; Castles and Miller (1993) term this the *globalisation of migration.*

As the volume of international migration has risen, so too has the complexity, with a greater variety of population fractions becoming involved for an increasingly diverse set of reasons. Many more countries have become both receivers of and providers of large numbers of migrants in recent years. Nevertheless, key geographical, demographic and legal trends are emerging, which allow us to begin to sketch out how international migration is evolving:

- There has been a decline in the number of people who migrate to another country and are immediately accepted for permanent settlement and given the appropriate citizenship rights. Few countries still give citizenship rights on settlement, and many countries are now revising their entry requirements to favour those with skills or those joining existing primary migrants (Salt 1987).
- Family reunion has come to dominate the immigration patterns of many advanced industrial countries. This has ossified flows of migration, since existing source areas have continued to provide the majority of immigrants. In Australia, the family reunion category now accounts for approximately half of all settler arrivals (Robinson 1993b).
- Many developed countries have sought to limit the total number of immigrants who are either eligible to seek entry or are granted it (Chapter 7).
- Countries are increasingly cooperating in immigration policy and forming themselves into political blocs, within which movement is encouraged, but into which immigration is restricted. Robinson (1996a) described how those countries which make up the Schengen group in the European Union (Belgium, Luxembourg, the Netherlands, France, Germany, Italy, Spain, Portugal and Greece) have a common list of 129 countries from which people are not allowed to travel to the European Union without acquiring a visa in advance. Such restrictions reduce flows of unskilled migrants from developing countries to the European Union.
- Where low-skill labour migration is still necessary, this has been increasingly closely regulated and usually defined as contract migration. Castles (1995) described how many of the so-called Asian Tigers are experiencing shortages of unskilled labour, and are having to turn to contract labour of the type previously used in Germany (the guest worker) or the Gulf (Chapters 7, 8 and 9*)*. This is producing new flows of migration (Figure 1.12) (page 30) and is making the South East Asian regional network of migration both much more complex and more globally significant (Skeldon 1994).
- Where disparities in economic wealth still exist (or are widening), unskilled migration continues to occur (Figure 1.13) (page 31). However, because of the tighter restrictions on legal immigration, this has become increasingly clandestine. Miller (1995) estimated that, in 1990, there may have been as many as 5.5 million illegal aliens in the United States, mostly from Mexico.
- As the emerging world economy continues to see capital and production capacity exported from the advanced industrial countries, so there has been a greater need for the spatial mobility of highly skilled personnel (Chapter 4). In 1988, Japan had 83,000 workers posted to overseas branches of Japanese companies. Salt (1995) has also shown how the circulation of highly skilled labour is not simply taking place between the developed and developing worlds but is also accelerating within the former.
- A growing proportion of movement has become temporary, with businesspeople, tourists and students all becoming more important. Castles and Miller (1993) demonstrated the importance of the latter, noting the 366,000 foreign students in the United States in the late 1980s, half of whom were Asian. Li *et al.* (1995) argued that such educational migrations often link developing with developed countries and are important in deciding subsequent migration strategies.
- As new global economic relations offer greater employment opportunities to women, so mi-

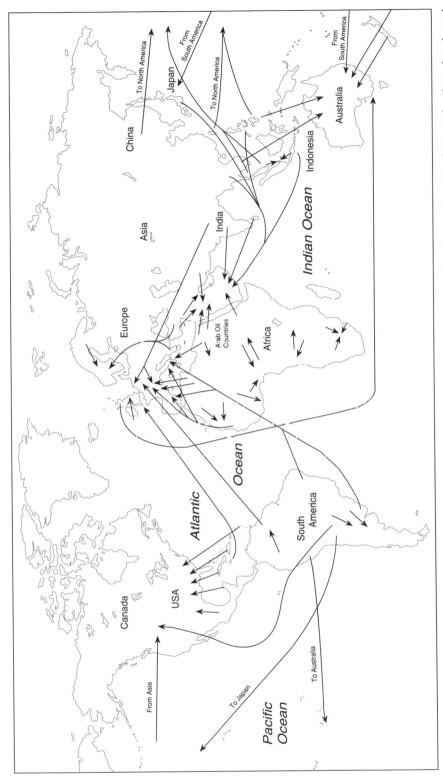

Figure 1.11 Global migratory movements since 1973. Note that the arrows have not been drawn proportional to the size of the flow passing through a channel since precise data rarely exist (source: Castles and Miller 1993, reprinted by permission of Macmillan Press Ltd.).

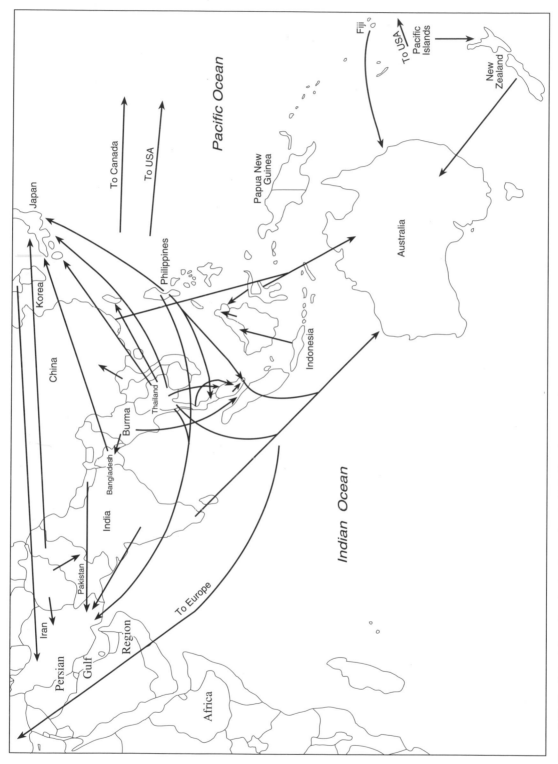

Figure 1.12 The Asian international migration network. Note that the arrows have not been drawn proportional to the size of the flow passing through a channel since precise data rarely exist (source: Castles and Miller 1993, reprinted by permission of Macmillan Press Ltd).

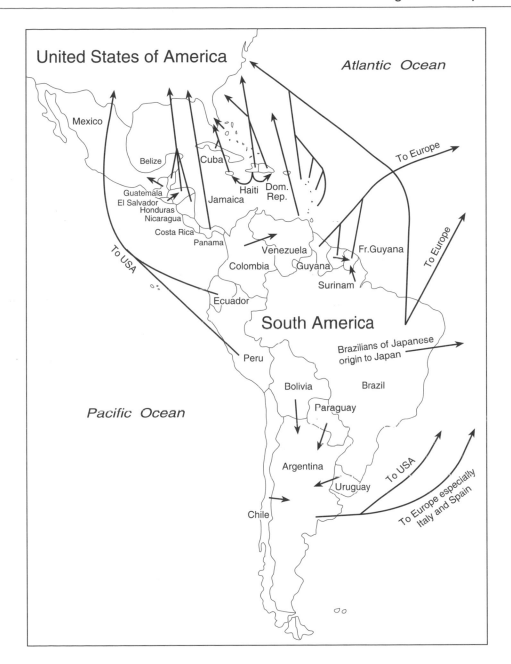

Figure 1.13 The American international migration network. Note that the arrows have not been drawn proportional to the size of the flow passing through a channel since precise data rarely exist (source: Castles and Miller 1993, reprinted by permission of Macmillan Press Ltd).

gration is becoming feminised. In turn, the polarisation of the labour market in global cities within the advanced industrialised countries has created employment niches for immigrant women to either service higher income groups or take de-graded manufacturing work (Chapter 4). Women must increasingly be seen as social actors in their own right, rather than marginal, tied migrants, with nearly half of all international migrants now being women (Campani 1995).

- International migration is increasingly being dominated by forced migration, whether the victims are driven by political, ethnic or religious persecution, or by development projects, environmental change, or state slavery (Chapter 8). Perhaps as many as 20 million of the 100 million international migrants in 1992 were forced migrants. Circumstances often produce 'flash migrations' of large volume but short duration. These are particularly common in Africa, making it a temporally volatile and geographically complex regional migration network containing some 47 percent of all global refugees (Figure 1.14). One example was the flight of one million people from Rwanda into Zaire in only five days in July 1994. While many of the asylum seekers originating in developing countries are simply displaced to adjoining developing countries, others seek entry to

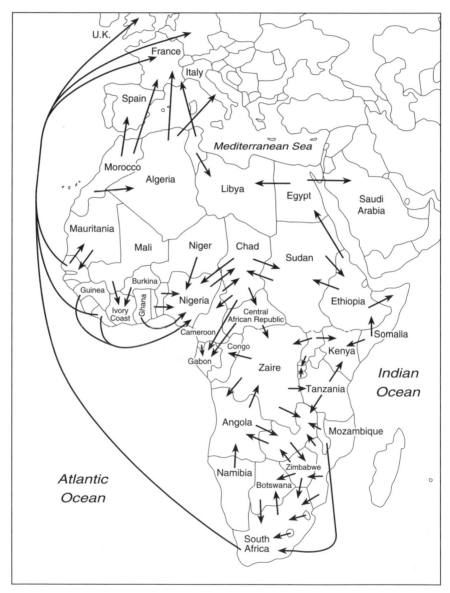

Figure 1.14 The African international migration network. Note that the arrows have not been drawn proportional to the size of the flow passing through a channel since precise data rarely exist (source: Castles and Miller 1993, reprinted by permission of Macmillan Press Ltd).

Table 1.10 Average annual number of asylum seekers in Europe by region of citizenship, 1983–1990 (000s)

Region of citizenship	1983–1984	1985–1986	1990
Sub-Saharan Africa	15.9	39.0	76.7
North Africa and West Asia	12.9	60.3	97.7
South Asia	26.4	58.0	65.9
East and South East Asia	6.0	4.0	13.7
China	0	0.9	1.9
Central America and Caribbean	1.1	1.3	0.9
South America	2.7	5.6	3.5
Total	88.0	240.9	412.9

Source: Zlotnik 1996.

Western Europe, North America or Australia (Table 1.10, above and Figure 1.11, page 29).

- Whereas migration was formerly targeted at a small number of economic 'honeypots', such as the Gulf, Western Europe or North America, increasingly international migration has been directed towards a larger number of alternative destinations. Hugo (1996) showed how migration within Asia increased sharply in the 1980s, as the more economically successful countries traded skilled workers and attracted cheaper workers from their less developed neighbours. Indeed, Asia is set to become the new centre of gravity of global international migration (Figure 1.12) (page 30).

Conclusion

Change is a pivotal theme in the empirical sketches presented in this chapter. Within the developed world, the rapid changes in metropolitan, rural and regional populations have meant that predicting future population redistribution is an extremely challenging task (Lutz 1991). Turnarounds in population trends have been highly unpredictable, and those identified above have usually taken academics by surprise. Change is also very much apparent in the developing world, where there is evidence that the prevailing tendency towards urbanisation is beginning to wane. Even in those cases where urbanisation continues rapidly, many rural–urban migrants circulate between rural and urban areas, and the importance of temporary movement, therefore, cannot be underestimated. In international migration, we have only recently passed a pivotal point in history. We can look back upon the period before the 1980s as an age of simplicity: patterns were dominated by stereotypical and easily understood migrations, such as the male-dominated guest workers, but in the 1990s such simplicity is giving way to a much more complex and varied set of migrations, taking in more countries, more social groups, more causal factors, and more people. Not surprisingly, what may be termed postmodern settings are creating postmodern migrations, which, in turn, require postmodern understandings (Chapter 3).

CHAPTER 2

Defining and measuring migration

Introduction

This chapter defines what the term migration stands for in theory, while emphasising that the practical definition adopted is often highly dependent upon the available data. Additionally, a number of terms that are commonly used in relation to migration are defined. Some influential sources of migration data are outlined and, although many of these sources are common to different countries, British migration sources are focused upon to show how the implied definition of migration and the adequacy of the different data sets varies considerably. The chapter also explains how migration is actually measured, beginning with the most simple descriptive methods and leading on to more advanced analytical methods that help us to explain the migration process. Although most of the methods described are quantitative, the importance of qualitative methods in migration research is illustrated with examples of qualitative methodologies. While certain techniques have become fashionable at certain periods, often matching the changing philosophical direction that geography has taken (Chapter 3), the relative strengths and weaknesses of the various methods need careful consideration. The choice of technique depends crucially upon the aims of the research project in hand.

Defining migration: the component parts

Most people are familiar with the term migration, but it becomes clear that there are definitional problems once a precise description is attempted.

Migration involves the movement of a person (a *migrant*) between two places for a certain period of time. The problem is defining how far someone needs to move and for how long. Migration must be distinguished from *spatial mobility*, which embraces all forms of geographical movement including flows of people over international borders at one extreme and trips to the local corner-shop at the other. Although it is impossible to define migration succinctly, in a way with which all researchers would agree, it is useful to identify some key components of any generally accepted definition. The first of these component parts concerns movement over space.

Migration over space

Migration is usually defined spatially as movement across the boundary of an areal unit. This suffers from the problem that within most sets of areal units (such as the Indian states or the German *Länder*) the size of the individual areas will vary considerably. Consequently, relatively long-distance moves will not be counted in one part of the migration system because they do not cross a boundary, while much shorter moves will be counted in a different part of the system because a boundary is crossed. Additionally, the decision about which areal units are adopted in the study will also influence which moves are identified as migrations. However, definition of the areal units to use is often critical, as population redistribution between these units often has policy repercussions. In Britain, for example, flows between local authority districts influence the funding that they receive from the government, as this funding is based partly on the total population who reside within their borders.

Where a boundary within a country (these are often defined for political reasons) is crossed, this is described as *internal migration*. Those moving into a particular areal unit are termed *in-migrants*, while those moving out are *out-migrants*. The total number of in-migrants to a particular area from all of the potential origins is the gross number of in-migrants, while the total number of out-migrants from an origin area to all of the potential destinations is the gross number of out-migrants. The difference between the size of the two groups is the resulting *net migration*. Those people who move between the same pair of origin and destination zones are usually referred to as individuals within a *migration stream*. Shorter-distance changes of residence that do not involve the crossing of a defined boundary are more generally termed residential mobility and, within urban studies, it is common to distinguish between inter-urban migration and intra-urban residential mobility (Roseman 1977). Finally, flows across national borders, such as between France and Spain, are described as *international migrations*. The movers into a country are *immigrants*, while the people moving out are *emigrants*. Again, we should not forget that this spatial definition is not directly related to the distance moved. A migrant from France to Spain may move only a few kilometres across the border, while an internal migration within France may involve a move over many hundreds of kilometres.

Migration over time

Migration is also defined temporally, as it is generally accepted that there will be some permanence to a move described as a migration. Households from Washington, DC who take short vacations in Miami over the spring break are clearly not migrants. Conversely, a couple who move from university in Manchester to employment in London and remain there for a period of some years certainly are migrants. Agreeing how long this time period should be for a migration to take place can be difficult. A Turk working in Germany during the winter but returning to Turkey in the summer may be more appropriately thought of as a temporary migrant or seasonal worker. An individual who regularly moves from Maidenhead in England to San Francisco for three-month periods associated with work may be regarded as a business visitor, while African households who more or less constantly move around within a broadly defined area are regarded as *nomads* rather than migrants. The last two examples are more commonly referred to as *circulation*, as the moves are not expected to be permanent and they are repetitive, involving the same, or similar, origins and destinations (Zelinsky 1971). Likewise, Gould and Prothero (1975) distinguished between three types of movement: daily, periodic and seasonal. These terms, which all relate to the permanence of the act of moving, demonstrate the difficulty of separating people into the dichotomous categories of migrant and non-migrant.

Cross-sectional studies focus on a migrant or a migration stream during one particular period in time. In contrast, other migrant analysts are concerned with *longitudinal* migration patterns. Such studies focus on the longer-term migration history of one or more individuals. For example, Cornelius (1991) traced Mexicans moving from their Mexican birthplace to Mexico City and then on to California.

It is common to refer to those who leave a particular place for a reasonable period of time, only to return at some later period, as *return migrants*. One example of this is the young adult who leaves home in search of employment but returns at a later date, either after failing to find work or after having earned sufficient money to be able to afford to return home. A second example are the retirees choosing to return to their place of birth after working elsewhere. The return may, or may not, be premeditated; in the first example, it is a response to unforseen circumstances, while in the second it may well be a carefully planned longer-term intention. Elridge (1965) distinguished between primary, secondary and return migrants based on the relationship between the migration event and the region of birth. Primary migrants are those who leave this region for the first time; secondary migrants are neither moving into nor out of their region of birth; and return migrants are those who are returning to their region of birth.

Much of the literature discussing return migration focuses on those who move from developing to developed countries for employment reasons (King 1986). These migrants move in order to accumulate savings, often so that they can invest in property or a business when they return home. However, for many, this intention to return may dwindle even though many of these individuals claim that they still intend to move home at a later stage. This failure to complete the intended migration route is referred to as the 'myth of return'.

Another temporal issue is that of *cumulative inertia*, first suggested by McGinnis (1968). Many argue that the longer an individual remains in a par-

When investigating return migration in the United States, identifying return migrants is problematic but they can be defined as inter-state migrants returning to their state of birth. Given the size of the American states, many of these migrants may not have been returning to their specific birthplace but, even so, the results are illuminating. The census question on migration asked respondents to provide their residential location five years previously, allowing return migrants to be compared with those moving away from their birth state (primary migrants) and those moving between two states, neither of which was their birth state (secondary, or onward, migrants). Between the late 1950s and late 1970s primary migrants represented a declining percentage of inter-state migrants, while return and onward migrants increased. Those living outside their state of birth were only twice as likely to move on to another state as back to the one they were born in. This indicates the importance of return migration within America:

If persons living outside their state of birth picked their next state of residence at random, the rate of onward migration would be 49 times the rate of return since there are 49 possible destinations.

(Long 1988: 105–106).

Inter-state migration in the United States

	Total inter-state migration	Primary migrants	Return migrants	Onward migrants
			Percentages	
1955–1960	100	49.9	17.1	33.0
1975–1980	100	42.0	19.2	38.8
	Migrants per 100 persons at risk (age standardised)[1]			
1955–1960	10.3	6.9	6.3	12.0
1975–1980	9.9	5.9	6.2	12.6

[1] Age-specific rates calculated from 1975–1980 base populations.

Source: Long 1988.

ticular location, the less likely that person is to migrate again (Goldstein 1954; Clark and Huff 1977; Huff and Clark 1978) because complex social ties develop over time. An alternative view is to recognise that the population is not a homogeneous migrant group, with much migration involving similar groups of individuals. Different sub-groups of the population have different migration propensities and there is a relatively small group who continue to move frequently (movers) and a larger group who rarely move (stayers). Regions with high proportions of stayers may appear to have numerous people experiencing cumulative inertia effects but this is misleading. Instead, movers and stayers have different migration probabilities, which do not alter considerably over time (Davies and Pickles 1983, 1985b; Pickles, Davies and Crouchley 1982).

One migrant stream may have an impact upon a second stream. The migration of an initial stream of people often encourages the migration of a second group; the innovators may be followed by family or friends at a different time, for example. This process is referred to as *chain migration*. Although this term is generally used in connection with groups of migrants, it can also be applied appropriately to individual migrants.

Migration and spatial networks

Defining migration based on spatial and temporal factors does have the drawback that the 'significance' of the move is usually determined by the distance and permanence of the move. Both of these assumptions treat the migrants' origin and destination as spatially fixed in space and time, such that, at any one time, a person occupies a single residence exclusively (Behr and Gober 1982). In certain types of migratory behaviour, such as the cyclical migration of elderly retirees, multiple residences or mobile homes are commonly used and more traditional definitions of migration would overlook these activities. Consequently, Behr and Gober (*ibid.*) suggest an alternative *spatial network* framework that traces individuals through time and space and, rather than asking people where they live, they are asked to identify places where the majority of their time is spent on a daily, monthly, seasonal and yearly basis. McHugh, Hogan and Happel (1995) integrate this perspective into a life-course model (Chapter 5) and argue that recurrent mobility between multiple residences is becoming increasingly prevalent for individuals at various stages in their lives (Figure 2.1) (page 38). This obviously blurs the distinction between migration and other forms of spatial mobility, which can be useful for many migration studies as it can be argued that 'there are no absolutely clean "breaks", natural or man-made, in

Box 2.2 Vacancy chains

The international migration of highly skilled professional and technical personnel highlights the importance of transfers within the same organisation's internal labour market (ILM). As many as 86 percent of the 8,754 long-term work permits issued in the United Kingdom in 1985 were given to professional, managerial and associated groups. Of these, 4,999, or 57 percent, were intercompany transfers. Although most of the moves within ILMs occur within the same country, complex vacancy chains, or 'chains of opportunity' (White 1970), may be identified; the chain identified here involved five countries and took eight months to complete. A new office opened in London and the post suited the skills of an employee in Milan due for promotion. The space in Milan was filled by an employee from Taipei who needed experience of European company operations, while the Taipei post suited a promotional position for an employee in Dubai. The move from Bahrain, where an office closed, to Dubai was suitable for an employee who needed more experience in the same region. This chain may have appeared to have been started by the office closure in Bahrain or the opening of the London office but, in fact, the promotion of the employee in Milan was the driving force behind this complex chain.

Source: Salt 1988.

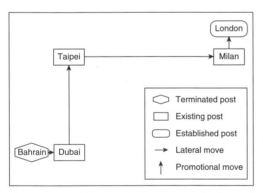

Movements within a single company's internal labour market (source: Salt 1988).

the spatio-temporal spectrum of mobility' (Zelinsky 1983: 36). Thus, an extreme example involves the transient homeless, for whom various locations, such as treatment centres or jails, act as centres of activity, rather than single residential locations (Spradley 1970). Even so, it is generally rare for this approach to be adopted, except in studies where the migrants involved are known to occupy multiple residences or activity centres (Chapter 9).

Migration and culture

Many have argued that migration, in contrast to mobility, involves a social or *cultural* change in the life of the migrant. For Bogue (1959: 489), migration describes moves that 'involve a complete change and readjustment of the community affiliations of the individual'. This behavioural perspective is useful if we are interested in identifying those people whose circumstances are considerably altered by the act of migrating. As Bottomley (1992) emphasises, the issue of cultural change is particularly relevant when dealing with ethnic minority migration. However, Fielding (1992a) argues more generally that migration is an important cultural event, distinguishing between migration that is 'exciting and challenging' and that which is 'rootless and sad' (Chapter 9). A less pessimistic distinction separates *innovative* and *conservative* migration; the former fits well with Fielding's first category, while the latter involves migration that is motivated by an aim to 'move geographically in order to remain where they are in all other respects' (Peterson 1958: 256).

The significance of migration for all those involved is enhanced considerably when it is regarded as a cultural experience rather than as a purely spatial or temporal event. Chambers (1994) highlights this emphasis by demonstrating the cultural importance that migration brings to bear on the attitudes and mores of contemporary society. He observes how:

migration and attempted relocation is a fragment, invariably caught in a press photo, on the news, in a television documentary, in immigration statistics, that nevertheless illuminates much of the landscape we inhabit.

(*ibid.* 1–2).

As state and cultural boundaries are crossed ever more frequently – as migration, mobility and travel blur – Chambers defines *migrancy* as an experience of constant change, as population movement where the points of departure and arrival are uncertain.

Migration and motivation

A distinction can be drawn between *forced* and *voluntary* migration. Forced migrants, such as *refugees*, are migrants who have little choice but to leave their

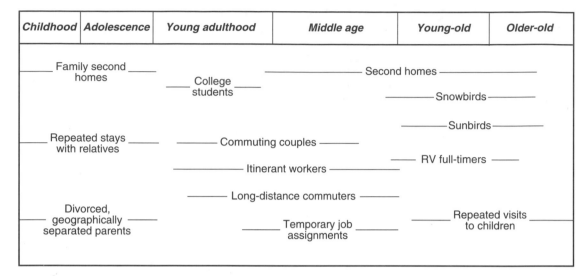

Childhood	Adolescence	Young adulthood	Middle age	Young-old	Older-old

Family second homes
College students
Second homes
Snowbirds
Sunbirds
Repeated stays with relatives
Commuting couples
RV full-timers
Itinerant workers
Long-distance commuters
Divorced, geographically separated parents
Temporary job assignments
Repeated visits to children

Figure 2.1 Multiple residences by life-course phase (source: McHugh, Hogan and Happel 1995).
Note: RVs are recreational vehicles or camper vans.

homes because of persecution, war or famine (Chapter 8). On the other hand, voluntary migrants are those who choose to move. Nonetheless, we should use these terms carefully, as few people move purely as a result of their own deliberations and an element of free will is apparent in many forced moves. The distinction between forced and voluntary should be seen as signifying relative levels of freedom, set in the context of personal characteristics and of the society in which the migrant lives.

In terms of labour migration, *speculative* migration and *contracted* migration can be distinguished (Silvers 1977). The former involves those who move with the intention of finding work at another place. Flowerdew (1992) argues that this type of migration is rare, despite the fact that neoclassical theories of labour migration (Chapter 3) assume that people move regularly to equalise imbalances in (un)employment or wage rates. In fact, a substantial proportion of labour migration is contracted, whether it involves moves within the internal labour market (ILM) of organisations or between organisations when individuals take up new jobs (Chapter 4).

Thus, although it appears simple at first to explain what the term migration refers to, various qualifications need to be considered when defining who migrants are and, once defined, distinguishing between the various categories of migrant. How far have they moved? How long did they move for? Are the social and cultural circumstances of the origins and desti-

nations considerably different? Do the movers believe themselves to be migrants? Was the migration forced or was the move essentially voluntary? Given this wide variety of factors, the definition used must be stated explicitly. The choice of this definition is not always determined by the researcher, as they may be forced to use data sets where the definition of migration has already been made for them.

Data sources

The data used in migration studies have often been collected such that deliberation over definitions is largely irrelevant. Here, primary and secondary migration data sources are considered and, although examples are drawn from various nations, special attention is afforded to British data sources to indicate the variety that are available in one particular place.

National censuses

Much migration research relies on data collected from surveys, and the most comprehensive and well-utilised of these are national censuses. The first population counts were carried out in Babylon, China and Egypt between 2800 and 2200 BC, but these were restricted to population sub-groups; women and children were rarely included (Brown

1976). Strictly speaking, these were not censuses according to the United Nations definition:

A census of population may be defined as the total process of collecting, compiling and publishing demographic, economic and social data pertaining, at a specified time or times, to all persons in a country or delimited territory.

(quoted in Brown 1976: 17).

It is only within the last 200 years that censuses have been carried out that accord with this definition.

Censuses are usually carried out decennially (every ten years) or quinquennially (every five years). In Britain the first census was held in 1801, while in the United States the first was in 1790. In other countries, such as Germany and the Netherlands, conventional censuses are not held and register-based sources are used to collate information about the resident populations (Langevin and Begeot 1992).

Those countries that do maintain conventional censuses do not necessarily carry them out at similar time intervals, and the comparability between questions is often poor (Redfern 1981). Some censuses ask for the respondent's place of residence at a previous fixed point in time (usually one or five years), although in some censuses (such as those in Australia, New Zealand and South Korea), the place of residence is requested for both points in time. Alternatively, the respondent may be asked where he or she lived prior to their present address, usually supplemented with a question on the duration of residence at the present address, allowing the precise time of the move to be determined. Having to remember when the move was made may introduce bias into the data, especially when 'Western methods of reckoning time are alien to the population under study' (Skeldon 1990: 18). Despite the early recognition by the International Statistical Institute in 1872 that international comparability would be advantageous, this has yet to be achieved satisfactorily (Dale 1993). There have been some advances along these lines, and work by the Statistical Office of the European Community (Eurostat) has succeeded in synchronising the timing and coverage of questions in some European states to varying degrees.

In recent British censuses migrants have been defined as those whose address on census night was different to their address one year previously. If the head of household (the first adult recorded on the census form) was a migrant, the household would be defined as a 'migrant household'; if the entire household at the destination moved as a single group they would be defined as a 'wholly moving household'. This definition is temporally rather than spatially based, as people who moved only short distances, perhaps along the same street, would be identified as migrants. It also ignores the fact that some migrants would have moved more than once during the one-year period. There may even have been a small number of return migrants who moved during the one-year period but resided at the same address at the beginning and end of the period. The question also relies on people remembering where they were living one year previously, which may have been difficult if the individual moved often. Finally, those migrants who were under the age of one on census night are inevitably ignored.

Further problems with the British census include the fact that it is only enumerated every ten years (with the exception of the 1966 census) and, because of the time required to produce the output, there are inevitably delays in data provision. These delays are most serious for the migration (and journey-to-work) data, where the origins and destinations need to be linked together. Also, despite the legal requirement to complete a form some people will inevitably be missed – approximately one million people from the 1991 census returns (Simpson and Dorling 1994). Omissions cause problems, not least because they are not randomly distributed either socially or spatially. Indeed, we might expect many young adults, who are typically the most mobile age group, to be most likely to have been missed, perhaps as a direct result of moving home at the time.

In spite of the problems experienced by any census, they provide an extremely useful migration data source. No other data source provides such a comprehensive coverage of the national population. In Britain, the resulting migration data are provided in a series of formats at a range of different scales. Although all moves are recorded, the output usually refers to moves that are between or within certain administrative boundaries (Figure 2.2) (overleaf). At the finest spatial scale (enumeration districts in England and Wales, postcode sectors in Scotland) the data refer to the migrants resident within the area. Flows between or within these small areal units are not produced. However, as noted above, the fact that a person has moved between areal units does not mean that she or he has moved further than someone who moved within an areal unit. This issue, and the 'multiple migrant' problem that results from the definition of migration used in the census, are demonstrated in Figure 2.3 (overleaf).

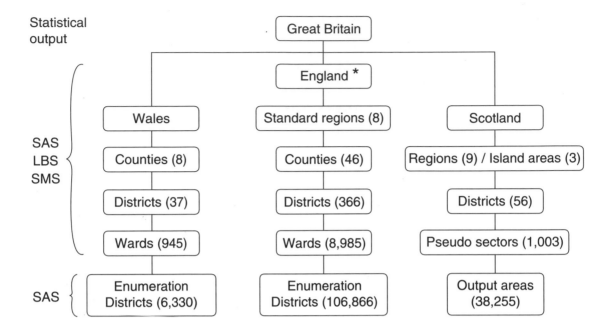

Statistical
output

* Note that politically England, Scotland and Wales are equivalent, but in most research exercises
  Scotland and Wales are treated alongside the standard regions.

Figure 2.2 Selected British census output areas (source: Denham 1993). © Crown copyright.

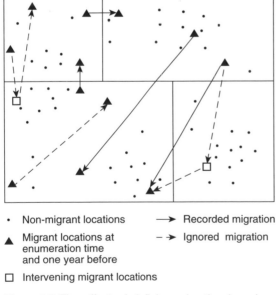

•   Non-migrant locations   → Recorded migration

▲   Migrant locations at   -→ Ignored migration
    enumeration time
    and one year before

□   Intervening migrant locations

Figure 2.3 The effect of defining migration based on
crossing areal units and recording moves at two points in
time.

In the British census, migrants are also disaggregated by age, sex, economic activity, occupation and many other variables. Much of the migration output is published in County Reports, but information is available at finer spatial scales in the Small Area Statistics (SAS) and the Local Base Statistics (LBS), both of which are held in computerised form. Even more useful for certain types of migration modelling are the Special Migration Statistics (SMS), which include inter- and intra-ward flows of migrants disaggregated by age and sex, and inter- and intra-district flows disaggregated by other socio-economic variables (Rees and Duke-Williams 1995). It is therefore possible to extract large matrices of migrant flows. Boyle (1995b) used the inter-district flows to examine migration into the remote Highland and Islands area of Scotland, demonstrating that this area continues to attract unusually high numbers of migrants from middle-class areas in southern England.

Longitudinal data on British migration can be derived from the Longitudinal Study (LS), which links together the 1971, 1981 and 1991 census records of approximately 500,000 individuals born on four secret dates in the year. For each individual, we have address information a year before each of the three censuses and information on their whereabouts at the time of each census (including the 1966 mid-term census because of the five-year migration question in the 1971 census). Not only does the source include information derived from the census but it also incorporates vital event information from other sources, such as births, deaths and cancer registration. Census and vital registration data can also be linked in other countries, such as Denmark, Finland, France and Norway (Dale 1993). Various authors have used the LS to relate migration to the changing socio-economic characteristics of the individuals over time. For example, Fielding (1989) analyses the relationship between occupational and geographical mobility between 1971 and 1981. However, we should not forget that the LS provides a very sporadic view of the migration histories of those individuals who are included.

One disadvantage of standard census data is that they are provided at the aggregate level and migrants are usually only disaggregated into groups based on one or two variables, such as age and sex. For the first time in the 1991 British census, individual data were provided in the Sample of Anonymised Records (SAR); for each individual we have information on every census variable, allowing much greater depth of analysis. A 2 percent sample of individuals in households or communal establishments and a one percent hierarchical sample of households were extracted from the census. The individual file is the most geographically detailed, providing records for 278 SAR areas, each of which has at least 120,000 people. Migrants therefore reside in one of 278 destinations and the distance that they moved is recorded, but unfortunately their origins are provided only at the regional level, of which there are ten. Although Britain is ahead of the rest of Europe in providing individual-level census output, Canada released similar data after its 1970 census and Australia after its 1980 census. In the United States the Public Use Microdata Samples (PUMS) have been available routinely since the 1960 census, providing a 5 percent sample of the resident population (Fotheringham and Pellegrini 1996). Cohen and Tyree (1994) extracted data from the PUMS on the 3,513 Israel-born immigrants in the sample and, using the ancestry and language questions, identified Arabs and Jews. While the Jews were more educated, and had better jobs and higher incomes, the differences between the two groups were not as extreme as they had been in Israel.

Population registers

The most detailed records of migration can be derived from national population registers, although, in most cases, little information about the migrant characteristics is available from these sources. They are maintained in most Eastern European states, Scandinavia, Germany, the Netherlands and Japan. They provide a continuously updated and reliable record of migration because people are required to re-register after making a residential move. These registers are maintained by governments as tools for policy planning, or for keeping track of people's movements within the state. They also play a valuable role in the production of demographic statistics. Ishikawa (1987) used population register data to analyse the 1960 and 1980 migration patterns between the 46 Japanese prefectures. The register has been updated continually since 1954, providing a highly reliable migration data source, and this allowed a detailed comparison to be made between migration in these two periods.

Medical records

One source of British migration data is derived from individuals' medical records. Most people in Britain are registered with a medical doctor known as a general practitioner (GP). Family health service authorities (FHSAs) administer the payment of GPs and this is based, almost entirely, upon the number of patients that are registered with them. There are 98 FHSAs in Britain, and their boundaries generally match the county and metropolitan district boundaries. When a patient migrates between FHSAs and registers with a new GP, their medical records are passed between the origin and destination FHSAs through the National Health Service Central Register (NHSCR), where a record of the move is made. From this source, it is possible to derive a continuous series of inter-FHSA migration data, broken down by age and sex. Considerable work has utilised migration data from the NHSCR, which is especially useful for analysing macro-level population redistribution (Stillwell, Rees and Boden 1992b).

Work is also underway to analyse the potential of using the records held by each individual FHSA (Shen, Boyle and Flowerdew 1994) to analyse local-level movement. Fortunately, historical records are kept of patients who change address within FHSAs and, theoretically, it is possible to derive long-term migration records for patients who have moved around within a single FHSA. Unfortunately, it is not compulsory for people to register with a GP, even though most of the population do so (Bone 1984). It is also difficult to predict the lag between a patient migrating and then attending a new GP practice. Young male adults are least likely to re-register quickly, as they rarely use the GP services, while others, such as the elderly or families with young children, are likely to re-register rapidly.

Other secondary data sources

Other national surveys, usually based on samples of the total population, are commonplace. In Britain, the General Household Survey (GHS) involves a sample of approximately 20,000 individuals. Such surveys have some benefits over the census. First, they are carried out regularly, providing more detail about changes over time; the GHS is carried out annually. Second, while most questions remain the same over many years, additional questions, specific to a particular research problem, may be included. Questions about the migration characteristics of people living in property bought from the government's public housing stock were added to the GHS in 1991, information that could not be derived from sources such as the census.

A range of other migration data sources also exist, but these are less commonly utilised in research, perhaps because of their reliability or coverage, or because they fail to capture a reasonable sample of the underlying population. These include housing registers, school records, or more obscure sources such as telephone directories or credit card companies (Bulusu 1991). Even electoral registers, which might be expected to be quite reliable indicators of population movement, are rarely entirely accurate. Smith (1993) showed that, in 1991, 7 percent of the eligible Britons who completed a census form were not recorded on the electoral register and that geographical variations in coverage were wide, with up to 20 percent being unrecorded in inner London. Indeed, difficulty in registering means that some of the most migratory individuals are frequently left off the electoral register (Halfacree 1992).

Different sources are available in different countries. In the Netherlands migration data can be derived from the CBS Housing Demand Surveys, which include approximately 55,000 respondents aged 18

and over who are not resident in institutions. Mulder and Hooimeijer (1994) used data from 1977, 1981, 1985 and 1989 to examine those people moving into owner-occupied housing for the first time, demonstrating that, after controlling for the age and life-course situations of the respondents, there has been an increasing propensity to enter owner occupation. University enrolment data are another potential source, which appears to have been used only rarely. Ishikawa (1987) illustrates the use of these data in Japan and, although these are only useful for examining student moves, they are a relatively reliable source of migration information for this group. In the United States, the Bureau of the Census has developed an annual record of inter-state flows based on federal income tax returns provided by the Internal Revenue Service (IRS) and it is thought that approximately 90 percent of the total population is captured in these records (Engels and Healy 1981). Barff (1990) used these data to investigate the migration response to the economic boom in New England during the 1980s. These time-series data allowed detailed migration responses to economic change to be analysed, demonstrating the utility of these data beyond census data in certain circumstances.

International migration data sources

Most of the sources described so far have been concerned with migration within countries, rather than international migration. Of course, a great deal of international migration is illegal and consequently goes unrecorded. To complicate matters further, the estimates of these flows usually vary substantially between the host and the receiving countries. In the United States, 17 million Mexicans were arrested and expelled between 1964 and 1989, but figures on the overall numbers of illegal migrants are hotly contested. Durand and Massey (1992) distinguish between *speculative* and *analytical* methods for producing estimates, the former being based on opinion or conjecture and the latter on quantitative procedures with verifiable assumptions. Speculative estimates are often far larger and attract much more attention in the press. Chapman (1976) estimated that 10 percent (6 million) of Mexico's entire population were living in the United States illegally, while analytical analyses suggested that the maximum possible number was 3.8 million in 1980 and a more realistic number would have been between 1.5 million and 2.5 million (Bean, King and Passel 1983).

More information is available about legal immigrants but it is also far from perfect. The British government issues work permits to those people who enter Britain for work, and this is a relatively reliable source of information on those moving in for employment. Acquiring information about the destinations of emigrants is far more difficult. Other sources on employees include the records held by recruitment agencies, some of which specialise in international recruitment, or the personnel records of transnational companies (TNCs). Salt (1988) examined the records of Honeywell, which has operations in 29 countries outside the United States, and showed the considerable extent of international movement among its employees.

Perhaps the most utilised source of information on international migration in Britain is the International Passenger Survey (IPS). This is a stratified random sample survey carried out at ports and airports. It provides coverage of all international routes, except those to the Republic of Ireland, but it is based on a small sample. Approximately 250,000 interviews are conducted annually, under 1 percent of which are emigrants or immigrants, based on the definition that they intend to leave or stay in the country for at least a year, having been present or absent for at least a year (Coleman and Salt 1992). Findlay (1988) notes that in 1985 the estimate that 108,495 Britons had emigrated was based on a sample of 578 respondents!

Reliable information about refugees is also difficult to obtain, because of the rapidly changing numbers involved and the limited resources available to document them. The United Nations High Commission for Refugees (UNHCR) stated that there were 18.2 million refugees worldwide in 1993 (UNHCR 1993). Annual data may be extracted from the *World Refugee Survey*, but it is acknowledged that one country's asylum seeker is another country's illegal immigrant and consequently these data must be treated with care.

Primary data collection

A problem with the migration data sources identified above is that they rarely provide information about the motivations of the migrants involved; why do different population sub-groups decide to move or stay? The most obvious way of collecting this type of data is to use specially designed surveys, although these inevitably provide smaller samples than most secondary data sources. In Britain, the Office of Population Censuses and Surveys (1983) carried out a large survey of housewives who had recently moved, focusing on movement between different housing tenures. This type of information is not available from standard sources, such as censuses, as from these we know the tenure characteristics only at the destination.

A typical example of survey work is the study of Williams and Sofranko (1979) on the motivations of in-migrants to non-metropolitan areas of the Midwest United States. The authors surveyed 708 households, asking them for their reasons for leaving their former residence and their reasons for choosing their current residence. These various reasons were placed into one of six categories (Table 2.1). Note the distinction drawn between 'push' and 'pull' factors (Chapter 3). The authors were able to conclude that quality of life motivations (Chapter 6) were more important in driving migration, especially for those who originated in metropolitan areas, than was suggested by the conventional economic emphasis in migration research (Chapter 4). Elsewhere, Halfacree, Flowerdew and Johnson (1992) present the results from Gallup surveys carried out in 1990 and 1991, which captured a total of 18,010 respon-

Table 2.1 A six-fold classification of reasons for moving.

Factors	Examples
Employment-related	Job transfers, moves on unemployment or underemployment, search for a better job.
Ties to destination area	Desire to return to area of birth or former residence, links in destination with family or friends.
Environmental 'push' factors	Negative attributes of origin area, whether general or specific.
Environmental 'pull' factors	Attribution of positive features to destination area.
Retirement	
Other reasons	Health, divorce, education, restlessness.

Source: Williams and Sofranko 1979.

dents. The migration questions were added to surveys that were already running, and consequently the costs of the venture were reduced. They provided information not only on the characteristics of migrants but also about the decision-making process that influenced the moves.

Primary surveys have been heavily used in the developing world, where standard data sources may be absent or their reliability may be suspect. Bascom (1994) considered Eritrean refugee repatriation from eastern Sudan using a field survey of 131 households resident in a United Nations refugee settlement adjacent to Wad el Hileau. These interviews were followed up in 1992–1993, when 35 of these households were interviewed for a second time. The repatriation decisions made by the refugees were determined, to a great extent, by the pre-flight social history of the refugees, but they were also influenced by the impacts of transformations in their social and cultural practices during the period of exile. These details about the migration process would be missed in conventional published migration data sources.

Other types of primary migration information may also be derived from migrant diaries, which respondents are asked to complete. Alternatively, secondary source material may also be gleaned from diaries kept by individuals for personal reasons, but which contain details about residential histories over time.

Qualitative data sources

Artistic sources of information are becoming increasingly important in migration research (King, Connell and White 1995), since the cultural significance of migration is apparent in literature, painting, music and other artistic media. Artistic sources express well the qualitative experience of migration rather than its statistical expression (Chapter 9). For example, Chandler (1989) uses literary sources for evidence of the importance of suburbanisation (Chapter 1) for the American population. This burgeoning body of writing dates back to the 1950s and is described as a 'subgenre' or 'a literature of the suburbs'. In the three examples Chandler chooses, suburbia is depicted as a place that is isolating and fragmenting as the influx of new people serves to repress the host communities. Thus, in Munro's *The Shining Houses* (1968), Mrs Fullerton is an old lady living in a ramshackle farmhouse that becomes surrounded by suburban dwellings. The attempts by the young suburbanites to move her on are used to high-

light some of the detrimental effects that suburbanisation has had on many rural communities. However, the experience of moving to and living in the suburbs has been debated hotly. In contrast to Chandler, Gerster criticises Australian fiction writers who have indulged in 'literary gerrymandering' by ignoring the importance of suburbia, which is home for the majority of Australians:

In fiction, suburbia is not only attacked by the pedlars of the bush mythology, it is habitually dismissed with cosmopolitan contempt by urban-orientated writers as a place fit solely for satire, if indeed it is a place worth writing about at all.

(1992: 19).

These sources of information provide valuable indications of the way in which migration processes are acknowledged and represented in society that might be difficult to glean from more traditional data sources.

Analysing migration quantitatively

Gross migration

A wide range of techniques have been used to analyse migration quantitatively and some of the more simple are discussed here. *Gross in-migration* (IM_j) is the sum of the people moving into place j from all of the other places in the migration system ($m_{1j}, m_{2j}...m_{nj}$), over a specified time period:

$$IM_j = \sum_{i \neq j} m_{ij}$$

Similarly, *gross out-migration* (OM_j) is the sum of the migration from j to all other places during this period:

$$OM_j = \sum_{i \neq j} m_{ji}$$

Net migration (NM_j) is the gross in-migration minus the gross out-migration:

$$NM_j = IM_j - OM_j$$

Migration rates

For many purposes it is useful to compare the relative in- and out-migration for different places, or for the same place over time. To facilitate such comparison, *in-migration rates* (IMR_j) and *out-migration rates* (OMR_j) are standardised by dividing the mi-

---

Box 2.3 Population change

The migration element of the population equation is the most difficult to estimate. Although we usually have accurate and relatively up-to-date records about births and deaths, it is more difficult to acquire migration data that is not outdated or incorrect. This problem, which was described in the previous section, is especially important in the developed world, because in nearly all areas population change can be attributed primarily to net in- or out-migration, as the spatial variations in births and deaths are generally slight. Population change in a particular area can be represented by a simple equation where the difference between one time period (*t*) and another (*t*+1), can be attributed to the births, deaths and net in- or out-migration:

$$P_{t+1} = P_t + B_{(t,t+1)} - D_{(t,t+1)} + IM_{(t,t+1)} - OM_{(t,t+1)}$$

Here, P_t is the population at time *t*, and *B* and *D* refer to the births and deaths, respectively.

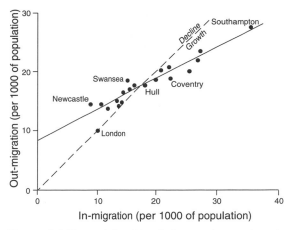

Figure 2.4 The relationship between in- and out-migration (source: Cordey-Hayes 1975, 'Migration and the dynamics of multiregional population systems', *Environment and Planning* A, **7**: 793–814. Pion Ltd London. Reproduced by permission).

gration totals by the population of the place (P_j). Thus, for in-migration:

$$IMR_j = (IM_j / P_j) \times 1,000$$

For out-migration:

$$OMR_j = (OM_j / P_j) \times 1,000$$

One difficulty with this measure is that the best 'at-risk' population to use is the population of place *j* in the middle of the specified time period. Hence, for calculating in- and out-migration rates using census data where the migration question refers to those who have moved in the previous year, an estimate of the population at the mid-point of that year is required. In practice, because of data constraints, the at-risk population is usually defined as the population in *j* at the end of the period. Also, while the at-risk population for the out-migration rate is conceptually reasonable, it is unrealistic for the in-migration rate because the denominator population includes those people already resident in *j* – they could not be in-migrants therefore. However, in most studies this problem is ignored.

Intuitively, we might expect that if the in-migration rate in a region is high, the out-migration should be low and *vice versa*. This assumes that the region is attractive, perhaps because the employment opportunities are good, while other regions are less attractive. However, using correlation coeffi-

cients, Cordey-Hayes (1975) showed that the relationship between in- and out-migration rates for city regions in England and Wales was positive when measured (Figure 2.4). One reason for this is that where the economic conditions are favourable people are both more willing, and more likely, to change jobs. Thus, places that are economically buoyant may both attract and lose migrants, who are moving from positions of strength.

Net migration rates (NMR_j) are calculated as the difference between the gross in-migration and gross out-migration rates:

$$NMR_j = IMR_j - OMR_j$$

These rates may be either positive or negative. Net migration rates have been extensively used to analyse patterns of population redistribution. For example, Champion (1987) uses such rates to examine counterurbanisation tendencies in Britain.

Demographic effectiveness

A measure that has been argued to be more useful than net migration rates is *demographic effectiveness* (E_j) (Thomas 1941). Unlike net migration rates, the measure is not based on the region's population size, as it is calculated entirely from migration data by dividing the net migration to region *j* by the total migration into and out of *j*, all expressed as a percentage:

$$E_j = 100(NM_j / TM_j)$$

Demographic effectiveness measures the extent to which migration is 'one-way' relative to the size of the flows. Plane (1994) also used a *system effectiveness (E)* measure, which summarises the effectiveness in the entire migration system, rather than being specific to a particular place:

$$E_j = 100 \sum_{j=1}^{n} |NM_j| / \sum_{j=1}^{n} TM_j$$

A *stream effectiveness* measure, which relates to each of the individual flows in a migration matrix, can also be calculated (Plane and Rogerson 1994). Such measures are useful for comparing different sub-groups or the same sub-group over a period of time. Boyle and Halfacree (1995) used these measures to contrast the redistribution patterns of service-class males and females between non-contiguous metropolitan counties and metropolitan remainders in England and Wales. Male migrants were shown to be more dispersed throughout England and Wales, while the redistribution of female migrants was more geographically focused, especially towards London (Chapter 4).

Gross migraproduction rates

A useful measure that can be calculated for different time periods or spatial scales of migration (Bonaguidi 1990) is the *gross migraproduction rate (GMR)* (Willekens and Rogers 1978). This measures the expected number of moves that an individual will make over his or her lifetime, assuming that they are exposed to the observed age-specific migration rates and that they survive to the oldest age group:

$$GMR = 5 \sum_{a=0} M_a$$

In this example, five-year age groups are used and M_a is the mobility rate of the age group a.

Gravity models

Defining the model

Considerable research has modelled aggregate migration flows across space. The *gravity model* is based on the law of universal gravitation derived by Isaac Newton when he was 23 years old. This law states that the gravitational force (attraction) between two objects is directly proportional to their masses and inversely proportional to the square of the distance between them. Zipf (1946) and Stewart

---

Box 2.4 The index of dissimilarity

The index of dissimilarity is not a measure of migration *per se* but it is commonly used to explain the geographical segregation or clustering of minority (often immigrant) groups within a total population (for example, Robinson 1980). It is only one of a variety of this type of measure, which also includes the index of concentration, but it is the most popular (Duncan and Duncan 1955; White 1986). The value provided is the fraction of the population of one of the sub-groups that would need to move so that the distribution of the two groups was evenly spread across all of the geographical regions in the study area. The index is calculated as:

$$D = \frac{1}{2} \sum_{i=1}^{n} |(P_{ig} / P_g) - (P_{ih} / P_h)|$$

In this equation there are n areas, and g and h are the two sub-groups of the population. Where the populations are similarly distributed across each area, the value returned is 0, while a value of 1 reflects complete segregation with no mixing of the two sub-groups at all.

Source: Plane and Rogerson 1994.

(1948) applied the basic equation to social interactions:

$$\hat{M}_{ij} = k \frac{P_i P_j}{d_{ij}^{b}}$$

In the equation, $\hat{M}_{ij}$ is the estimated number of migrants moving between i and j, K is a constant, P_i and P_j are the populations of the origin and destination and d_{ij} is the distance between them. The exponent b is a *distance decay* parameter, which in Newton's model was 2 but in models of migration may vary considerably. The larger this parameter, the steeper the distance decay effect. The model provides estimates of the aggregate flows between each pair of places in a migration system, with the origin population representing the potential migrants at i and the destination population representing the possible opportunities at j. The distance is a surrogate for a number of variables, including the potential migrant's knowledge about a destination, the physical cost of moving between the two places and the psychological cost of moving away from a familiar place.

These models have generally been operationalised as ordinary least squares (OLS) multiple regression models based on the normal distribution. Parameters for the origin and destination populations are also estimated, because it is rare for migration to be exactly proportional to population size. This means that the estimated flows between i and j will not equal the estimated flow between j and i, which would have been the case in the basic gravity model. The variables are logged: a_0 is the constant, a_1 and a_2 are the population parameter estimates, and b is the distance decay parameter estimate:

$$\ln(\hat{M}_{ij}) = a_0 + a_1 \ln(P_i) + a_2 \ln(P_j) - b \ln(d_{ij})$$

The distance decay is represented here by a *power function* (the distance variable is logged). Another popular distance decay function is the *negative exponential*, where distance would not be logged. Fotheringham and O'Kelly (1989) suggest that the power function is appropriate for analysing population migration, while the negative exponential is appropriate for analysing residential mobility.

Additional variables relating to the socio-economic conditions at the origins and destinations have been incorporated into these migration models. Others have suggested original alternatives to the basic gravity model principle. Plane (1984) devised an *inverse gravity model*, where the distances that matched the observed migration were output. It is then possible to create warped migration spaces that reflect the locations of the American states relative to the migration into a particular destination (Figure 2.5).

Alternative modelling approaches have been developed to analyse flow data. Wilson (1967, 1970)

advocated a mathematical approach (in contrast to the statistical approach) that is *entropy-maximising*. The estimated flows constitute the most probable configuration of flows, given a set of constraints defining the system, and this approach is often referred to as *spatial interaction modelling*. Rather than using origin and destination populations as the mass terms, the total in- and out-migration for each of the areas is used instead (Senior 1979).

Flowerdew and Aitkin (1982) argued that statistical models should be based on the Poisson distribution, rather than the normal distribution, because the data involved are counts. There are also problems calibrating OLS models based on the normal distribution when some observed flows are zero, as it is not possible to calculate their logarithm. Usually, a very small constant is added to these zeros but choosing the constant size is arbitrary and altering it slightly will produce considerably different results. Also, the estimated migrant total in OLS models does not necessarily equal the observed number, as is the case in Poisson models. Flowerdew (1991) also favours Poisson models over maximum-entropy models. Most importantly, Poisson models provide a goodness-of-fit measure, allowing different models to be compared, and it is simpler to include extra explanatory variables.

The models described above are *unconstrained*, meaning that the total observed and estimated (from the model) number of migrants in the whole migration system will be equal (in the Poisson and the maximum-entropy approaches) but the total estimated and observed flows into or out of each individual area will not be equal, unless by chance. Constrained models can be fitted where the estimated total out-migration from each of the origins is forced to match the observed out-migration (*production-* or *origin-constrained*); the estimated in-migration to each of the destinations is forced to match the observed in-migration (*attraction-* or *destination-constrained*); or both the origin and destination observed and estimated flows are equal (*doubly constrained*). It is also possible to extend these models so that *origin-* and *destination-specific* distance decay parameters are derived rather than a single distance decay parameter for the whole migration system (Stillwell 1991).

Scale issues

One problem with the specification of the models outlined above is the *spatial structure effect* (see the

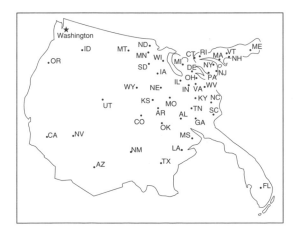

Figure 2.5 Migration space based on in-migration to Washington (source: Plane 1984).

Box 2.5 Intervening opportunities

The intervening opportunities concept was introduced by Stouffer, who argued that there was:

no necessary relationship between mobility and distance [and] the number of persons going a given distance is directly proportional to the number of opportunities at that distance and inversely proportional to the number of intervening opportunities.

(1940: 846).

The concept dealt with socio-economic rather than physical space, and the intervening opportunities were measured within a circle with a radius equal to the distance between the origin and destination pair (circle A). As Stouffer pointed out, these opportunities could take many forms; in many cases the total number of migrants is used. Empirical tests of the models proved successful (Bright and Thomas 1941; Isbell 1944), but the method was later altered and the intervening opportunities were calculated within a circle that had a diameter equal to the distance from i (St Louis) to j (Denver) and that passed through i and j (circle B). This refinement was made in order to account for the uneven distributions of opportunities within circle A, as larger flows would be expected in certain directions from i than in others. Choice of shape is arbitrary, with Rowlingson and Boyle (1992) showing how geographical information systems (GIS) can be used to compare alternative shapes. A second refinement is the competing

migrants concept. According to this principle, migration from i to j is inversely proportional to the number of competing migrants from elsewhere (or the total number of out-migrants leaving places that were as close to, or closer to, j than i) and these were measured within circle C. Jansen and King (1968) used this approach to show that inter-county migration flows within Belgium conformed to Stouffer's hypotheses *within* the French and Flemish linguistic regions.

Source: Stouffer 1940, 1960.

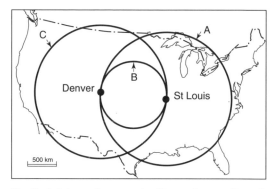

Stouffer's intervening opportunities and competing migrants concepts (source: Stouffer 1960).

Regional Studies debate: Curry 1972; Johnston 1973, 1975; Curry, Griffith and Sheppard 1975; Cliff, Martin and Ord 1974, 1975, 1976). This problem occurs when the distance decay effect in the migration system varies (as shown by origin- and destination-specific distance decay parameters), with peripherally located areas having higher parameters than centrally located places. Fotheringham (1991) maps origin-specific distance decay parameters against a simple measure of origin accessibility (or centrality within the migration system), showing how these are more negative for origins that are least central (Figure 2.6) (page 50). This locational effect can be controlled for by incorporating an accessibility measure into the standard gravity model (Fotheringham 1983, 1984).

A fundamental problem with many geographical modelling applications is the modifiable areal unit problem (MAUP), and this is no less the case in migration models. In fact, the problem becomes more complex when multivariate models are fitted, rather than simple bivariate models (Fotheringham and

Wong 1991). Much of the research into the MAUP has used migration model applications (Masser and Brown 1975, 1977; Openshaw 1977; Batty and Sikdar 1982a–d). There are two aspects to the MAUP; the *scale* and the *zonation* problems. If we assume that we have migration flow data for a set of areas it is possible to fit simple gravity models to these flows and a set of parameter estimates, which describe the underlying migration process, will be derived. The scale problem occurs when we use data for the same broad study area, but we disaggregate the area into a different number of zones. Almost certainly, the parameter estimates at this scale of analysis will be different. The zonation problem occurs when we retain the same number of zones in the migration system, but they are derived differently and once again the parameters will vary. It is difficult to decide which of the models is the most appropriate to base conclusions upon, particularly as the number of alternative configurations is enormous. Despite the fact that the problem has been acknowledged for many years, a solution seems some way off (if it is

Box 2.6 Implementing gravity models of inter-county migration using OLS and Poisson regression

The flow data used here are the number of young male apprentices from the 33 Scottish counties moving to Edinburgh between 1775 and 1799. The most prominent among these migrants were baxters (bakers), cordwainers (shoemakers), goldsmiths, hammermen (metal workers) and tailors. The apprentice data, the populations of the origin counties and the distances between these counties and Edinburgh are provided below. For comparative purposes, the resulting model parameters, estimated flows and standardised residuals are provided for unconstrained OLS and Poisson regression models.

Inter-county apprentice migration in Scotland, 1775–1799

Origin county	Apprentice migrants	Origin population	Distance moved	OLS estimates	Poisson estimates	OLS standardised residuals	Poisson standardised residuals
Midlothian	225	56,000	21	46.5244	173.11981	26.1661	3.9430
West Lothian	22	18,000	24	16.7734	45.05569	1.2762	−3.4348
East Lothian	44	30,000	33	15.7460	32.72293	7.1202	1.9714
Kinross	3	7,000	33	5.3911	8.85317	−1.0298	−1.9672
Fife	41	94,000	36	32.3847	73.79342	1.5139	−3.8175
Peebles	9	9,000	41	4.8082	6.52574	1.9117	0.9686
Clackmannan	2	11,000	41	5.5740	7.81480	−1.5138	−2.0801
Selkirk	5	5,000	52	2.2467	2.15237	1.8369	1.9410
Lanark	23	147,000	54	25.7261	40.91390	−0.5375	−2.8006
Berwick	11	31,000	56	7.7756	9.24765	1.1563	0.5762
Roxburgh	9	34,000	67	6.4983	6.48060	0.9814	0.9897
Stirling	13	51,000	71	8.0862	8.09505	1.7280	1.7240
Perth	26	126,000	78	13.8260	14.49511	3.2741	3.0218
Dumbarton	0	21,000	79	3.6303	2.80977	−1.9053	−1.6762
Angus	5	99,000	85	10.2814	9.45951	−1.6471	−1.4500
Dumfries	3	55,000	86	6.5620	5.42168	−1.3905	−1.0400
Renfrew	1	78,000	92	7.7334	6.29240	−2.4213	−2.1098
Ayr	2	84,000	110	6.3824	4.34474	−1.7347	−1.1249
Kirkcudbright	0	29,000	110	2.9161	1.67129	−1.7077	−1.2928
Kincardine	1	26,000	125	2.2556	1.10839	−0.8360	−0.1030
Bute	0	12,000	132	1.1839	0.48438	−1.0881	−0.6960
Aberdeen	3	123,000	156	5.2191	2.60449	−0.9714	0.2451
Wigtown	0	22,000	157	1.4563	0.54634	−1.2060	0.7392
Banff	4	36,000	159	2.0567	0.82443	1.3550	3.4974
Moray	2	27,000	174	1.4694	0.51072	0.4377	2.0839
Nairn	0	8,000	175	0.5951	0.16887	−0.7715	−0.4109
Argyll	1	72,000	179	2.9099	1.15013	−1.1196	0.1400
Inverness	7	74,000	234	2.0515	0.61221	3.4549	8.1640
Ross and Cromarty	4	55,000	274	1.3262	0.31882	2.3219	6.5195
Caithness	1	23,000	274	0.6978	0.14569	0.3618	2.2382
Sutherland	0	23,000	283	0.6674	0.13462	−0.8169	−0.3669
Orkney	0	29,000	366	0.5551	0.08839	−0.7451	−0.2973
Shetland	1	22,000	491	0.3020	0.03362	1.2703	5.2703

The model parameters from the OLS model were:

$$\ln(\hat{M}_{ij}) = -0.01122 + 0.7365 \ln(P_i) - 1.38 \ln(d_{ij})$$

The Poisson model parameters were:

$$\ln(\hat{M}_{ij}) = 2.777 + 0.8983 \ln(P_i) - 2.445 \ln(d_{ij})$$

Relatively similar conclusions would be drawn from these model parameters. The distance decay parameters were steep, indicating that the apprentices were prone to moving over short distances, while the population parameters were close to 1, suggesting that out-migration is approximately proportional to population size at the origin. The values for both the OLS and the Poisson models can be estimated simply from these model parameters – here the calculations for Midlothian are provided:

$$\hat{M} = \exp(-0.01122 + 0.7365 \ln(56000) - 1.38 \ln(21)) = 46.5$$

$$\hat{M} = \exp(2.777 + 0.8983 \ln(56000) - 2.445 \ln(21)) = 173.1$$

Box 2.6 (cont.)

The OLS model accounted for 63.7 percent of the variation, while the Poisson model accounted for 88.3 percent. The Poisson model performed especially well when the size of the flows was large; the flow from Midlothian was predicted especially well. A further benefit of the Poisson model is that the estimated number of migrants always equals the observed number (468), while a total of only 251.6 migrants were estimated using the OLS model.

Source: Lovett, Whyte and Whyte 1985.

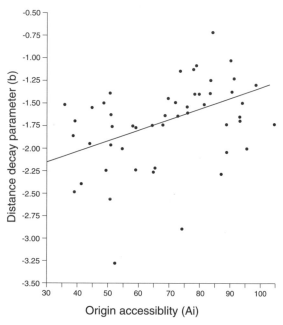

Figure 2.6 The relationship between origin-specific distance decay parameters and origin accessibility (source: Fotheringham 1991).

achievable at all) and many researchers appear to ignore the implications of this problem when undertaking aggregate data analysis.

Micro-level models of migration

In individual-level migration models, the focus moves from the characteristics of places to the behaviour of individuals. The decision-making process is explicitly modelled and the underlying assumption is that migration improves the individual's or the household's position. This modelling strategy takes account of individual differences in knowledge, contacts, economic and social conditions and previous experience, but this type of model also allows the incorporation of regional characteristics.

Discrete-choice models are one type of individual-level approach, rooted in the idea of utility maximisation (Chapter 3). Individuals are expected to choose the region that offers the highest utility, measured using a number of regional attributes, such as cultural characteristics, local amenities, climate and economic circumstances. Individuals perceive opportunities differently, and therefore the socioeconomic status of the individual is usually built into the model. These models are often fitted as logit models, where the dependent y variable is a 0 or 1, depending on whether or not the individual makes a decision to move.

Flowerdew and Halfacree (1994) used data from a commissioned survey to examine the characteristics of long-distance (between county) compared with short-distance movers. The modelling approach enabled them to determine whether or not highly correlated variables had independent associations with the likelihood of migrating. They found that occupation and the age at completing higher education, which were apparently related to long-distance migration when examined alone, were not significant once the remaining variables had been controlled for in the analysis.

Event history analysis

Event history analysis relies on longitudinal rather than on cross-sectional data and is used to explain the transition of individuals through their life-courses using variables that relate to specific points in time. The continuous observation period (usually a person's lifetime, or a good proportion of it) for each individual is scattered with changes from one discrete outcome to another and these occur in a particular sequence. The outcomes may be migrations from one place to another that are related to a sequence of life-course changes (Chapter 5). Supplementary data on the individual's personal or locational characteristics are used to explain what influences the changing outcomes, and how these outcomes occur at different times for different subgroups. A common form of these models is Cox's

(1972) *proportional hazards model*, where the *hazard rate* measures the probability, per unit of time, that an event will occur among those at risk during the study period.

These types of model can be extremely complex but they can be an improvement on cross-sectional models, which may suffer from omitted variables because of data inadequacies (Pickles and Davies 1991). Davies and Pickles (1985a) argue that the concept of the housing career, and its inevitable relationship with migration, is a useful one as it represents an important form of consumption that may be expected to vary in some systematic fashion with passage through the stages of life (Rossi 1955). Hypotheses about these relationships can be tested more adequately using longitudinal, rather than cross-sectional information. Thus, Davies and Pickles (1985b) used data from the first ten interviews of 887 households participating in the Michigan Panel Study of Income Dynamics, 1968 to 1977. They challenged the conventional wisdom about life-cycle effects on housing careers as there was no evidence that changing financial circumstances had a role in the life-cycle variation in residential mobility:

Neither financial stress often associated with the child-rearing years nor surplus of income to needs often associated with later middle age appears to have any significant effect on residential mobility. There is not even any evidence that financial hardship restricts a household's ability to move in order to alleviate overcrowding.

(*ibid*.: 214).

They argue that more theoretical insights are required to explain these surprising empirical findings.

Forecasting migration

From a planning perspective, various types of resource allocation and service provision need to be decided on the basis of future population estimates. *Cohort component models* subdivide the population into age and sex groups (cohorts), although more advanced models may also disaggregate the population into ethnic groups. The expected change in the size of cohorts is then estimated using information on births, deaths and migration. Survival rates for each cohort, and age-specific childbearing rates for the females, are used to refine these estimates but projections at the local level are dominated by the migration element, which accounts for the majority of population change. These types of model can be highly complex. Rogers and Willekens (1986) and Shen (1994) provide an introduction to them, and Isserman *et al.* (1985) focus on a migration example.

Markov models are used to predict destination choice probabilities of individuals where matrices of migration transition probabilities are constructed, based on the flows between pairs of areas in the migration system, and these are then used to estimate the flows in future time periods. The underlying proposition in Markov models is that the outcome is directly dependent on the outcome in the previous time period. The method assumes that the probability of moving does not change over time and that the probability of an individual migrating is based entirely upon where they reside. It ignores previous migration histories and assumes that each individual within the same region will be governed by the same migration probabilities. These assumptions are ultimately unrealistic but for short-term forecasting these models provide useful indications of population change.

Analysing migration qualitatively

Interviews

The quantitative approaches outlined in the previous section explicitly aim to generalise about migration. More recently within geography there has been a move towards more qualitative methods that delve into individual idiosyncrasies and responses. Increasingly, methods of this type are combined with quantitative techniques, as a multiplicity of techniques often improves the research project. Interviews are one of the most commonplace of the qualitative methods.

Interviewing usually involves face-to-face verbal discussions but can also include group interviews, questionnaire studies and/or telephone surveys. Questionnaire responses provide the basis for valuable quantitative data, particularly when they have a structured design, the responses from which can be coded easily. However, qualitative data can also be collected using unstructured questions, designed to allow respondents to elaborate their answers to less explicitly expressed questions.

A popular qualitative technique, which provides a greater breadth of information, is in-depth ethnographic interviewing, where an open-ended research strategy is used (Cook and Crang 1995).

Ethnographic research has a strong emphasis on investigating particular social phenomena rather than constructing and testing hypotheses about these situations. The data tend to be unstructured and the aim is to provide substantial detail about a small number of cases, rather than a smaller amount of information about a large body of respondents (Atkinson and Hammersley 1994).

In-depth interviews vary between situations and researchers but certain questions about how the research is conducted are common to most studies.

Box 2.7 In-depth interviewing

In a detailed example of an in-depth interview with a Russian immigrant to Israel, who moved in 1990, open-ended conversations were carried out over a six-month period and details about the respondent's migration experience and the considerable psychological impact this had had upon her were obtained. This admittedly time-consuming approach offers a great deal towards the understanding of the migration experience. Natasha (a pseudonym) was Jewish and she described how daunting the initial experience was:

It was horrible, horrible. I suddenly realised that I really left this place for ever and ever, never to return. But the full realisation took perhaps three months more to sink in.
(quoted in Lieblich 1993: 102).

This illustrates how important the timing of the interview is in relation to the migration event. Natasha also explained how difficult it was to be part of a minority group:

I have become a poor person. And I am still a member of a minority group. This is funny: in Russia we were told: you're different, you're Jews; here we are told: you're different, you're Russian.
(quoted in Lieblich 1993: 123).

Natasha was involved in a deeply emotional experience; the loss of home and the adoption of a new society, culture and language are difficult to cope with, especially when the decision to move is not made personally. As the researcher points out:

I understood Natasha's situation at the beginning of the academic year as that of a person exerting a tremendous struggle to cope with the crisis of immigration.
(*ibid.*: 96).

Clearly, it is difficult to get such detailed responses using more structured research methods.

Source: Lieblich 1993.

How does the researcher 'get into' the situation from which the information is sought? How are particular respondents identified? How does the researcher present himself/herself to the respondents? A distinction is often made here between trying to appear similar to or different from the respondents. How are the results from the interviews collected and analysed? And from our perspective, how do we identify relevant migrant groups? Douglas (1985) argues that researchers need to be creative in conducting interviews as it is important to adapt to the (dis)advantages of everyday situations. Postmodern interviewing strategies are more concerned with the way that researchers themselves influence the data collection and the techniques of reporting the results (Marcus and Fischer 1986). One approach, which counters this problem to some extent, is *polyphonic* interviewing. Rather than summarising the various findings from one or more respondents, a tactic which may capture some of the (sub)conscious biases of the researcher, the voices of the subjects are recorded with minimal interference from the researcher and the sheer variety of different responses is acknowledged instead of being smoothed over.

Participant observation

Participant observation is the main ethnographic technique where the researcher plays an established role in the studied scene. Although it is difficult to experience migration *per se*, because of the length of time that may be needed to examine a community before migration occurs and the difficulty of experiencing that migration personally, it is possible to observe migrant groups at the destination. Of course, researchers may influence the study results themselves; the very act of becoming involved in a situation will inevitably alter the composition of events (Adler and Adler 1987). Even so, participant observation can provide original insights into the impact of migration, upon both the destination areas and the migrants themselves and, of the qualitative techniques that are available, it is perhaps the least intrusive.

Often, researchers wish to investigate particular circumstances where it is difficult to participate and in these cases *non-participant observation* may be used. One example of this type of research was carried out on a Bangladeshi group in Rome (King and Knights 1994), where the second author spent some years living in Rome teaching Bangladeshis Italian. Learning the local language is integral to the

Box 2.8 Polyphonic interviewing

Case studies of elderly women in Wales demonstrate that autobiographical material may be difficult to generalise from but is extremely useful for putting 'flesh on the bones of analysis'. In this extract, the strategies adopted to deal with life in a new and unfamiliar place are highlighted:

Martha, from Ynyswen in the Rhondda Valley, left Wales in February 1929, to follow her husband who had gone to the Midlands about four months before in the hope of getting work when Bute Pit closed. His brother had got work in the Austin factory, but he had to settle for a job as a baker's roundsman, despite his qualifications and experience as a pit deputy.

Martha had a hard time settling in to two rooms in Halesowen, and soon had to cope with a young baby, and her husband losing his job because of pneumonia. She found the city a lonely place after her life in Wales in a family of ten, with 'crowds of cousins and friends'.

Martha's husband got a job at the Corona works, where he joined the male voice choir that was possible because of the large numbers of Welsh employees. Martha's company were her children, and then she joined the Cooperative's Women's Guild which, she wrote, was the 'beginning of my education'. During her long involvement with the Guild she held every office within it, and eventually became a magistrate, serving for eighteen years on the Birmingham Bench. Through her Party activities Martha chaired conferences, visited Budapest, Warsaw, Paris, Vienna, East Germany, Switzerland and Milan, and went on a Peace Caravan with Dora Russell to Moscow.

Leaving school at $13\frac{1}{2}$ with no further opportunity for education Martha looked back on her life as having been both interesting and rewarding. She has been recently nursing her husband who has pneumoconiosis, his legacy from pit work 56 years earlier.

Source: Bartholomew 1991: 175.

naturalisation process and the classroom is a valuable environment for learning about the trials and tribulations of life in a new place.

Migrant histories

A technique that is especially appropriate in migration studies is a *life-history* approach (Chapman 1985; Taylor 1986; Thrift 1985a). Individuals are traced through their lives and their biographies are built up, emphasis being placed upon the migrations that have shaped and have been shaped by these lives (Courgeau 1985). This approach has been common in developing world studies, where other data sources may be less reliable. In a developed world context, Forrest and Murie (1990, 1991, 1992) reconstructed histories from in-depth interviews undertaken with owner-occupiers of housing in Bristol, England. Such illustrative material was meant to complement rather than replace the insights gained from more quantitative studies and to generate ideas to be explored further through subsequent surveys.

These methods are, however, subject to problems of memory lapse and the definition of migration used needs careful thought. Auriat (1993) makes the interesting point that respondent's errors in dating events are usually in years rather than months, because the time of year when an event occurred is easier to remember. Using a sample of 500 couples in Wallonie, Belgium, she compared their recall of migration events with the records in the Belgian population register (Auriat 1991). Higher rates of mobility result in more confused recall and migration into places where the duration of residence was short were more likely to be forgotten. She also identified gender differences in recall, with females performing significantly better than males, perhaps because of the different roles they take in the migration process.

A *biographical* approach also has many advantages in migration research (Chapter 3). This can be distinguished from the life-history approach because it stresses the structural and societal context of a respondent's migrant history; it emphasizes how this is rooted in their everyday existence, rather than stressing the migration events as occurring at specific points in time (Halfacree and Boyle 1993). Thus, it uses qualitative techniques in a philosophically different way from the relatively narrow focus of much of this work.

Ethical, cultural and gender caveats

There are a number of important ethical, cultural and gender questions attached to migration research. Although these concerns apply to both quantitative and qualitative research, the literature addressing these issues is more substantial in the qualitative field. Individuals have a right to privacy and certain research strategies may invade this excessively. While observational techniques may be amongst the least obtrusive they can suffer from two types of privacy invasion: venturing into private places and the researcher misrepresenting himself/herself as a 'member' of the observed group. Researchers often aim to be inconspicuous in the research setting and an anticipated consequence of this is that information may be gained that would not have been forthcoming in a more structured research setting. Deviant behaviour often occurs in private settings, avoiding the gaze of those with different moral perspectives, and it can be argued that social scientists have no right to invade these situations and that subjects of study must be informed before the research commences:

it is unethical for a sociologist to deliberately misrepresent his identity for the purpose of entering a private domain to which he is not otherwise eligible.

(Erikson 1967: 373).

Others are less worried, arguing that important social concerns may be misunderstood without disguised research and that covert strategies mirror the deceitfulness of everyday life (Punch 1986).

Problems also occur when research is carried out by researchers from a different culture to that of the respondents. Much migration research in the developing world has been carried out by white researchers from the developed world whose sensitivity to the social and cultural differences between themselves and the respondents may be inadequate. These problems may be compounded by language difficulties because, even when the researcher and respondent are fluent in the same language, there may be different ways of saying things that are not understood by both parties. Stanfield argues strongly that ethnic differences are important and that we must:

call into serious question the vast warehouse of knowledge that researchers of European descent have been accumulating and legitimating as ways of knowing and seeing.

(1993: 183).

Rather than merely encouraging researchers to be more sensitive to ethnic and cultural difference, Stanfield suggests that indigenous 'ethnic' models of qualitative research must be developed and more research on ethnic groups should be carried out by members of those groups using techniques that are more appropriate to the cultural norms of those societies.

It has also been argued that research techniques often fail to account for gender differences. As Denzin (1989: 116) points out, 'gender filters knowledge' because the sex of the respondent and the researcher makes a difference. Interviews necessarily take place in a cultural setting and in most, if not all, settings masculine and feminine identities are differentiated. For example, Kane and Macaulay (1993) show that male respondents offer significantly different responses to male and female interviewers on questions that relate to gender inequalities. Interviewing can also be seen as hierarchical, with the respondent subordinate to the interviewer and this problem is exacerbated by gender differences. A

Box 2.9 Migration life-histories

Migrant life-histories can be built up from secondary data sources. Historical data from Sweden were used to test whether the rapid demographic urbanisation (primarily driven by rural–urban migration) reflected in early census figures provided a misleading impression of the nature of urban growth in nineteenth-century Sweden. Many migrants circulated between urban and rural areas – as seen today in many developing countries where urbanisation is rapid – and the impression given by cross-sectional census data, that urban areas were gaining net migrants from rural areas, was false. It may be erroneous to classify these migrants as either urban or rural. Only by examining the migration life-histories of individuals can these types of question be answered.

One set of longitudinal data examined was derived from *husförhörslängder* (household registers), compiled for every Swedish parish from the 1780s, including the names and addresses of household heads, the age, place of birth, last and next residence, relation to household head and occupation of each member of each household. Attention was given to 87 residents in the town of Vadstena and some neighbouring rural parishes in 1855. The life-paths of these individuals were traced through the registers of all of the other parishes that they had resided in. Most occupational groups migrated regularly:

The transition from bourgeois to industrial capitalism was neither produced nor contained within the urban system. Moreover, the pattern of links between the towns and rural areas also changed fundamentally. Vadstena's plebian population was transient and circulatory.

(Langton and Hoppe 1990: 159).

However, the contrasting migration histories of master craftsmen and farmers is clearly evident, with the former changing residence much more often both within and out of Vadstena.

Source: Langton and Hoppe 1990.

Box 2.9 (cont.)

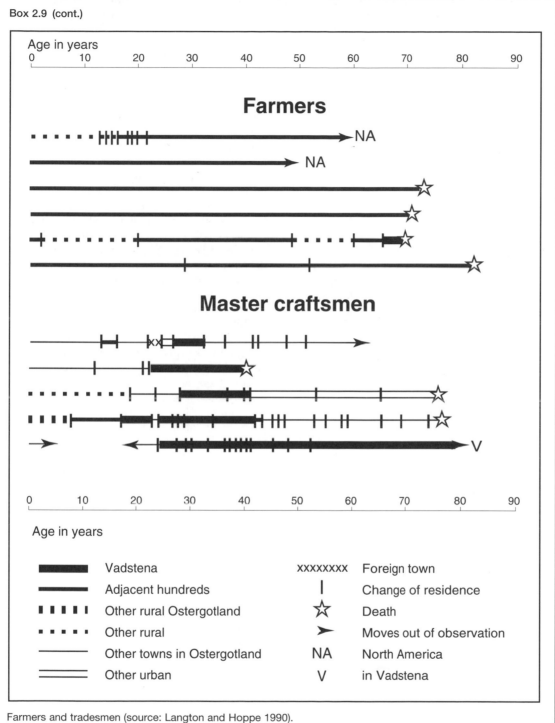

Farmers and tradesmen (source: Langton and Hoppe 1990).

common plea is that a closer relationship between the interviewer, and the respondent should be developed so that status (and gender) differences are minimised. In short, interviewers should:

show their human side and answer questions and express feelings. Methodologically, this new approach provides a greater spectrum of responses and a greater insight into respondents.

(Fontana and Frey 1993: 370).

Others, however, worry that the answers provided may be influenced by such a stance. Lastly, from a strongly feminist perspective, Stanley and Wise argue that:

future research should be *on* and *for* women and should be carried out *by* women. Such research is, at least in part, 'corrective'.

(Stanley and Wise 1993: 30).

This somewhat extreme stance does at least remind us that many academic researchers are white males and, consequently, the results may be different from those gathered by females or those from other ethnic groups.

Conclusion

Defining migration explicitly has been shown to be problematic. In reality, many studies rely on the definitions of migration used in large-scale surveys or censuses and, consequently, deliberation over who is and who is not a migrant is not seriously considered.

For those designing their own survey, the definition is important as it will influence the conclusions drawn. It has also been shown that the range of migration data sources is wide – there are others, which have been ignored in the space of this chapter – but that none of them is perfect. Researchers, therefore, must decide which of the myriad of migration data sources are appropriate for the specific research project.

A wide variety of research techniques may be implemented in migration studies. Much of this chapter has considered migration as a quantifiable event described, or modelled, using stark numbers. However, migration is also a cultural event that impacts upon both the people and the places which are involved. Consequently, facts and figures about migration do not necessarily provide the whole picture and qualitative analyses are helpful in highlighting some of the social and cultural processes of change that are involved. Indeed, using a combination of qualitative and quantitative methods may provide a more rounded picture in the research endeavour.

This chapter drew attention to some of the problems with particular research techniques, be they quantitative or qualitative. These problems cannot always be solved but migration researchers must be aware of the drawbacks of the methodology they employ. Such a rather pessimistic conclusion should be tempered, however, as a wide and impressive range of research has been carried out over many years, in a variety of locations, and a considerable amount has been learnt about migration processes and patterns. Much of this work is addressed in the remainder of this book.

CHAPTER 3

Contrasting conceptual approaches in migration research

Introduction

We saw in the previous chapter the different ways of defining and measuring human migration and the corresponding difficulties this presents. However, there is another important obstacle facing anyone studying migration, namely how the act of migration should be conceptualised from a philosophical standpoint. Indeed, at least until a recent upsurge in interest in theoretical questions within population geography more generally (Findlay and Graham 1991), migration research has typically neglected and overlooked the philosophical debates that have raged, often quite furiously, in geography regarding the 'correct' way in which patterns and events should be studied and explained (Pooley and Whyte 1991a; White and Jackson 1995). Research has been much more concerned with questions of methodology and technique (Chapter 2), part of a move suggested by Findlay and Graham (1991) that takes population geography closer to 'spatial demography' (Woods 1982) than to the rest of geography. Nonetheless, these philosophical debates *have* influenced the way in which migration research has been carried out, and migration studies have taken on board, to a greater or lesser extent, all the main philosophical currents that have been experienced by geography and the other social sciences. It is therefore important that we describe the range and variety of conceptual approaches that have been adopted as a result.

Migration research has been dominated by a series of dichotomies with respect to philosophical and methodological approaches (White 1980). In particular, there are those pieces of work that concentrate on individual migrants and their decision making, and those that focus on groups: *micro-*

analytical versus *macro-analytical* traditions, respectively (H. Jones 1990; Morrison 1973). Thus, Lewis (1982) describes 'objective' levels of analysis concerned with 'broad patterns of migration flows' and the 'normative' and 'psycho-social' levels of analysis that deal with the norms underpinning migration and the decision-making process of the migrants themselves, respectively. Similarly, White (1980) distinguishes between 'objective' and 'cognitive' approaches, the latter concentrating upon the individual migrant's cognition or perception of the relative merits of the origin and destination.

An alternative dichotomy, one opposing *determinist* to *humanist* conceptual approaches, is used in this chapter. The former, which are by far the most commonplace in the literature, play down the role of the individual in actively deciding whether or not to migrate by assuming migration to be an almost inevitable response to some rational situation, whether this rationality is judged from the supposedly objective perspective of an outside observer or from the subjective perspective of the potential migrant. In contrast, humanist approaches stress the importance of seeing the individual migrant as an active decision-maker, whose ultimate decision to migrate may or may not be rational from any particular perspective. For the determinist the migrant has little choice whether or not to migrate, given the broad environment in which he or she exists. For the humanist, an individual has a very real choice whether or not to migrate and this decision can only be explained through in-depth study of the person involved.

Although it starts off with a conventional dichotomy in this manner, the chapter goes on to describe some more integrated approaches to conceptualising migration. These approaches argue that a balance needs to be struck between regarding migration as an inevitable response to particular circumstances

Box 3.1 Contrasting philosophical approaches to migration research

Different philosophical traditions inform contrasting ways in which migration research is undertaken and the results analysed and interpreted. Four traditions in particular are noted in the literature:

Positivism

A philosophy proposed by Comte in the early nineteenth century, predominantly concerned with distinguishing science from religion and metaphysics. It seeks to establish the scientific status of geography, since its 'basic approach . . . is to assume that geography is a strict science and then to proceed to examine the substantive results of such an assumption' (Bunge 1962). Five key steps in the determination of scientific status are noted:

1 Scientific statements are grounded in direct, immediate, empirical experiences; observation is privileged over theoretical statements.
2 Observations must be repeatable; made possible through the use of a uniform 'scientific method', involving hypothesis testing and controlled experiments.
3 Science progresses through the formulation of laws, the result of empirically verified theories.
4 Laws take the form 'if A, then B', being purely technical and descriptive, without moral or ethical judgement; the scientist is a neutral observer.
5 Scientific laws are to be progressively integrated into a coherent body of eternal knowledge and truth.

Behavioralism

This approach concentrates upon individuals rather than the aggregate analysis of positivism. It recognises that positivist laws of human behaviour might not apply at the individual level; in particular, individuals often act in ways that appear irrational to the observer. Behavioralism argues that attention must be given to the individual as a decision maker. Thus, behavioralism in geography reflects a 'psychological turn' within the discipline, emphasising the role played by individual psychological factors and processes – notably cognition, knowledgeability and motivation – in mediating between the environment and spatial behaviour. Nonetheless, in spite of its different scale of analysis, behavioralism shares positivism's assumptions of generalisability and law-like behaviour, frequently relying upon statistical techniques and models. In a way, behavioralism represents positivism applied at the level of the individual.

Structuralism

A philosophy which argues that just looking at the observable and the empirical, as in positivism and behavioralism, tells us little about what causes human behaviour. Instead, we should delve beneath the level of surface appearances and expose the essential logic that binds together human behaviour into regular patterns. The enduring and underlying structures that are responsible for this order can only be exposed through theoretical intellectual development. Human action is seen as being highly constrained and determined by forces outside one's everyday control. This suggests similarities with positivist efforts to recognise laws of human behaviour but, unlike the positivists, the crucial objective of structuralists is to expose the underlying forces moulding behaviour and, in the case of structural Marxism, with its strong political ambitions, to overthrow and replace oppressive structures, such as capitalism, in the quest for human liberation.

Humanism

This radical departure from the other philosophical traditions gives central importance to the operation of human awareness, consciousness and creativity. In many ways, humanism within geography represents a backlash against the impersonality and coldness of the other traditions. Humanism stresses the essential and irreducible subjectivity of both the observer and the observed, thus going beyond the formal individualism of behavioralism, the search for laws within positivism, and the determination by hidden structures that characterises structuralism. Humanism demands that individuals – their behaviour, thoughts, feelings, priorities – must be studied and valued in their own right. The world that geographers study is nothing more (and nothing less) than the sum of all the engagements of these individuals with their environments, understood in the broadest sense. Humanism encourages the use of qualitative research methods.

Postmodernism

A relatively recent and extremely diverse field of thought that is united by scepticism towards the explanatory claims of other (modern) theories such as structural Marxism, humanism or positivism. It emphasises the need to accept and celebrate a wide range of voices, explanations and perspectives. Postmodern style is characterised by an emphasis on appearance over functionality; postmodern method seeks to throw doubt upon existing interpretations through techniques such as deconstruction; and the idea of a postmodern era suggests that we have moved into a set of 'new times', where post-Enlightenment certainties of progress have been progressively undermined and rejected. Postmodernism has been seen by many as re-stating the importance of place and geography in the contemporary world, as the sheer 'messiness' and complexity of everyday life can only be contemplated successfully within its localised context.

Source: Cloke, Philo and Sadler 1991; Johnston, Gregory and Smith 1994.

Box 3.2 An alternative classification of differing conceptual approaches to migration research

The classification scheme used in this chapter for describing the different ways in which migration has been conceptualised in the academic literature is not the only classification available. Given below is an alternative classification scheme, tailored to migration in the developing world. The first four conceptualisations are grouped by Shrestha as 'conventional' perspectives, in contrast to the radical alternative, the 'neo-Marxist' perspective.

Conceptual perspective	Independent variables	Comments
Economic behavioural	Wage/income differentials; employment opportunities; utility maximisation, expectation	These studies assume a two-sector rural–urban economy and argue that migrants make a rational decision by moving in the direction where they get or expect to get the highest benefits; generally apply partial equilibrium, single-equation models (Todaro 1969; Harris and Todaro 1970; Greenwood 1969; Sjaastad 1962).
Eco-demographic	Population pressure; carrying capacity	Assume overpopulation and argue that population pressure in an area leads to out-migration (Simkins 1970; Dahal, Manzardo and Rai 1977; Grigg 1980).
Spatial	Distance; spatial attraction; place utility/dis-utility	Migration is an inverse function of distance and other intervening factors. Evaluation of place utilities and dis-utilities plays an important role in the migration decision (Beals, Levy and Moses 1967; Masser and Gould 1975; Olsson 1965; Wolpert 1965).
Anthroposociological	Group, kinship or ethnic network; modernisation factors; adjustment and assimilation	Some kind of security assurance in the destination facilitates migration; the stronger the kinship, group or ethnic networks, the greater the propensity to migrate (Abu-Lughod 1961; Mangin 1970; Zelinsky 1971).
Neo-Marxist dependency	Socio-economic structure; modes of production; colonial–capitalist penetration	Although migration may appear to be 'voluntary' and 'rational', it is conditioned and manipulated; it is directly related to colonial–capitalist penetration (Amin 1974; Portes 1978a; Omvedt 1980; Gregory and Piche 1978; Meillassoux 1981; Van Binsbergen and Meilink 1978).

Source: Shrestha 1988, reprinted by kind permission of Arnold.

and seeing it as a completely individual action. While such an integration is both theoretically and methodologically challenging, it is, as we show, seen by many as being the way ahead for migration research in the future.

Determinist accounts of human migration

Empirical laws of migration

Ravenstein, gravity models and the Newtonian legacy

Migration became an issue worthy of serious academic study in the developed world in the wake of the Industrial Revolution (Courgeau 1989), when the academy and the idea of 'knowledge' became organised along lines familiar to us today. A landmark was reached towards the end of the nineteenth century, when three papers were published by a German-born former cartographer for the British War Office named Ravenstein. These papers elaborated eleven principal laws of migration, derived from analysis of census data, which was beginning to be used as a source for migration research. Ravenstein's laws provided the hypotheses upon which much future migration research and theorising was built, although little attention was paid to his work until it was revived by the historical geographer Darby in 1943 (Grigg 1977). Ravenstein remains one of the pre-eminent geographers with regard to research on migration in nineteenth-century Britain. For example, in the mid-1960s, Lee

(1966) developed his laws into a series of hypotheses concerning the volume of migration, the development of stream and counter-stream migration, and the characteristics of migrants. Ravenstein's ideas remain very influential in contemporary work (Massey *et al.* 1993; Zolberg 1989), even though, as Woods (1982) notes, his laws are based largely on empirical observation and lack explicit theoretical grounding.

Attempts have been made to verify and explain some of Ravenstein's laws and to give them some theoretical credence through the use of the gravity

---

Box 3.3 Ravenstein's 'laws of migration'

Ravenstein's laws, better seen as hypotheses, were published in the *Geographical Magazine* of 1876, and the *Journal of the Statistical Society* in 1885 and 1889. They were derived from British census place of birth tables for 1871 and 1881, supplemented in the final paper by similar data from censuses in North America and Europe. Eleven major laws were identified:

- The majority of migrants go only a short distance.
- Migration proceeds step by step.
- Migrants going long distances generally go by preference to one of the great centres of commerce or industry.
- Each current of migration produces a compensating counter current.
- The natives of towns are less migratory than those of rural areas.
- Females are more migratory than males within the kingdom of their birth, but males more frequently venture beyond.
- Most migrants are adults: families rarely migrate out of their county of birth.
- Large towns grow more by migration than by natural increase.
- Migration increases in volume as industries and commerce develop and transport improves.
- The major direction of migration is from the agricultural areas to the centres of industry and commerce.
- The major causes of migration are economic.

All of the laws, excluding the fifth one, have been shown to be more or less accurate for the nineteenth century and many of them are true today, attesting to the lasting importance of Ravenstein's work.

Source: Grigg 1977.

---

model (Chapter 2), whose basis lies in the extension of the laws of physics to the human sphere. Specifically, Newton's Law of universal gravitation states that: 'two bodies in the universe attract each other in proportion to the product of their masses and inversely as the square of the distance between them' (cited in H. Jones 1990: 189). This is a classic determinist law, as one would expect from physics, where inanimate masses have no say over their behaviour. In the 1930s and 1940s, this law was applied to humans and social physics was born. While the basic gravity model linking migration flows to distance and population size remains sensible at first sight – one might expect distance to deter migration and population size to relate both to numbers willing and able to move and to opportunities available for such migrants – work in the years after 1949 soon revealed how it was much more than just distance and numbers of people that determined migration flows. Consequently, the gravity model itself has been adapted over time (Chapter 2).

Demographic change and the mobility transition

Another famous attempt to produce general statements about migration was undertaken by Zelinsky in the 1970s and 1980s. Zelinsky (1971) sought to generalise the transitions that take place in rates and scales of migration as a society is transformed over time. Migration, through its link to personal mobility, was fundamentally related to social change, especially modernisation, and the resulting *mobility transition* was drawn up with inspiration from other contemporary theories and models, notably the well-known demographic transition model. Thus:

There are definite, patterned regularities in the growth of personal mobility through space-time during recent history, and the regularities comprise an essential component of the migration process.

What is attempted ... is the application of the principle of the spatial diffusion of innovations to the laws of migration ... set within the same sort of temporal structure that has been developed for the demographic transition. ... The exposition is almost entirely at the descriptive level; no serious effort is made to plumb the processual depths.

(*ibid.*: 221–2, 220–1).

The resulting mobility transition model identified five phases. In Phase 1, a 'pre-modern traditional society', there was limited migration and circulation (Chapter 2), all of which was highly localised. This situation changes over time until Phase 5 is reached in what would now be described as a 'post-industrial society'. Here, there is a general decline in migration as it is absorbed into and displaced by circulation and the communications media.

Zelinsky and others subsequently revised the basic model in the light of criticism (Woods 1993; Zelinsky 1983, 1993), with a recent simplified version illustrated in Figure 3.1. Overall, the model has been shown to be quite effective at describing events, especially in the developed world, when adjusted to local cultural conditions. However, it is flawed, as indeed is the demographic transition model, in its rather vague conception of 'modernisation' as being the driving force behind change. In particular, its positivist universality and political naïvety have been noted. Thus, a recent revisitation of the model described it as 'a child of its time ... before geographers became far more critically aware of the political connotations of theory and model construction' (Woods 1993: 214; also Cadwallader 1993). As with many other such empirical models linking migration change to modernisation/social change (for example, Pryor 1975; Lewis and Maund 1976), it can be argued that the mobility transition model's highly generalised and succinct character, although extremely useful as an educational device, tends towards the trivial when we attempt to understand and explain migration patterns and behaviour in detail. Consequently, whilst Skeldon (1990) used Zelinsky's model as an inspiration for his recent account of internal migration in the developing world, he recognised that we are still a long way from any 'grand theory' of demographic change.

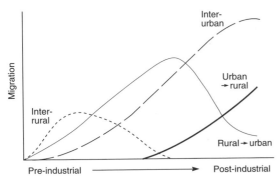

Figure 3.1 A representation of the mobility transition model (reprinted with permission from H. Jones (1990), *Population Geography*. © 1990 Paul Chapman Publishing Ltd, London).

Capital maximisation approaches

Neoclassical insights

Most early theories of migration were rooted strongly in *neoclassical economics* (Greenwood 1975, 1985; Massey 1990). In particular, emphasis was given to paid workers responding in economically rational terms to wage differentials across space, thereby moving from low-wage to high-wage areas. This followed Hicks's (1932) reasoning that the existence of wage differentials was the main cause of migration. Migration was thus 'labor reallocation in response to market need' (Ritchey 1976: 364), the scale of which could stretch from the local area to the globe as a whole. The effect of such reallocation would be the creation of an over-supply of potential workers in the high-wage areas, which would cause wages there to drop in line with neoclassical theory. Ultimately, this would generate migration elsewhere and so the cycle would start again. A typical example of this reallocation approach in the United States literature was Rogers's (1967) study of California, where wages and other labour force variables were used to predict inter-metropolitan migration flows.

These ideas were extended to work on migration in the developing world. Crucial here was the contribution of Todaro (Harris and Todaro 1970; Todaro 1969, 1976), which emerged from work on the causes and effects of rural-to-urban migration in East Africa. Todaro extended the wage differential model of migration to include a measure of the *probability* of finding employment. Thus, wage differentials came to be weighted by unemployment rates. This reasoning, of course, suggested a less smooth relationship between wages and labour supply than the original economic theory implied but it remained rooted in the neoclassical tradition (Molho 1986).

Unfortunately, even sophisticated wage-based models were limited in their scope and applicability, even when just considering labour migration. For example, the trend towards geographical economic equilibrium in wage levels clearly has not come about, whether at the global scale or even within individual nations (Wood 1982). Thus, in his review of migration in 1975, Greenwood concluded that differences in wage differentials had been shown to be a poor determinant of migration. It was therefore unsurprising that research extended outwards to investigate other factors driving migration.

Human capital theory

A key extension of the neoclassical perspective on migration came with the development of *human capital theory* (Milne 1991). A principal figure here was Sjaastad (1962), who concerned himself with the idea of weighing up the costs of a migration against its returns. This approach argues that potential migrants base their migration decision on 'an assessment of the anticipated future stream of benefits (both monetary and psychic) as a consequence of migration' (Molho 1986: 398). The approach remains wedded to the idea of maximising income but the definition of 'income' is extended beyond the narrow confines of wages (McNabb 1979; Borjas 1989). Crucially, migration streams need not be dominated by flows from low-wage to high-wage areas, since a whole variety of social, environmental and economic factors can drive migration. Thus, not all migration is seen as being driven by labour market concerns as was implied by neoclassical approaches. Moreover, the idea of costs allows such things as the expenditure on information gathering and the psychological consequences of leaving a 'home' environment to be incorporated into the model.

In addition to acknowledging the diversity of costs and benefits associated with migration, the human capital approach recognises the need to take a broader temporal perspective on migration. It became apparent that the time horizon for expected returns from a migration will be neither instantaneous nor uniform. An appreciation that this return period varies with age and other characteristics – captured in the idea of migration as an investment – allows variation in migration behaviour amongst different population groups (Molho 1986); for example, elderly people will search for more immediate returns than their younger brethren, as they are likely to have fewer years in which to capitalise on their investment (Hart 1975). Overall, the human capital approach regards migration as a holistic investment decision for an individual based on long-term as much as short-term benefits.

The human capital approach has been used extensively in migration research (Cadwallader 1986; Da Vanzo 1980), notably in the United States, where Cebula's (1979; Cebula and Vedder 1973; Gallaway and Cebula 1972) work is typical. Here, investment in migration is regarded as depending upon three general sets of forces: expected real income differentials, expected amenity differentials, and expected differential benefits and costs from state and local government policies (Cebula 1979: 40). From these three sets of forces, Cebula constructed a complex model describing the likely probability that an individual would migrate from area A to area B.

A principal problem with the human capital approach, however, is the way it treats the *information* processed by the potential migrant. The approach suggests that relevant information concerning such factors as wages and amenity is received and processed by the individual in a relatively unproblematic manner. In other words, individuals acquire *perfect* information – accurate and complete – which they can then process to decide whether or not to migrate in a rational manner with respect to human capital gains. This is clearly an unrealistic assumption and its attempted resolution leads us into more behavioural theories of migration, which give greater attention to the individual.

Behavioural decision-making approaches

The behavioural model

Behavioural approaches to migration are critical of the spatial analysis perspective of the previous approaches. Instead of concentrating on the immediately observable and measurable, quantifying migration and presenting the resulting patterns of flows, they stress the importance of noting the *mechanisms* behind individual acts of migration. Specifically, they investigate how psychological processes of cognition and decision making mediate between the environment and the individual. They also represent a response to the concern with the *scale* at which migration research was being undertaken. In particular, there was the danger of the 'ecological fallacy', whereby incorrect inferences about individuals are made from aggregate studies (Golledge 1980). Just measuring the attributes of different places when modelling migration flows ignored the significance of the perception and processing of information by the individual. Investigating this processing and perception is fundamental:

> Basic to this approach is the idea that spatial preferences are subjective evaluations, and the perceived attractiveness or perception of residential desirability of alternative locations is a critical element in the decision-making process of migration and a critical determinant of migration and its direction.
>
> (Ritchey 1976: 397).

Wolpert was a key figure in the development of behavioural approaches. In a series of papers in the mid-1960s (Wolpert 1964, 1965, 1966), he developed the concept of 'place utility', defined as

the net composite of utilities which are derived from the individual's integration at some time and space ... may be expressed as a positive or negative quality, expressing respectively the individual's satisfaction or dissatisfaction with respect to that place.

(1965: 60).

Individuals will move to a place that offers a higher overall place utility than either the origin or alternative destinations. Crucially, what takes this understanding beyond that of the human capital approach is that exactly *how* these relative utilities are expressed is problematised. They may be the actual characteristics, as the neoclassical tradition would suggest, or they may be represented by the subjec-

tive and possibly inaccurate perceptions of the individual in their behavioural environment (Figure 3.2).

The importance that the behavioural approach attaches to perceptions emphasises how migration may be far from rational and optimal to the external observer. Wolpert expressed the irrationality of much migration by drawing on Simon's (1957) concept of the 'satisficer'. Here, people are seen as being satisficing rather than maximising agents, with migration decisions relating to personal aspirations of satisfaction rather than optimality. These aspirations change constantly but are linked to factors such as previous experiences and the behaviour of people of a similar socio-economic position. Level of satisfaction and probability of migration are also associated with the extent to which a person is established in

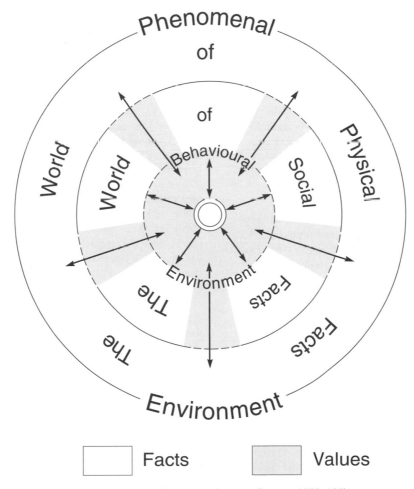

Figure 3.2 The behavioural and phenomenal environments (source: Gregory 1978: 136).

one place and the concept of cumulative inertia (Chapter 2). Migrants' limited and bounded knowledge was recognised by Wolpert through his emphasis on migrants tending to relocate to familiar places. Individuals have what is termed an 'action space', which consists of the places that are well known and that tends to bound the range of probable destinations. Again, this action space is highly dynamic, expanding over time as more information is gathered about more places. Important for shaping this action space are ties with places, such as the presence of family or friends, and experiences of those places.

The concept of *stress* is also important to the behavioural model. The satisficing human will tolerate a degree of discomfort or stress at his or her current location up to a given threshold. However, when the stress crosses this threshold the individual will either raise the threshold or deem the current residential location to be unsatisfactory and seek an opportunity to migrate. Finally, stress is not identical to 'dis-utility', where the place utility of the current location is perceived as being lower than that of an alternative location. With stress, only the unsatisfactory nature of the current location is at issue, and the act of comparison is not necessarily made. Stress is created by such things as changes in household structure, dwelling characteristics or environment, or by the tolerance threshold moving down as aspirations rise.

The model applied

The behavioural approach tends to conceive of migration as a sequential decision-making process. This represents the legacy of Rossi's (1955) pioneering study of intra-urban migration in Philadelphia and its link to changes in the life-course (Chapter 5). Rossi divided the migration decision-making process into three relatively discrete stages: the decision to leave the current location; the search for a better location; and choosing from alternative destinations. This sequence is explicit in Zuiches's model:

In such a model, the decision to move is analytically divided, first, into a phase of evaluation of one's current residence, in which a threshold of dissatisfaction may be reached, bringing the household to consider the possibility of a move. The second stage involves the search-and-selection procedures and includes a comparative evaluation of alternative sites. It is at this stage that locational preferences play a role in the selection process and that, finally, a decision to move is made.

(Zuiches 1980: 184).

A flavour of the resulting models is given in Figure 3.3. First, there is Clark's (1986) modified version of Brown and Moore's (1970) migration model which included the concept of stress (Figure 3.3a). This stress was seen as originating in the needs and expectations of the household, on the one hand, and in the environment, on the other. In contrast, both Roseman's (1971) model (Figure 3.3b) and Molho's (1986) contribution (Figure 3.3c) assume that a level of intolerable stress has been reached and focus on the subsequent search process. The final model (Figure 3.3d), taken from Lee (1966), is a more simplified representation of migration. As with the place utility concept, the origin and destination can be compared in terms of their overall combinations of *pull factors* (attractions) and *push factors* (repulsions). However, any simple comparison is complicated by the presence of 'intervening opportunities', obstacles such as family obligations at the origin or the high costs of moving, which may prevent migration occurring. All four models emphasise the problematic nature of any migration and the need to overcome a series of barriers and obstacles.

Capital maximisation approaches are preoccupied with capturing the economic dimension of migration. In contrast, behavioural models begin to consider the extent to which some migration is brought about by individuals attempting to improve their quality of life beyond economic considerations. In particular, such an appreciation comes from the frequent use of questionnaire surveys as a research tool (Chapters 2 and 6), where respondents themselves tell the researcher why they have moved. However, there are continuities between the behavioural and the capital maximisation perspectives, which locate behavioralism within the determinist category (Cloke, Philo and Sadler 1991). Indeed, behavioralism can be seen as being as positivist as a capital maximisation approach. First, both treat migration as a relatively systematic response, whether to the prospect of higher wages, the potential to increase human capital or the need to overcome stress. Thus, they both anticipate the ability to produce generalisations of migration behaviour (Golledge 1980), which potentially can be expressed as statistical regularities or as the empirical laws of Ravenstein, Zelinsky and others. Second, both behavioural and capital maximisation approaches concentrate on quantifying the causes of the migration. The former do this in terms of costs and benefits, tightly or loosely defined, while the latter quantify in terms of levels of stress, place utility or, from survey

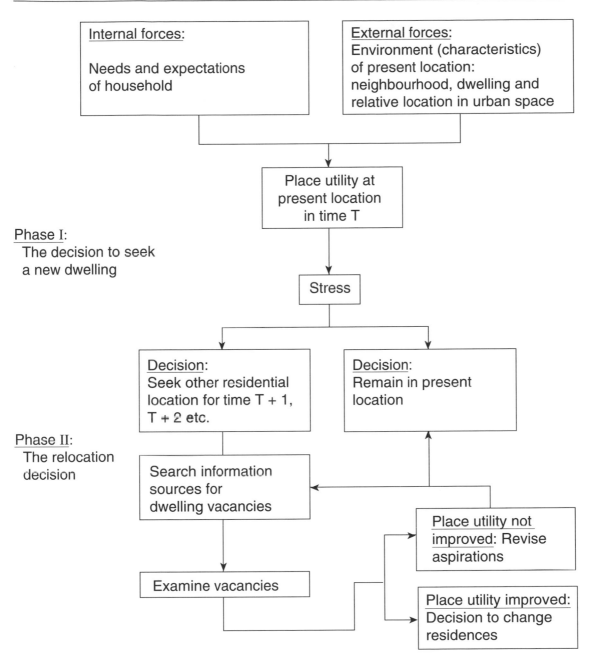

Figure 3.3 Behavioural models of the migration process. a. Clark's (1986: 49, after Brown and Moore 1970) model of the 'residential location decision process'.

work, tables of reasons for moving. Third, all the approaches, in effect, naturalise migration, regarding it in very benign terms (Halfacree 1995a); migration is something that is always there and is a 'good thing' because it leads to better living conditions, higher wages and more pleasant residential environments. Migration is thus placed 'beyond' politics. This third criticism can be addressed through examination of structural perspectives on migration.

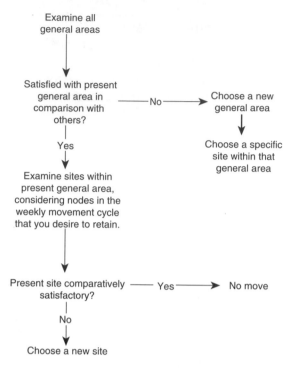

Figure 3.3 Behavioural models of the migration process. b. Roseman's (1971: 592) 'generalized locational decision schema'.

The structuralist critique: migration and power

The structural model of society

For *structuralists*, explanation of a phenomenon such as migration is located at the structural level, conceptually beneath the level of everyday appearances and beyond the confines of everyday categories and concepts. The real forces driving and structuring society are hidden and have to be brought into the open through academic work (Sayer 1984). Thus, in contrast to the positivist emphasis on observation and experience, for the structuralist the causes of migration are hidden from the gaze of the observer and require theoretical endeavours to determine what they are. Consequently, positivist and behaviouralist work is regarded as reductionist through its neglect of the macro-scale conditions within which decision making is undertaken. This work thus presents an 'unwarranted veneer of free choice' (Wood 1982: 305):

Quite clearly, individuals migrate for a number of different causes. ... Nothing is easier than to compile lists of such 'push' and 'pull' factors and present them as a theory of migration. The customary survey reporting percentages endorsing each such 'cause' might be useful as a sort of first

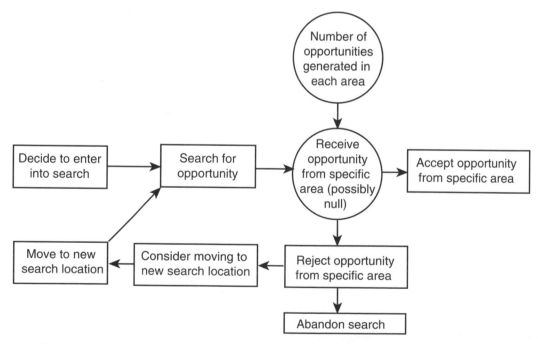

Figure 3.3 Behavioural models of the migration process. c. Molho's (1986: 404) 'migration decision-making framework'.

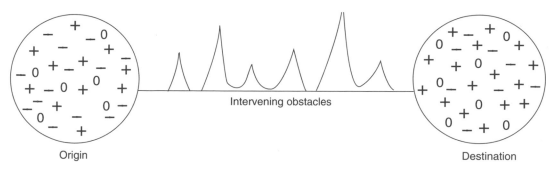

Figure 3.3 Behavioural models of the migration process. d. Lee's (1966: 48) 'intervening obstacles' model.

Box 3.4 Push and pull factors influencing migration

So-called 'push' and 'pull' factors have a long pedigree within migration research. The original formulation envisaged a combination of push factors from the origin and pull factors from the destination bringing about any one migration. These factors were summarised by Bogue as follows:

Push factors

- Decline in a national resource or the prices it commands; decreased demand for a particular product or service; exhaustion of mines, timber or agricultural resources.
- Loss of employment due to incompetence, changing employers' needs, or automation or mechanisation.
- Discriminatory treatment on the grounds of politics, religion or ethnicity.
- Cultural alienation from a community.
- Poor marriage or employment opportunities.
- Retreat due to natural or humanly created catastrophe.

Pull factors

- Improved employment opportunities.
- Superior income-earning opportunities.
- Opportunities for specialised training or education.

- Preferable environment or general living conditions.
- Movement as a result of dependency on someone else who has moved, such as a spouse.
- Novel, rich or varied cultural, intellectual or recreational environment (especially the city for rural populations).

From the perspective of the 1990s or from within specific societal contexts we may, of course, wish to modify these factors quite considerably. Indeed, any citing of just push and pull factors is now generally considered as being far too simplistic to explain observed migrations. We can recognise the presence of both push and pull factors in both origin and destination. In addition, attention must also be paid in particular to 'intervening obstacles' that can impede particular migrations, such as family obligations at the origin, the costs of moving, legal constraints and personal anxiety about migration. Understanding these obstacles suggests a need to adopt more humanist and structuralist perspectives when investigating the migration decision-making process.

Source: Bogue 1969; Lewis 1982.

---

approximation to the question of who migrates. In no way, however, does it explain the structural factors leading to a patterned movement, of known size and direction, over an extensive period of time.

(Portes 1978a: 5).

As Woods (1982: 152) notes, the empirical perspective in general 'avoids the question, "why does migration occur?" '

Adopting a structural perspective means that migration should be seen as a fundamentally *social* phenomenon – migration is a product of society. Given that the form of society is historically specific, so too is the form and character that migration takes. There are no pre-given 'laws' of migration. Migration should not, therefore, be treated as an unproblematic, natural and eternal form of behaviour. Additionally, the structuralist approach regards the organisation of everyday life through a much more critical lens than does positivism, which it regards as being a defender of the status quo within society.

Marxian structural approaches

In *structural* Marxism, the essential logic binding together human actions to produce regular forms of behaviour, responses and outcomes is that of the *mode of production* or the way in which specific societies organise their productive activity. In the present day, worldwide, the predominant mode of production is capitalism. Production is organised around the *bourgeoisie*, who are the owners of the *means of production* or the items required for production, such as land and tools. As owners of the means of production, the bourgeoisie extract surplus value from the *proletariat*, who have only their labour power to sell from the three key factors of production (land, labour and capital). The concept of *surplus value* refers to the unrewarded portion of the work undertaken by the proletariat for the bourgeoisie, which is translated into the capitalist's profit.

Applying a Marxist perspective suggests that the causes of, and the driving forces behind, migration should be located within the 'hidden' logic of the capitalist mode of production. Any explanation of migration cannot rely solely on either measuring characteristics of origin and destination locations, as suggested by the capital maximisation model, or concentrating on the actions and priorities of individuals, as suggested by the behavioural model. Primary attention must be paid to how the capitalist economy operates and evolves over time, and the way in which this dynamic structure has implications for the character and prevalence of migration. Hence, while Marx himself may have had little specific to say about migration, he

> examined migration in terms of its production roles in that migrants were a factor of production, whose destiny was determined by the capitalist demand for labour.
>
> (Shrestha 1988: 181).

Similarly, for Harvey:

> In search of employment and a living wage, the labourer is forced to follow capital wherever it flows. ... Wage differentials ... provide the means to co-ordinate workers' moves to capital's requirements. ... The more mobile the labourer, the more easily capital can adopt new labour processes and take advantage of superior locations. The free geographical mobility of labour power appears a necessary condition for the accumulation of capital.
>
> (1982: 381).

Marxian structuralism applied

H. Jones (1990) considers Marxian approaches to the study of internal migration within the developed world (Fielding 1982; Rees and Rees 1981) to be less sophisticated than those applied to studies of developing world and international migration. One developed world example, however, comes from Jones's (1986) own work on migration patterns in Scotland. Historically, Jones argues that the spatial inequalities generated by capitalism in the eighteenth and nineteenth centuries provided the spur to initiate new migration patterns. In the lowlands, a more capitalist approach to agriculture did not, as has often been thought, lead to high levels of out-migration as marginal producers were driven off the land. This was because the new agriculture created many jobs and also because of the rise of small-scale commodity production in rural areas. This, in turn, led to a significant redistribution of population through short-distance migration to the emerging industrial centres.

More dramatic migration patterns were apparent in the Highlands and Islands, where the clan-based pre-capitalist production system was replaced by capitalist relations. Large-scale agricultural change culminated in the infamous 'clearances' of the early nineteenth century, when sheep replaced people on the landlords' estates. Tenants were evicted to barren coastal smallholdings, or crofts, whose lack of productivity necessitated migration for temporary employment in seasonal work, the Army or the fishing industry. These people provided what in Marxist terms is regarded as a *labour reserve*, whose reproduction costs were only partly borne by the capitalist system, through wages, as the crofts provided the rest of the livelihoods. Eventually, however, the landowners began to pay for some of the crofters' upkeep and the crofters thus became a net burden. Consequently, with the rise of an unsympathetic urban capitalist class, the 'Highland problem' emerged, resulting in large-scale evictions and long-distance emigration to countries such as the United States.

Looking at labour migration in developing countries, especially that from rural to urban locations, Shrestha (1988) outlined another structural model of migration. Conventional models of migration are regarded as superficial and as giving too much importance to free choice and economic rationality. Instead, migration must be seen as being linked to changes in the *social relations of production* – the way in which participants in the productive process relate to one another – and the uneven development of the economy. As regards the former, attention is paid to the changing socio-economic structure of the countryside. For example, moderate land reforms have prompted landowners to consolidate their lands and become more capitalist in their operations. This marginalises many peasants, who are presented with three options: stay put and make the best of deteriorating conditions; revolt against the system; or migrate elsewhere. The latter is frequently chosen, as the *uneven development* of capitalism (Frank 1969) makes certain locations, especially big cities, appear more attractive for making a living. Generally, the development of capitalism sees the traditional mode of production undermined in the countryside, with agriculture becoming more capitalist and mass-produced commodities displacing locally produced goods. Government investment and attention becomes oriented to the 'modern' city and migration 'becomes a channel through which the surplus value of migrants' labour is driven or drawn from the periphery into the capitalist core' (H. Jones 1990: 221).

Shrestha went on to define three major issues for examination as regards migration and development. First, there is the question of *who* migrates, attention being concentrated on the migrants' class positions. Specifically, members of the subordinate class are less likely to be able to make strategic choices compared with members of the dominant class. Second, we must ask ourselves *why* migration occurs, with explanation moving beyond migrants' subjective responses through linkage to their class positions. Third, concern must be given to the socio-economic *impacts* of migration for the migrants, delineated largely in class terms. Hence, in conclusion:

For the large majority of migrants, migration is merely a holding action – a transitory spatial escape from the harsh realities of the regressive agrarian social structure to the different but equally prohibitive urban economic terrain dominated by capital.

(Shrestha 1988: 198).

Marxian perspectives on international migration

Zolberg (1989) has argued that much of the most stimulating recent research on international migration has come from a structuralist perspective, being historically specific, critical of the status quo and global in orientation (Castles and Kosack 1973). Two specific traditions within this work (Massey 1990; Massey *et al.* 1993) are concerned with labour migration (Chapter 4). First, there are studies stemming from Piore's (1979) examination of migrant labour in industrialised societies. Piore argued that international labour migration is a chronic feature of modern capitalism, since there is a permanent demand for immigrant labour to fill vacancies for unskilled jobs. Such jobs are increasingly distinguished from the jobs of the non-immigrant populations due to the institutionalisation of a *dual labour market*. The two labour markets are linked, with a capital-intensive primary sector and a labour-intensive secondary sector. Secondary jobs are poorly paid and insecure and display little opportunity for advancement. It is thus difficult to recruit people into this sector, except from poorer nations. Piore emphasises the demand-led character of labour migration, in contrast to the supply-led explanations of neoclassical theories: capitalism's need comes first.

Second, Wallerstein's (1974, 1979, 1983; Taylor 1989) *world system theory* focuses on the spatial growth of the capitalist economy. By around 1900, he asserts, this economy covered the entire globe. If we recognise such a global geographical structure, economic variation between countries reflects the uneven character of capitalist development. International labour migration is a response to this unevenness and its geographical segmentation into core and peripheral areas, defined by the sorts of economic activities that go on in each. Migration is 'a natural outgrowth of disruptions and dislocations that inevitably occur in the process of capitalist development' (Massey *et al.* 1993: 445) and 'follows the political and economic organization of an expanding global market' (p. 447). While much world systems analysis focuses on migration from peripheral to core locations and the exploitation of these migrants in the core areas, appreciation of a *global economy* helps to explain other flows. For example, Findlay *et al.* (1996) looked at skilled expatriate workers in Hong Kong and linked them to the rise of world cities and global restructuring, and to the demands for integration with international markets that these shifts necessitate. In short, migrants provide a linking mechanism for global capital.

A final example returns us more directly to the work of Marx. Miles (1987) is concerned to explain why capitalism, as a globally dominant mode of production, has not totally eliminated all forms of 'unfree labour', notably the migrant worker (Chapter 7) whose

status as a wage labourer was only formal because both the state and the contract of recruitment prevented them from freely disposing of their labour power on the labour market.

(*ibid.*: 167).

He begins his study with a Marxian emphasis on how migration must not be seen solely as spatial mobility, since such a conception

abstracts people from their material context because the significance of migration does not lie in spatial mobility *per se*, but in the position in the relations of production occupied before and after such movement.

(*ibid.*: 6).

People, therefore, relocate with respect to these relations of production, with movement from one 'class site to another, both within and between social formations' (*ibid.*: 7). The book then demonstrates how migrant labour has proved crucial to the development of the capitalist economies of Western Europe and elsewhere. For example, in the post-1945 period, labour recruitment from abroad has been considerable in countries such as Switzerland, France, Belgium, Britain and Germany (Chapters 7 and 9), sourced from (former) colonies, the European periphery and North Africa. Consequently, by the 1980s, although still concentrated predominantly into a limited range of low-skill manual occupations 'a larger proportion [of migrant workers] are employed in key sectors of the economy and constitute ... a relatively permanent part of the European proletariat' (*ibid.*: 159).

Theoretically, the need for these workers is explained in terms of the fluctuating labour requirements of capitalism. Besides a general tendency towards boom–bust economic cycles, Marx identified a contradiction within capitalism between a desire to employ as little labour as possible (notably by displacing people by machines) and a desire to employ as much labour so as to expand the accumulation of surplus value or profit. Resulting fluctuations of labour demand necessitate the presence of a 'reserve army' of labour, to be exploited in the workplace or thrown out of work at will. In short, Miles showed that capitalism has increasingly turned to migrant labour rather than to domestic sources (for example, former agricultural workers or women previously working only in the house) to provide this latent surplus population. Migrant labour contracted from the developing world is especially attractive in this respect as it seen as being highly flexible with respect to economic cycles, as having some of its production and reproduction costs 'externalised' through falling on the migrant's original society, and as providing little political disruption to the coherence of the host society (*ibid.*: 212–13, 1982; Chapter 9). Hence, Miles (p. 212) con-

cludes that migration is 'an integral feature of the development and expansion of the capitalist mode of production.'

An alternative structural approach

Structures other than capitalism can also be put forward as the driving force behind migration. One such structure is that of *patriarchy*, which represents the subordination of women to men through 'a system of social structures, and practices in which men dominate, oppress and exploit women' (Walby 1989: 214). Patriarchy need not be reducible to capitalism but represents a separate basis for exploitation (Mitchell 1975). In looking at patriarchy's link to migration, we must acknowledge how migration frequently occurs at the household rather than the individual level, so that the former is the appropriate unit of analysis (Boyd 1989; Wood 1981, 1982). However, the significance of gender extends beyond the household, with

migration decision making processes ... shaped by sex-specific family and friendship sources of approval, disapproval, assistance and information.

(Boyd 1989: 657).

With patriarchy, these shaping factors adopt a clear structural character.

Women in the developed world tend to show a greater reluctance to migrate than men, a relationship that holds even after controlling for factors thought to influence willingness, such as occupational status and home ownership (Markham and Pleck 1986). Just considering the case for married or cohabiting women, a large number of reasons have been proposed to explain why there should be an independent association between sex and willingness to move (Halfacree 1995a). On the one hand, there are internal perspectives, which focus on the individual woman and her family, discussing migration decisions in terms of issues such as net gains in human capital or gender roles within the family. On the other hand, there are external perspectives, which locate the potential migrant and her family within their broader social context. It is from the latter perspective that an understanding of women's relative immobility in terms of the structure of patriarchy can begin. Differences in the willingness and propensity to migrate by sex can be seen to follow the logic of the patriarchal model. For example, in delineating work into 'women's jobs' and 'men's jobs' (Bradley 1989), patriarchal forces have meant that, in general, women have much less to gain financially or in career terms than men because 'their' jobs tend to be lower paid and less career-oriented. In addition, the unequal power relations in the household, which reflects the domestic dimension of patriarchy, give women less say about, and control of, the migration decision than men. Overall, married women's reluctance to move relative to their spouse must be traced back to the gender-specific logic of patriarchy.

Structuralist insights suggest a highly determinist conceptualisation of migration, with individuals almost excluded from the picture. Although politically it comes from a very different tradition, this approach thus reveals considerable similarities with the neoclassical positivist tradition discussed earlier. Counter to all these determinist perspectives, however, are humanist accounts of migration, which question the systematic and quantifiable character of migration.

Humanist accounts of human migration

The humanist critique of determinism

The *humanist critique* argues that potential migrants must be regarded first and foremost as individuals, with the decision on whether or not to migrate being irrevocably that of the individual. In contrast to deterministic approaches, migration cannot be predicted by measuring the characteristics of individuals or of different places. Instead, we must engage with what Davis (1945: 248) termed 'the motives that migrants carry in their heads', concentrating upon *personal characteristics*, such as beliefs, aspirations and obligations, always acknowledging that such characteristics never bear a one-to-one relationship with the likelihood of migrating. Consequently, humanist approaches to migration research involve intensive qualitative methodologies, such as in-depth interviewing or constructing life-histories (Chapter 2).

The importance of producing humanist accounts of migration to complement the plethora of quantitative work into patterns and generalities of migration has been noted in the literature. For example:

Heavily quantitative studies using large data sets tend to produce an impersonal, dehumanized approach in which flows replace individual people and the motives for migration are assumed rather than proven, often being interpreted in a simplistic and generalized way to a point where they have little meaning. … In the process of aggregating the data, individuals, with their hopes, fears and aspirations, become lost.

(Pooley and Whyte 1991a: 4–5).

Lack of data and often poor-quality material have proved important in explaining this relative underdevelopment of humanism within migration research (Courgeau 1989; Fielding 1992a; Taylor and Bell 1994). This is perhaps unsurprising given the very intensive character of qualitative methods. Thus, key reviews of theories of migration tend to ignore the contribution, actual and potential, of humanist approaches (W. Clark 1982; Greenwood 1985; Massey 1990; Massey *et al.* 1993; Molho 1986; Woods 1982). In spite of what Lewis (1982) suggests, however, the lack of humanist work should *not* be associated with the problems of acquiring a representative sample, since humanist work is not about seeking generalisation; it concentrates instead on describing and relating individual experiences, experiences that may or may not be shared by others. It attempts to gain subjective understanding rather than producing statistical description. Thus, for Miles and Crush (1993: 87), 'one life-history text is equally as "valid" as a historical text as 100.'

Behavioralism in migration research, as with behavioralism more generally within geography (Cloke, Philo and Sadler 1991), acted as something of a bridge between the 'people-less' conceptualisations of positivism and structuralism and the highly 'peopled' conceptualisations of humanism. Survey work, associated with behavioralism, goes some way towards accepting the humanist critique of much of migration research, in that it actually involves asking the migrants themselves why they have moved house. However, such an approach often tends towards positivism rather than humanism in the way that it feeds into quantitative modelling approaches and strives to generalise across a population. The survey approach does not fully take on board the humanist emphasis on the uniqueness of the individual. Moreover, humanism requires much more detailed work in teasing out the causes of the migration than is usually captured by surveys, although the latter provide a useful first step.

Humanist insights

Migrant history work

A more in-depth analysis of individual migration is apparent in much of the work on migration done within historical geography (Langton and Hoppe 1990; Pooley and Whyte 1991b). On the one hand, limitations in data sources, such as their levels of coverage and completeness, encouraged historical geographers to consider alternative data sources but, on the other hand, there was a feeling that quantitative methods tended to underplay the personal and qualitative aspects of migration. A key effort in this area of research has been to build up *migrant histories* (Chapter 2), where individuals are traced through their lives and their biographies are built up, emphasis being placed upon the migrations that have shaped and have been shaped by these lives (Bartholomew 1991; Chapman 1985; Courgeau 1989).

In justifying their use of migrant histories, Forrest and Murie argue that they avoid both the 'problems of the averages of aggregated situations, where such averages may not reflect the actual situation of any one case' (Forrest and Murie 1990: 193) and the 'arid and depersonalized' (Forrest and Murie 1991: 64) perspectives of much survey work. However, they did not favour an extreme humanist 'every case is different' model (Forrest and Murie 1990: 207) but saw the need for case studies to illustrate and inform the impersonality of more general accounts of migration trends and patterns. This argument points towards the attempts at integrating different philosophical approaches, discussed later in this chapter.

In contrast to Forrest and Murie, from a developing world context, Gulati makes an impassioned case for adopting a more strictly humanist philosophy:

Perhaps my friends are rational agents maximizing utility in a perfectly competitive market, but I do think one should ask them about their lives before making those assumptions. I am sure they would appreciate our listening to their stories before we assume behavioural characteristics for them and proceed to solve models and make policy decisions for them.

(1993: 10).

Gulati's work should not, however, be regarded as somehow idiosyncratic or uncoordinated, since, as qualitative methods demand, she draws out some general conclusions. The crucial point is that individual stories are expressed and respected in their own right. A similarly impassioned plea for the use of migrant life-histories is made by Miles and Crush (1993), as they reflect upon their own research into the experiences of former migrant workers in southern Africa, notably Swazi women.

Pooley and Whyte (1991a) have noted how the emergence of novel sources of migration data has often been associated with key developments in migration research. Such an opportunity is now being presented to some degree for humanist work by the emergence of *longitudinal data sets* (Chapter 2). Such data sets facilitate the migrant history approach by providing, systematically, large amounts of information, which can be used to track the migration behaviour of individuals through time. The value of longitudinal data in this respect is currently being realised (Greenwood 1985), with Nash (1994: 388) describing them as 'the most interesting current research frontier in migration studies'. Nonetheless, longitudinal data sets may not be used predominantly to raise the profile of humanist work in migration research. Much of the use to date has been concerned with modelling migration behaviour (Davies and Flowerdew 1992). Moreover, the actual data provided by the longitudinal surveys tend to be rather restrictive, largely limited to the types of statistical information typically associated with censuses and other social surveys.

Migration and culture

The use of migrant histories can be extended to a recognition that migration is a highly *cultural* experience for all those involved (Bottomley 1992; Fielding 1992a), a recognition that has been best achieved within anthropology (Moon 1995; Taylor and Bell 1994). Migration tends to have meanings for a given society or for a sub-section of that society that cannot be reduced to clear-cut economic or social factors alone. For example, in their in-depth interviews with professional people in Hong Kong regarding ideas about emigration, Li *et al.* (1995) stressed the centrality of concerns over retaining Chinese ethnic identity with any migration. Ethnicity was itself strongly associated with place, and potential migrants raised issues such as whether experiences at the destination would undermine this cultural identity. Elsewhere, on a religious level, Abu-Sahlieh (1996) describes the Islamic conception of migration, expressed in Islamic law in terms of issues such as the obligations to and of migrants, or when emigration should take place. Furthermore, locating migration within cultural patterns and practices helps to naturalise it amongst those involved. This was noted by Lewis (1982: 123) with respect to the migration of young adults from the Welsh district of Colwyn on the English border, where migration was 'the "done thing" … a part of family life …

irrespective of … educational attainment or desired occupation.' Overall, therefore, migration and culture can be linked in *cultures of migration* (Chapter 9).

As with the emergence of longitudinal data, an appreciation of migration as being highly cultural need not promote more strictly humanist conceptualisations of migration itself. Fielding (1992a: 202) observes that culture is a 'property of individuals and groups'. Hence, we can envisage the cultural dimension of migration in very individualist terms or we can see it as more of a group-based phenomenon. The latter approach takes us back towards more structural conceptualisations. Nevertheless, adopting a cultural perspective on migration does suggest the need for a more ethnographic and qualitative approach to research. For example, Table 3.1 delineates some of the intersections between migration and culture. To the five points listed in the table, we can add how 'people from "another place" often fail to accept without question the taken-for-granted "realities" presented [to them]' (Bottomley 1992: 43), and how this draws attention to issues of domination, resistance and socio-cultural transformation. Exploring where the individual migrant (or nonmigrant) fits within these intersections clearly gives plenty of scope for the play of individuality.

Taking things a step further, the study of migration can be used to investigate the human condition (White 1995). As White notes, migration and the associated experiences of dislocation and marginality can undermine the stability of our cultural identities and our sense of place in the world (Chambers 1994; Gilroy 1993). This is apparent in the coverage of the experience of migration and the expression of what it is like to be a migrant that is contained in literature (King, Connell and White 1995) and other artistic media.

Although a humanist conceptualisation of migration has been relatively poorly developed in the migration literature, it has become more popular in recent years. Where humanism has emerged it has often appeared more in the form of a critique of existing conceptualisations than as a coherent alternative. This is typical of the impact of humanism in geography more generally (Entrikin 1976). Thus, we have seen that the use of survey work typically feeds into developing behavioural models, longitudinal data invigorates positivist work, and migration as culture can be seen as coming from as much a structural as a humanist direction. This cross-fertilisation of philosophical perspectives takes us on to efforts to produce more integrated accounts of migration.

Integrated accounts of human migration

Calls for integration

The contrasting conceptual approaches in migration research and their attendant methodological strands are not necessarily mutually exclusive. The various traditions have learned much from one another and it is rather too strong to suggest, even with respect to 'microeconomic' and 'historical–structural' studies, that 'they are two academic communities that speak utterly different languages for which there is no common idiom' (Wood 1982: 312). Nonetheless, there is a common feeling that 'disciplinary parochialisms' (Massey 1990: 4) have caused a fragmentation of migration research and reduced communication between different traditions. As Massey (p. 19) concludes:

> What is needed most is not better methods, better data, or different theories, but a simple realization that the fragmented way social scientists have pursued the study of human migration in the past is not likely to bear much additional fruit. What is needed now is for migration researchers in all fields to read more of one another's work and to spend more time thinking about how to integrate their theories and findings.

Awareness of the fragmented state of migration research at all geographical scales is widespread and has prompted calls for integration (Massey *et al.* 1993; White 1980; Woods 1982). There are a number of efforts to produce more integrated accounts of migration and a more holistic, encompassing conceptualisation of the migration process. Such work reflects the recent revival in interest in theoretical issues in migration research (Findlay and Graham

Table 3.1 Intersections between migration and culture.

- The varied cultural characteristics of different groups of migrants.
- How the cultural characteristics of different places influence migration.
- How migration shapes culture.
- The way in which cultural change provokes different migration responses.
- How migration can change and mould cultures.

Source: Fielding 1992a: 202–3.

1991; Halfacree and Boyle 1993). On its own, Massey's advocacy of greater interaction between the different traditions may be insufficient. Instead, developments within social theory can be taken on board in order to produce novel conceptualisations of migration.

Broadening conventional theories

Extending capital maximisation insights

Studies of labour migration rooted in the neoclassical tradition have taken on board the human capital and other criticisms of the early post-Hicks work. For example, Stark's (1991) study of migration in the developing world adopts two central premises emerging from this criticism. First, he sees a need to acknowledge that 'there is more to labor migration than a response to wage differentials' (p. 3) and, in a critique of the social physics tradition, migration is seen as being 'fundamentally dissimilar to the flow of water, which will always be observed in the presence of height differentials' (p. 3). A greater humanist or, at least, behavioural element must come into this work. Second, labour migration is not to be seen as individual optimising behaviour but as reflecting the importance of relationships of mutual interdependence and the family context.

This second point draws attention to the issue of *risk*, which has been used to extend the human capital approach more generally. Thus, risk and uncertainty have been incorporated into human capital models in a variety of ways that assume a strategy of risk avoidance (Hart 1975). For instance, measures of risk – such as the likelihood of becoming unemployed – can be built into the models through the use of measures of probability (Milne 1991). From a more behavioural angle, it can also be noted that individuals vary in respect of their willingness to take risks in order to obtain benefits – they vary in their willingness to gamble – which feeds into the specification of the migration as an investment decision (Molho 1986). This variation in attitude to risk on the part of the potential migrant reflects both individual and contextual factors: personality and knowledgeability.

Recognition of the role of risk features strongly in the work of Stark, as mentioned above (Katz and Stark 1986; Stark and Levhari 1982), who is associated with the *new economics of migration* school (Stark and Bloom 1985). For this school, migration can be interpreted as being guided by a family-based risk-reduction strategy rather than as an individual income-maximising action (Massey 1990; Massey *et al.* 1993); it is fundamentally built upon risk avoidance. Thus, in the context of urban-to-rural migration in the developing world, Stark and Levhari (1982: 192) argue that

an optimizing, risk-averse small-farmer family confronted with a subjectively risk-increasing situation manages to control the risk through diversification of its incomes portfolio via the placing of its best-suited member in the urban sector, which is independent from agricultural production.

Box 3.5 Stark on labour migration

Stark's *The Migration of Labour* represents an important recent contribution to the field of migration research in the developing world. The extract produced below, taken from the chapter on 'The new economics of labor migration', emphasises the need to adopt a more humanist perspective on the 'economics' of this migration:

Whereas owners of production inputs or commodities, such as bricks or bottles of wine, can ordinarily ship them away (so as to maximize profits or utility) while themselves staying put, owners of labor must usually move along with their labor. Furthermore, owners of labor have both feelings and independent will. Indeed, *most aspects of human behavior, including migratory behavior, are both a response to feelings and an exercise of independent will*. These simple observations divorce migration research from traditional trade theory as the former cannot be construed from the latter merely by effecting a change of labels.

(Stark 1991: 24, our emphasis).

Yet, qualifying this potentially individualistic understanding of migration, Stark also emphasises the importance of taking family relations into account:

Just as it is clear that neither a brick nor a bottle of wine can *decide* to move between markets, so should it be equally clear that a migrant is not necessarily the decision-making entity accountable for his or her migration. Migration decisions are often made jointly by the migrant and by some group of nonmigrants. Costs and returns are shared, with the rule governing the distribution of both spelled out in an implicit contractual arrangement between the two parties.

(*ibid.* 1991: 25).

Note, however, that this second quote does not make explicit mention of the forces of capitalism in also determining migration and, hence, leaves Stark open to criticism from a more structuralist angle.

Source: Stark 1991.

Furthermore, the role of *information* is central to a risk avoidance strategy and has thus been considered further than in the original human capital model. In particular, the intervening role of a migrant's level of information about the relative opportunities in a specific place, and the accuracy of this information, can turn upside-down more objectively defined migration outcomes. Thus, in the example of labour migration, an asymmetry of information between employers and employees (Katz and Stark 1984, 1986; Kwok and Leland 1982) – either the employers or the employees have 'more of the relevant information' (Stark 1991: 169) – can be incorporated into models to produce varied migration outcomes (*ibid.*). Overall, therefore, we see a greater appreciation of the way in which migration is a highly contextual phenomenon rather than the isolated, objectively rational act that it was initially thought to be.

Extending behaviouralism

Turning to other developments from behavioralism, Speare, Goldstein and Frey (1975) put forward a model that sees migration in behavioural terms but that integrates the stress-threshold model with the cost–benefit considerations of human capital approaches; response behaviour is combined with goal-seeking behaviour. In behavioural style they break migration into stages: the emergence of a desire to move; the selection of a potential destination; and the ultimate decision on whether or not to move to that destination. The first stage represents the emergence of dissatisfaction, due to factors such as life-cycle change and environmental deterioration (Graves and Linneman 1979), while the selection stage involves action spaces and place utility comparisons (Huff and Clark 1978; Phipps and Carter 1984). However, the third stage owes much to human capital approaches, with the operation of a complex cost–benefit analysis.

Elsewhere, Cadwallader (1989) has proposed a model that tries to integrate the macro and micro traditions, albeit from an explicitly positivist–behavioural perspective (Figure 3.4). At the top of the figure, the 'objective' variables (O_h, O_i, O_j), such as comparative wage and unemployment levels at the origin and destination, can be combined in positivist fashion to model observed migration behaviour (M) through stage (i). However, the mechanisms in the 'black box' that bring about this link are to be explored further through behavioural concerns. Thus,

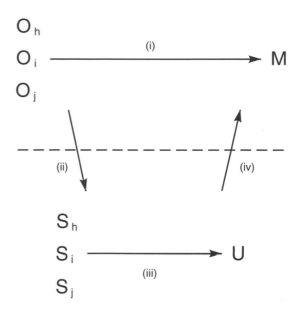

Figure 3.4 A conceptual framework for analysing migration behaviour (source: Cadwallader 1989: 496, reprinted by kind permission of Arnold).

through stage (ii), the objective variables are transformed into their 'subjective' counterparts (S_h, S_i, S_j), such as the migrant's perceptions of comparative wage and unemployment levels. These, in turn, combine to form an overall measure of attractiveness for the destination (U) through stage (iii). Finally, in stage (iv), investigation must be made of how the overall utility function generates the actual migration observed and modelled initially (M).

Extending place utility

The use of place utility, too, has been developed and broadened substantially since its introduction by Wolpert. Development of *multi-attribute utility theory* (MAUT) concerns itself with obtaining multiple goals simultaneously (Baron 1988; Moon 1995). When applied to migration it therefore recognises that the decision to move typically involves the resolution of a range of concerns and not just obtaining the net composite of utilities suggested by Wolpert. Unfortunately, as Lin-Yuan and Kosinski (1994) make clear, MAUT does not appear to work for individuals, in particular due to their adopting satisficing and non-rational behaviour. Thus, it has been suggested that individuals adopt a range of decision-making strategies (Svenson 1979), demonstrated by Lin-Yuan and Kosinski (1994) with respect to Chinese immigrants to Canada.

In spite of these developments, place utility remains firmly within the behavioural tradition and retains the sequential tradition inaugurated by Rossi's model (Woods 1982). In particular, we can be critical of how the approach suggests a very formal model of individual decision making. This formulation is part of the 'intellectual fallacy' within academic work, whereby

the human agent [is] reduced to a cognitive drone, to a string of internal programmes responding to an external environment. People's action is governed by some 'inner', on-board computer which assimilates all the available knowledge, works out the angles and decides on an appropriate course of action.

(Thrift 1986: 87).

Thrift's observations echo Gulati's concerns, outlined earlier, as to the dehumanising character of much of the work. A higher utility somewhere else may not lead to migration, not because rational decision making does not always lead to appropriate behaviour but because decision making is not as systematic as the intellectual fallacy has us believe (Woods 1982).

Advocates of a modified place utility approach partly recognise the problem of the intellectual fallacy. For example, Lin-Yuan and Kosinski (1994: 51) note as one of the failings of MAUT its requirement for 'large amounts of information and rather complicated computation'. They also note that the less rigorous (in terms of data and decisions) decision-making strategies they illustrate remain 'stylized characterizations of the decision-making process' (Goodman 1981: 133) and that the decision makers themselves may not 'consciously follow certain decision rules or use those rules to check their behaviour' (Lin-Yuan and Kosinski 1994: 59). Nonetheless, they do not go further to question the sequential tradition itself and to set migration decision making in the broader context of the individual's biography, as discussed below.

Extending structuralism

The Marxian structural model of migration has also been subjected to critique and modification. Marxian approaches have been criticised for their 'timeless functionality' (Cohen 1987: 144), in which migration for labour market reasons is seen as a permanent feature of capitalism, neglecting the historical specificity of this mode of production (Zolberg 1989). Bach and Schraml (1982: 337) argue that 'the logical connection between the development of capitalism and migration has been overplayed'. In response to such criticism, as we saw with respect to Miles's (1987) exploration of capitalism's link to 'unfree' labour, much work has been done to show how and why this logical connection exists.

In addition to work detailing the linkage between capitalism and migration, reference can also be made to the presence of other social structures, each with its own essential logic, such as the patriarchy discussed earlier. Nevertheless, acknowledging the importance of these other structures can still be seen as insufficient, since it is the determinism of structuralism that is the real problem. Hence, much Marxian work now tends to adopt the label 'political economy' (H. Jones 1990), trying to take on board a more humanist feel and reflect how labour power is not the same as other means of production, as it has a 'human element'. Thus, in terms of labour migration in the developing world, the destruction of the peasant economy is not an inevitable one-way process but is a contested experience, where the individual can, at least in some circumstances, 'make a difference'.

Migration systems and networks

Defining migration networks

Research on international migration has increasingly stressed the importance of *networks* in moulding and structuring patterns of migration (Findlay 1992; Nash 1994). Such an emphasis was apparent in Marxian studies of international migration, as outlined above. Network studies concentrate on linkages between sender and recipient countries (Kritz, Lim and Zlotnik 1992), linkages that help to sustain patterns of migration over time. A pioneering emphasis on the importance of such connections was Thomas's (1954, 1973) investigation of the relationship between migration and economic growth within what he termed the Atlantic economy (Skeldon 1990: 144–6). For Thomas, the nations each side of the Atlantic – North America and Western Europe, notably Britain – should be regarded as a single economy. However, this was by no means a static economy, with the fortunes of Britain relative to North America fluctuating over the late nineteenth-early twentieth-century period. Drawing upon the economist Schumpeter's linking of migration to business cycles within the economy caused by factors such as technological innovation and the relative success of different businesses, Thomas argued that migration could be explained through observing 'patterns traced by other series reflecting activities bearing some relation to migration' (1973: 26). Specifically, when business cycles, such as the building cycle, were on the upswing in Britain, migration was dominated by internal flows to the buoyant urban areas, but when there was a downswing emigration to the rival upswing economies of North America predominated. Moreover, these flows should be seen not just in terms of labour power but also in terms of capital and commodities. Hence, a migration network was apparent.

Massey (1990), also drawing on macro-economic theory, argues that networks promote positive feedback mechanisms of 'cumulative causation', whereby the establishment of links between origins and destinations has the effect of promoting greater levels of migration, sustaining and enhancing the flows. One way in which this can take place is through employment growth and migration being mutually supportive: employment growth stimulates in-migration and the resulting population growth – the new arrivals tend to be young, well-educated or trained, and highly productive – stimulates more employment growth, prompting more in-migration, and so on.

Social networks also serve to link origins and destinations, seen most clearly when migration is treated as being a family or household experience (Mincer 1978; Sandell 1977). In Gulati's (1993) study of the women 'left behind' in Kerala, India, she emphasised the sustained, if often difficult to maintain, connections that the women kept with their husbands working overseas. Crucial here were the monetary remittances sent back home (Chapter 4), which were used to support the family and to invest in property. Ultimately, these linkages led to return migration by the husbands, although in other situations the 'guests' may stay in their new home areas (Schaeffer 1991). Similarly, Trager (1988) explored migration and family interdependence in the Philippines, while Lewis (J. Lewis 1986) and Van Westen and Klute (1986) stressed the importance of remittances and return migration for regional development in the sending nations. Generally, family, friendship and community linkages underlie many of the contemporary patterns of international migration, going beyond simple economic relations between countries to extend to being information

conduits and to providing a cultural cushion for the migrants from negative aspects of the host society (Boyd 1989; Gurak and Caces 1992). Again, a pattern of cumulative or at least sustained causation is apparent. Migration flows become self-sustaining as information flows, patterns of assistance and obligations develop between the two locations.

Migration as a system

The presence of these varied socio-economic linkages in a wide range of migration situations suggests that we must not consider migration just from the perspective of its origin or destination. There are interrelationships between *all* of the components involved in the migration process, since the networks that link migrants and non-migrants across time and space tend to assume fairly regular and relatively permanent structures. In short, we can conceptualise migration as a *system* linking the sending and receiving areas (Castles and Miller 1993). For example, we have systems such as Arab and Asian labour migration to the Gulf states, whose significance was exposed starkly in the Gulf War of 1990–1991, when the return home of two million migrant workers within four months of the invasion of Kuwait led to the loss of hundreds of millions of dollars in remittances (Connell 1992). Other examples of international-scale systems include the links between Francophone Africa and France (Garson 1992), the Caribbean nations in a global context (Simmons and Guengant 1992), and the Southern Cone of Latin America (Balán 1992). Additionally, at both the inter- and intra-national scale, there was the example of the notorious migration system established by the apartheid-era South African mining companies as a form of labour control (Chapters 4 and 7).

An early call to focus on such systems and to integrate macro- and micro- perspectives, through the use of systems theory, came from Mabogunje (1970) in his model of rural-to-urban migration in West Africa. The model he constructed was meant to be calibrated and quantified so that any change in one aspect of the system could be traced throughout the whole. There are four main components to the model (Figure 3.5) (overleaf), which are typical of systems work generally (Bennett and Chorley 1978): the environment; the migrant; control sub-systems; and adjustment mechanisms (Table 3.2) (page 79). When the migration has taken place, it may cause either positive or negative feedback of information

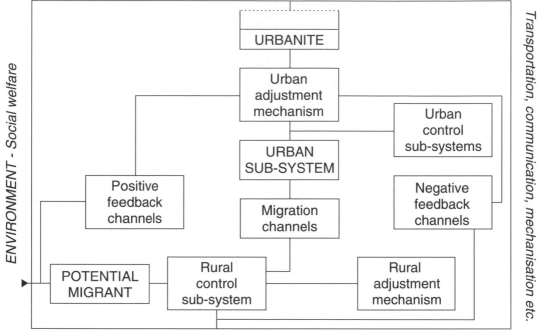

ENVIRONMENT

Economic conditions; Wages, prices, consumer preferences, degrees of commercialisation and industrial development

ENVIRONMENT

Governmental policies, agricultural practices, marketing organisations, population movement etc.

Figure 3.5 A rural–urban migration system (from 'Systems Approach to a Theory of Rural Urban Migration' by Akin L. Mabogunje, Geographical Analysis, Vol. 2, No. 1 (January 1970), reprinted by permission. © 1970 by Ohio State University Press. All rights reserved).

between the migrant and the origin area, respectively encouraging or discouraging further migration.

In spite of Mabogunje's detailed elaboration of a systems approach to migration and the regular occurrence of his model in migration texts, it is not a model that has been widely applied. In practical terms, this is because of the immense difficulties faced by its application, especially as regards data demands. Furthermore, systems approaches can be criticised for the way they isolate and draw boundaries around the systems in such apparently precise ways. 'Real world' systems are unlikely to be as precisely delimited as the theory would have us believe. Finally, systems theory can be criticised for the positivism of its emphasis on modelling flows and interactions between the various components of the system and, from a humanist angle, for the overall

mechanistic metaphor on which the concept of 'the system' rests. In particular, the formalisation apparent in systems theory neglects the more social element of networks, captured by Boyd (1989) in her understanding of migration as a 'social product'.

In spite of these difficulties and criticisms, interest in a systems approach remains. For example, Poot (1986) applied a systems approach to inter-urban labour migration within New Zealand. He argued that the novelty and power of his approach to the study of aggregate flows lay both in the quality of its variables, concentrating upon male workers and precisely defined labour market areas, and in the appreciation of interdependencies between flows. The latter, in particular, was characteristic of the systems' emphasis upon context. Kritz and Zlotnik (1992) also draw upon Mabogunje's ideas in their advocacy

Table 3.2 Main components of Mabogunje's systems model.

Component	Details
Environment	This is the context within which the migration system sits. In Mabogunje's case it was one of rapid economic development and an end of rural isolation, which was resulting in an integration of the rural and urban economies.
Migrant	Within the system itself, the first core component is the migrant, who in Mabogunje's Africa was being encouraged to leave the rural environment for the city. The 'energy' with which the migrant was propelled on his or her quest was seen to vary with the individual but was related to factors such as awareness of potential opportunities.
Control sub-systems	These components control the flows (migrations) within the system. For Mabogunje, the rural control sub-system comprised such things as the family and the community, which could either encourage or discourage the migrant. In the urban arena, the opportunities for housing, employment and general assimilation were stressed.
Adjustment mechanisms	Although bounded, the system itself is not static as a result of a migration. Adjustment mechanisms operate in the rural area to deal with the loss, while other mechanisms operate in the city to incorporate the newcomer.

Source: Mabogunje 1970.

of a systems approach to the study of international migration. Their migration systems comprise groups of countries exchanging significant numbers of migrants, whose importance is enhanced in this era of enhanced 'temporary' migration and global interdependence.

Rejection of systems theory does not mean a rejection of the significance of migration networks. Indeed, work originating in structuralism has embraced the adoption of systematic perspectives. Through its very metaphors of structure and layers of reality, structuralism emphasises linkage (Boyd 1989; Massey 1990). With its strong emphasis on historical specificity, structuralism also helps us to broaden system concepts, emphasising the role of macro-economic forces in structuring the system. This has most comprehensively been explored in the work on dual labour market theory and, of course, Wallerstein's world systems, as outlined earlier.

The structuration of migration and the biographical approach

Structuration theory

Structuration theory, usually associated with the work of Giddens (1984), represents an explicit attempt to overcome the determinist–humanist divide that bedevils social science in general. Giddens's theory hinges on a well-known quote from Marx:

Men make their own history, but they do not make it just as they please; they do not make it under circumstances chosen by themselves, but under circumstances directly encountered, given and transmitted from the past. The tradition of all the dead generations weighs like a nightmare on the brain of the living.

(Marx 1852, quoted in Tucker 1972: 437).

Drawing upon this quote, Giddens argues that we should attach equal importance to *human agency*, or the actions of individuals, and *social structure*, as understood by structuralists. Indeed, the two cannot and should not be separated, as they continuously (re)produce each other: 'agency produces structure produces agency produces structure in a never-ending recursive process' (Thrift 1985b: 612). This process Giddens terms the 'duality of structure'. Structures for Giddens exist only because people make them exist: thus, there would be no capitalism if we did not reproduce its structures every day through engaging in wage labour, uneven development, commodity buying and selling, and so on. Structures do not, therefore, determine behaviour as structuralism suggests. However, Giddens is not a strict humanist advocate either, because the presence of structures within which society is organised means that people are likely to conform to their norms and expectations, whether willingly or not. The individual is not as free, at least in the everyday, 'real world' context, as humanists would have us believe.

While Giddens's ideas are complex and wide-ranging, structuration theory has reached migration research. It has been applied by Halfacree (1995a) to conceptualise the gendered character of labour migration in the United States. Looking at a typical

American family, Halfacree argues that the main actors involved – the wife, husband and employers – are likely to engage with the structure of patriarchy when deciding upon and undertaking a migration. The consequence of recognising this subordinate position for the woman is that her priorities with respect to a migration are played down compared with those of her husband. Typically, this results in her becoming a 'tied' migrant, where the migration is primarily for the benefit of the husband. Thus, the structure is used by the human agents to bring about and justify a particular type of migration. Moreover, in becoming a tied migrant, the wife also helps to perpetuate the original structures of patriarchy through the duality of structure. In the domestic sphere, the status of the wife as a co-provider is undermined, while within the labour market her position is weakened, supposedly legitimating the sex-typing of occupations by reinforcing the perceptions of employers and others that women are uncommitted employees. Thus, by 'obeying' the structure in the way that they do, the migrants help to perpetuate it as an effective representation of the 'real world' when applied to migration.

If structuration theory were a humanist theory, then it would suggest that women themselves are partly to blame for the perpetuation of patriarchy. This would be unfair and superficial, as well as being highly controversial! Reproduction of the patriarchal structure is likely to be an *unintended consequence* of the migration. The household migrated for economic betterment; in a way it was just bad luck that the package involved sustaining patriarchy. The migrant woman must also be seen within her social context: as part of a household and of wider society. She is under considerable pressures to conform to the norms and expectations – to the *power relations* – of that society, which are typically in evidence around her all of the time. It takes considerable struggle and effort to break completely with a highly entrenched patriarchal society.

Interest in structuration theory has also been shown by those working on migration in the developing world. For example, Forbes (1984) stressed the need to overcome the agency–structure divide and, in their comments on a conference on international migration, Gilbert and Kleinpenning (1986) noted favourably an increased recognition of the presence of differing responses to similar structural socio-economic realities. They applauded the intentions of structuration theory, namely the way in which it 'overtly recognises the role of individual re-

sponse to the environment of constraint and opportunity' (p. 5). Elsewhere, Goss and Lindquist (1995) used structuration theory to interpret international labour migration from the Philippines. Accepting that the uneven development of capitalism provides the structural context for the existence of international labour migration, they argue that attention must be paid to the problematic character of this migration actually taking place. Potential migrants must have the appropriate structural resources and rules to engage with the migration agencies and brokers, who, in turn, reflect the institutionalisation of this migration. Such engagement is a skilled accomplishment and also serves to reproduce the institutionalisation itself.

The biographical approach

Structuration theory's emphasis upon the contexts and cultures in which migration takes place brings us to Halfacree and Boyle's (1993) advocacy of a *biographical approach* to the study of migration (White and Jackson 1995). This call comes, on the one hand, from a critique of the inadequacy of past ways of conceptualising migration and, on the other hand, from a recognition of the considerable value that many of these conceptualisations demonstrate. The biographical approach complements the structuration perspective and is intended to be used to illustrate in more depth the contextual processes through which the structuration of migration takes place. Crucially, it goes beyond the humanist bias suggested by many previous uses of the term biographical to allow explicit recognition of the structural constraints and enablements moulding migration.

Migration, Lin-Yuan and Kosinski (1994: 50) have noted, 'is generally viewed as goal-directed behaviour', exemplified most clearly in the behavioural models of migration illustrated in Figure 3.3 (pages 65–67). The biographical approach starts by questioning this assumption and qualifying it heavily through developing the critique introduced by Thrift's intellectual fallacy. Instead of stressing the purposeful and calculating character of migration, the biographical approach emphasises its location within the individual migrant's entire biography. In particular, three key issues are stressed (Halfacree and Boyle 1993).

First, migration has to be seen as an action in time, whereby the reasons for moving are not to be understood just at the instant prior to the move, as a result, for example, of comparing place utilities. Instead these reasons also relate in some way to the migrant's past and anticipated future. The reasons for moving are seen as being part of the migrant's whole life – their biography – and thus are unlikely to be appreciated fully just by asking blunt questions such as: 'Why did you move?' Instead, there is a need for in-depth qualitative work, enquiring around the subject and building-up a picture of the migration decision from a variety of angles, demonstrating how and where it fits into a person's life (Stubbs 1984; Vandsemb 1995).

Second, the biographical approach argues that specific migrations, because of the way they are embedded within our biographies, are likely to exhibit varied causes. Thus, multiple reasons will be involved in explaining specific moves. While all of these reasons will not be of equal significance, they all contribute to explaining the detailed form that a specific act of migration takes. This is not to suggest a neat sequence of discrete steps behind each move – for instance, arguing that reasons for leaving and reasons for coming are distinct (Williams and McMillen 1980) – as this would be to fall prey to the intellectual fallacy. Instead, because the act of migration is immersed within the migrant's biography, causally it is likely to be a very mixed-up and chaotic decision. Hence the difficulty in representing the decision-making process accurately.

Third, and chiming with the calls of Fielding (1992a), migration must be seen as a very *cultural* event. Individual migrants are embedded within and are part of varied cultures. These cultures introduce them to and socialise them into the normative behaviour and responses of the structures described by structuration theory. Consequently, as with structuration, the biographical approach should not be seen in highly humanist terms but as stressing the actions of contextualised individuals (single people, households, larger groups, and so on).

Explicit use of a biographical approach is in its infancy, but there are numerous pieces of work and ideas that are sympathetic to the approach. From the material described earlier, we can draw attention to the work on the cultural dimension of migration, the interest in longitudinal data, the investigation of housing histories, and structuralism's turn towards political economy. Davies (1991; Pickles and Davies 1991) even displays a biographical sensitivity in his attempts to model migration behaviour. He stresses the need to understand migration in terms of both long-term 'dynamic optimisation behaviour' and more short-term issues, without recourse to utility

maximising, satisficing or 'other particular formalised representations of decisionmaking' (*ibid*.: 480). Elsewhere, and generally reflecting the stronger anthropological tradition within developing world migration work (Skeldon 1995), Chapman has cautioned against the un-biographical formalised approach to understanding migration that has characterised much work on Pacific island migration:

Counting people who left island communities produced a numerical skeleton that was replicated by tabulating their personal characteristics and demographic profiles. A complex social process was reduced to a mechanical sequence of discrete events, abstracted from the broader structural contexts of environment, history, culture, society, economy, and polity. . . . the ferment of island mobility does not merge easily into a scholarly tapestry woven with the threads of dichotomized thinking and dualistic models.

(*ibid*.: 267, 287).

The biographical sensitivity suggested by Chapman has emerged in work exploring the narratives of everyday life within which migration is situated. A narrative can be defined as 'a sequence of events told in words, and the events are ordered chronologically' (Vandsemb 1995: 412–13). Reflecting, first, a strong humanist impulse and, second, an attempt to allow alternative voices to that of the (male, Eurocentric) mainstream to be heard (McDowell 1992), narrative stories are seen to 'give the numbers a human face' (Vandsemb 1995: 412) and to provide an opportunity to facilitate the presentation of the 'lost geographies' (Miles and Crush 1993: 84) of formerly marginalised people. Moreover, since '*social* life is itself storied' (Gutting 1996: 483, our emphasis), investigation of narratives takes us beyond a humanist preoccupation with the individual to 'illuminate both the logic of individual action and the effects of structural constraints within which life courses evolve' (Vandsemb 1995: 414).

Wrapping up migration within a series of narratives of daily existence requires us to conceptualise the action in terms of *identity* rather than behaviour: migration is an 'expression of people's sense of being at any one point in time' (Gutting 1996: 482) rather than some more-or-less automatic stimulus–response. Within this strong emphasis upon context, there is a need to disentangle the relative importance of different narrative strands to any one migration decision, whilst at the same time appreciating how any migration only occurs within the totality of an individual's narrative portfolio. Thus, using the example of Turkish labour migrants in Germany (Chapter 9), Gutting interprets one

couple's residential history through narratives of 'return' (to Turkey), 'real life' (within Germany) and
'family'. However, to date, narratives of migration
have been explored most in the developing world,
such as in Vandsemb's (1995) tale of 'Amma' and
her relationship to spontaneous migration in Sri
Lanka or in Crush's (1995) account of labour migration to South Africa's gold mines.

Conclusion

An international group of migration researchers recently observed, in a review of theories of international migration, that 'popular thinking remains
mired in nineteenth-century concepts, models, and
assumptions' (Massey *et al.* 1993: 432). We have seen
plenty of evidence to support this accusation in the
continued use of the ideas of Ravenstein, Marx and
the neoclassical economists. The sustained importance of these ideas does also, of course, reflect the
persistence of often highly regular trends in migration behaviour. Nonetheless, this chapter has
provided evidence for renewed interest in debates
over the conceptualisation of human migration, debates that can help us to re-evaluate past work.
Attempts at combining the insights provided by determinist and humanist accounts of migration have
stressed the need for migration researchers and students to accept and develop a more *pluralist* understanding of the migration process. Consequently, the
thematic chapters in the book should be viewed as
complementary perspectives rather than as referring
to clearly distinct types of migration. Adopting a
more holistic perspective suggests, for example, that
migration for quality of life reasons may also involve
employment and housing considerations.

Although the chapter finished with a discussion of
structuration theory and the biographical approach,
this is not meant to suggest that such an approach is
the way in which migration should be researched.
Neither is the new approach meant as another of
the 'fads' of migration research (Gilbert and Kleinpenning 1986). Indeed, with the rise in prominence
of postmodern thinking this would be a misguided
assertion. *Postmodernism* rejects the assumption
that one 'true' approach to research or one 'true'

way of understanding is possible (Ley 1994) and requires migration studies to be sensitive to diversity,
difference and context (Graham 1995). Postmodernism is highly sceptical of the sweeping claims
and theories that were seen as characteristic of modernism, such as Marxism. These grandiose declara-
tions are regarded as elitist and unduly divisive:

> The modernist search for a unified theory of society and
> social knowledge ultimately created a menagerie of inter-
> nally-consistent yet mutually-exclusive conversations in
> social theory.
>
> (Dear 1988: 267).

Instead, postmodernism offers us a 'vision ... of
many voices talking ... without any one being dominant' (Findlay and Graham 1991: 156) and the consequent need to accept the validity of multiple
methodologies (Vandsemb 1995).

Postmodernism invites us to choose our theory
and methodology to fit the task at hand. If we are
concerned with quantifying migration flows we probably adopt positivist modelling methods. If, on the
other hand, we wish to understand how information
is processed and used through a migration then we
turn to behavioralism, and to understand the idiosyncrasies and emotional character of migration we
draw upon humanism. Standing back from the decision-making process and considering the more fundamental question of power in society and how it
links with migration beckons us towards structural
perspectives, such as political economy. Trying to
put this all together and gain a rounded understanding of migration suggests adopting a structuration–
biographical approach. This more eclectic strategy in
migration research clearly requires work to overcome local theoretical traditions, which have seen,
for example, micro-economic approaches associated
especially with North America and historical–
structural approaches the province of African and
Latin American work (Wood 1982). Lastly, and
bringing migration research back down to earth, a
distinction between the relative freedom of different
groups of people means that we must not overlook
the issues of power and contrasting resources, which
tend to be sidelined in the postmodern celebration
of relativity, and the way in which these power relations feed into the resulting patterns of observed
migration.

CHAPTER 4

Migration and employment

Introduction

More work has probably been published on labour migration in capitalist societies than on any other form of human migration. The aim is to understand the factors that encourage and discourage the movement of economically active people between countries and broad regions within countries. This concern with labour migration is not surprising, as persistent inequalities in the distribution of unemployment are indicative of imperfections in the labour market.

The fact that regional inequalities in wage rates and unemployment exist suggests that labour may not react in a 'rational' manner, at least in terms of the economic theory described in Chapter 3. The findings from behavioural studies, which identify and explain some of these labour migration anomalies, are therefore also discussed in this chapter. Additionally, the freedom of movement assumed in neoclassical models has also been questioned. Various examples discussed here suggest that the migration of individuals is bound up with 'hidden' structural factors, which guide the migration decisions of individuals to greater and lesser degrees. Extreme political control over migration in the developing world, the legacy of colonial links between the developed and the developing world, the corresponding migration channels that result, and the discrimination inherent in the segmented labour markets of the developed world, all further complicate the relationship between migration and employment.

International labour migration

Neoclassical interpretations

As explained in Chapter 3, a great deal of the work on both international and internal migration has adopted a neoclassical framework and, although there are different interpretations of this theory, certain underlying concepts are emphasised. Of particular relevance to migration are the theories of *utility maximisation* and *market equilibrium*. Within these theories, individuals and households, as sellers of labour services, and employers, as purchasers of labour services, are treated as equals, and it is assumed that there is a strong reliance on wage rates to produce adjustments to changing labour market conditions (Fischer and Nijkamp 1987). Furthermore, inequalities in wages and unemployment are important targets of public policy and many governments have introduced measures designed to reduce these differences. Understanding the processes that influence labour adjustments through migration to labour market inequalities is therefore crucial.

Such neoclassical interpretations have also been applied to international migration. For example, rising oil revenues in the oil-rich Arab countries can be associated with large increases in immigration experienced by these countries (Findlay 1994a). Birks, Seccombe and Sinclair (1986) used time series data to relate the issuing of migrant work permits to levels of government expenditure in these states between 1975 and 1984. The results indicated a strong relationship in most cases. In Oman there was an almost perfect correlation, while in Kuwait the relationship was weaker but still statistically significant, with about 56 percent of the variation in permit issues explained by this expenditure. Birks, Seccombe and Sinclair also anticipated that any reduction in oil prices and consequent declines in economic activity would reverse the immigration trend and result in an export of 'surplus' immigrant workers. To some extent this prediction came true, as the slump in oil prices in 1986 was matched by a reduction in the migrant stock in Saudi Arabia, for example (Findlay 1994a). While the volume of return migration did not match the 36 percent fall in oil prices, the crisis provided the political context for the introduction of new migration policies; the Saudi Arabian government announced the objective of reducing its immigrant workforce by 600,000 people and restricted immigrants from moving between employers within the country. Despite the geographical variations in the strength of this relationship, therefore, it does appear that neoclassical assumptions are valid in extreme migration situations such as these.

Reinterpreting international labour migration

Recently, neoclassical interpretations of international migration have been questioned on a number of counts (Portes and Rumbaut 1990; Sassen 1988). In particular, the assumption that individuals are free agents who 'search for the country of residence that maximises their well-being' (Borjas 1989: 461) has been regarded as simplistic and incapable of explaining or predicting international migration. It is rarely the poorest people from the least developed countries who move to the richest countries. Moreover, a narrow neoclassical model does not explain why certain groups of migrants are attracted to one country rather than another; for example, why should Turks be especially attracted to Germany, or Algerians to France?

In contrast to the neoclassical perspective, it can thus be argued that prior links between countries, based on colonial, political, trade, investment or cultural ties (Castles and Miller 1993), are better at explaining patterns of international migration, and the idea of individual migrants making free choices outside these constraints is unrealistic. Likewise, according to world systems theory (Chapter 3), it is the structure of the world market and the penetration of capitalist economic relations into peripheral developing societies that stimulates out-migration. This penetration began with assistance from colonial regimes and has continued with neo-colonial gov-ernments and multinational firms, which perpetuate the power of supportive national elites (Massey *et al.* 1993).

In the context of mass migrations to and from Western Europe, Fielding (1993a) has emphasised how such migration is linked to the economic organisation of these countries. Fielding identifies three levels of this organisation, noting how changes in this set-up impact upon both the levels and direction of mass migration flows. First, there is migration linked to the often rapid cyclical variation in the economic prosperity of different parts of these core economies – migration associated with the business cycle. Second, at a longer timescale, changes in the overall organisation of production within each European country themselves have migratory consequences. Notably, there is economic restructuring, involving a shift from an economic organisation based upon regions or individual countries to one based upon a more international structure – the *new international division of labour* (see below). Third, at an even more engrained and slow-changing temporal scale, there is mass migration associated with the general economic 'health' of Western Europe compared with many other regions of the world, notably the developing countries. Broadening Fielding's perspective, we can thus appreciate the importance of the economic structure of the world to the explanation of international migration.

Such an emphasis on global economic structure can generate apparently paradoxical migration streams in the context of neoclassical expectations. For example, Sassen (1988) noted an emigration of people from various developing countries – where industrial growth rates were high and foreign investment was increasing – into the United States, where the economy was marked by growing unemployment and increasing inflation. This was explained through reference to multinationals and the changing organisational structure of the world economy. First, *multinational companies* (MNCs) have organised the large-scale mobilisation of young females into the industrial workforce in many developing countries, disrupting traditional patterns of waged and unwaged work structures and bringing about a cultural distancing of these women from their origin communities. In such instances, young males face fewer opportunities for employment as women take over many of the jobs, and this may stimulate male emigration. Additionally, long-term employment for women in export factories is unlikely, as there is a preference among employers for the women to be young and more exploitable. Once laid off, and having been 'westernised' by cultural work practices, the option of emigration becomes an attractive one for many of these women. It is the strong foreign corporate presence, therefore, that becomes crucial in establishing a cultural link that may influence the emigration of all of these unemployed persons.

Second, the world economy is now managed in many respects from a small number of *global cities*, where wealth and the highly educated population are concentrated (*ibid.*). While the globalisation of these cities generates highly paid employment opportunities, a corresponding service sector that supports the increasingly technical infrastructure and offers low-paid opportunities is also expanding, causing a bifurcation of the labour market. Thus, despite the high levels of US unemployment, new opportunities are arising in the global cities, such as New York and Los Angeles, for employment in low-paid jobs that are often rejected by native-born Americans. At the same time, traditional manufacturing industries have moved away from employing unionised labour towards the use of sweatshops and industrial home-work, particularly in the garment industry, and immigrant labour is the cheapest workforce available. The growth of this immigrant labour force in the global cities has encouraged a reconcentration of small-scale, labour-intensive manufactur-

ing into these cities, reinforcing international migration flows.

Skilled international labour migration

Skilled transients (Appleyard 1985) are replacing settler migrants as the most important international labour migrants, their outstanding characteristic being their willingness to migrate between countries regularly (Richmond 1968). Modern industries increasingly rely upon the use of human expertise to add value to their operations (Salt 1992a) and, if such expertise is not available locally, employers are forced to search abroad. Findlay and Garrick (1990) have suggested that three main channels are responsible for moulding skilled transient migration during the 1980s (Figure 4.1).

First, large numbers of skilled employees move within the *internal labour market* (ILM) of MNCs between the various branch and company headquarter locations. The new international division of labour describes the integration of developing countries into the global economy, as MNCs spread their operations unevenly across the world. Head offices tend to remain in prestigious global cities, while the majority of the more menial production work is carried out in poor countries, where unskilled labour is cheap (see above). Consequently, companies have

Figure 4.1 Channels of skilled migration (source: Findlay and Garrick 1990).

Box 4.1 Linking international and internal labour migration

There is growing recognition of the positive effects that immigrant workers can have on destination economies (Borjas 1990; Spencer 1994b), as skilled immigrants bring and generate savings and add entrepreneurial skills to the labour force. Unskilled immigrants may also take on jobs that natives refuse (J. Simon 1989) or jobs in areas where there is a spatial mismatch between the labour force and employment opportunities (Salt and Kitching 1992). On the other hand, more negative interpretations of the economic impact of immigrants centre upon concerns about the anticipated detrimental effect on employment opportunities and wages for the native population (Friedberg and Hunt 1995). Recently, there has been a growing literature focusing specifically on linking immigration and internal labour migration (Pryor 1983; Walker, Ellis and Barff 1992). The American states can be divided into three categories, based on their dominant immigration and migration dynamics, as shown in the table below. While it is not surprising that most immigrants gravitate to a single 'port of entry', it is significant that these states are also not attracting large numbers of internal migrants.

State	Contribution to 1985–1990 change (thousands)	
	Migration from abroad	Net inter-state migration
High immigration states		
California	1449	174
New York	614	−821
Texas	368	−331
New Jersey	211	−194
Illinois	203	−342
Massachusetts	156	−97
High internal migration states		
Florida	390	1071
Georgia	92	303
North Carolina	66	281
Virginia	149	228
Washington	102	216
Arizona	80	216
High out-migration states		
Louisiana	30	−251
Ohio	69	−141
Michigan	74	−133
Oklahoma	32	−128
Iowa	17	−94

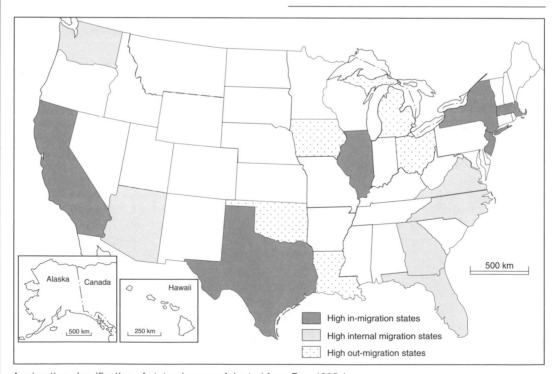

A migration classification of states (source: Adapted from Frey 1995a).

commitments in many areas and the most highly qualified personnel need not be available locally. Workers are therefore encouraged to move between operations, furthering their careers as they do so (Salt 1988). Their migrations are eased by consider-able relocation packages, which include financial in-centives and advice on settling into the new area, increasingly utilising specialist relocation agencies. Second, as rapid economic growth has proceeded apace in some parts of the developing world, de-mand has grown for skills that are rarely available in the indigenous labour market. External expertise is required, and the international contracts gained by companies in the developed world result in the mi-gration of workers, often for relatively short periods of time (Appleyard 1989). Third, international re-cruitment agencies have been employed by various institutions in the developing world to recruit appro-priate professional and managerial employees. They are responsible for filling skill gaps and have been especially influential in the recruitment of skilled labour to the Middle East (Figure 4.2). Findlay and Stewart (1986) estimated that over 40 percent of the 110,000 United Kingdom expatriates in the Middle East in 1981 were recruited through these agencies, while Salt (1992b) suggested that highly skilled pro-fessionals and managers accounted for more than half of the labour migrants moving in and out of Britain in 1989 (Table 4.1) (overleaf).

The role of remittances

For many international (and internal) migrants, links with the origin community are maintained

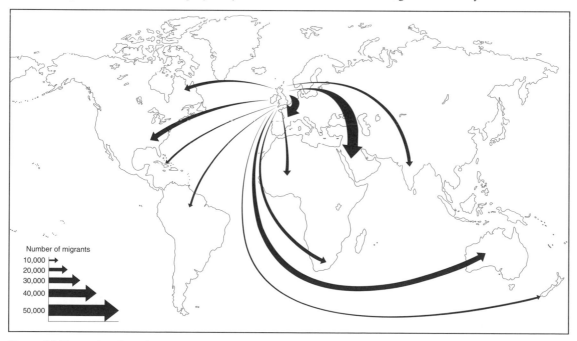

Figure 4.2 The emigration of professional and managerial British citizens, 1980–85 (Reprinted from *Geoforum*, **19**, Findlay, A, From settlers to skilled transients: the changing structure of British Internationl migration, pp. 401–10, © 1998 with kind permission from Elsevier Science Ltd, The Boulevard, Langford Lane, Kidlington, OX5 1GB, UK).

Table 4.1 Professional and managerial flows in the United Kingdom, 1989.

	Professional and managerial percentage
All labour migrants entering the UK	61
British	69
Non-British	54
All labour migrants leaving the UK	59
British	57
Non-British	63

Source: International Passenger Survey 1989.

through the practice of sending remittances home. In some nations, these remittances represent a substantial proportion of the gross national product (GNP) (Table 4.2). Russell (1986) distinguishes between three major components of the remittance process: the decision to remit; the method used to remit; and the use that is made of the remittances in the origin community (Figure 4.3).

Various factors are liable to influence the decision to send remittances, including exchange rates, the relative differences between interest rates in the origin and receiving countries, and the availability and quality of institutional transfer mechanisms, such as banks. The socio-demographic characteristics of the migrants will also influence remittance payments. Women appear to be more frequent remitters than males, although they often lack the earning capacity to send as much as men (Connell and Brown 1995; Shankman 1976). Additionally, Oberai and Singh (1980) found that lower caste migrants in India were more likely to remit than those from higher castes, while Serageldin *et al.* (1981) suggested that less educated people and those of lower occupational status

Table 4.2 Overseas remittances in relation to gross national product

Country	Year	Total remittances (million $)	Remittances as percent of GNP
Bangladesh	1992	1020	4.39
India	1990	2337	0.79
Indonesia	1992	184	1.48
Pakistan	1992	2772	5.88
Philippines	1993	398	0.75
Sri Lanka	1993	560	5.59
Thailand	1990	26	0.03

Source: International Monetary Fund 1994.

Box 4.2 Skilled international migration: the case of Scotland

Scotland is a source of skilled international labour migrants. The emigration rates for professional and managerial workers in 1980–85 were higher only in London, where a substantial number of MNC head offices are located and the emigration rates would be expected to be high. This is demonstrated in the following table:

Regions	Professional and managerial emigrants (per 1000 persons*)
Greater London	11.8
Scotland	*9.0*
Rest of South East	8.2
North West	6.0
North	5.8
South West	7.2
East Anglia	6.6
Wales	5.9
Yorkshire and Humberside	4.9
West Midlands	4.7
East Midlands	5.1
Northern Ireland	3.5
Total UK	14.0

* Rates were calculated relative to the population in the economically active age groups, 15–64 for men and 15–59 for women.

This finding may be related to the industrial history of Scotland, especially since relatively high numbers of engineers were involved in these outflows. There is also evidence that the domestic employment opportunities are perceived as being particularly poor by Scottish professionals and managers, but this does not explain why they choose to move outside Britain. More convincing are the arguments that, first, Scotland is host to a large number of foreign-owned branch plants – many American MNCs opened their first European branch plants in Scotland – and migration within the ILMs of these companies is likely to have been influential, and second, Scotland, more than the regions of England, maintains a distinct cultural identity. This Anglo-Scottish cultural divide may encourage Scottish workers to look further afield than London and the South East, which are relatively more attractive to potential migrants from other English regions.

Source: Findlay 1988; Findlay and Garrick 1990.

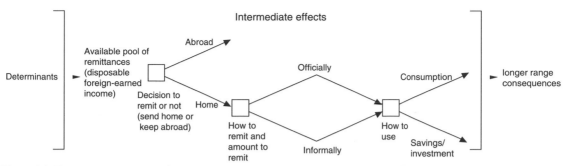

Figure 4.3 The remittance system (Reprinted from *World Development*, **14**, Russell, S., Remittances from international migration: a review in persective, pp. 677–96, © 1986, with kind permission from Elsevier Science Ltd, The Boulevard, Langford Lane, Kidlington, OX5 1GB, UK).

were also more likely to remit, although other studies failed to replicate these results (Gilani, Khan and Iqbal 1981). What is certain, however, is that migrants do not send remittances for purely altruistic reasons. Instead, Stark (1992) talked of 'tempered altruism or enlightened self-interest', as the migrant's longer-term intention to return home means that remittance payments are an investment in both their family's and their own futures.

In assessing the impact of remittance payments, it is relatively difficult to isolate the effect of remittances from other sources of income. In many cases, the consumption of luxury items outweighs investment in more productive enterprises in the origin

Table 4.3 Remittance expenditure in Pakistan, 1979.

Expenditure	Average expenditure per migrant ($)	Percent
Consumption	*1820*	*62.2*
Recurring consumption	1668	57.0
Marriages	69	2.4
Consumer durables	83	2.8
Real estate	*633*	*21.6*
Construction/house purchase	355	12.1
Home improvement	66	2.3
Commercial real estate	167	5.7
Agricultural land	45	1.6
Investment/savings	*379*	*13.0*
Agricultural investment	97	3.3
Industrial/commercial investment	240	8.2
Financial investment/savings	42	1.4
Residual	*93*	*3.2*
Total	*2925*	*100.0*

Source: Parnwell 1993; Stahl and Arnold 1986.

communities (Table 4.3). This apparent lack of investment has prompted many authors to argue that economies that become remittance-dependent are experiencing an undesirable and unsustainable stage in their development:

Despite the foreign exchange and balance of payments advantages, do remittances help the development process or, like drug dependency, do their existence and current uses primarily feed the need for more foreign exchange and exacerbate the balance of payments process, thus increasing the need for ever more remittances and the accompanying dependency on the receiving countries?

(Kritz, Keely and Tomasi 1981: xxv).

The implication from such observations is that dependence upon remittance should be discouraged, with the small economies reverting to subsistence (Forsyth 1992; Tisdell and Fairburn 1984). In contrast, Bertram and Watters (1986) argued that remittances, foreign aid and the spread of state bureaucracy have become dominant factors in many small islands' *migration–remittances–aid–bureaucracy* (MIRAB) *economies*. Moreover, these economic structures were regarded as being durable and sustainable, as households received considerable sums for consumption needs.

Brown and Connell (1993: 616) argued that interpretations of the role of remittances tended to represent 'mechanistic and static understanding of the relationship between external and internal dimensions of the development process'. This was because they assumed that migrants' remittances were used exclusively for consumption needs, rather than as a source of investment. Fieldwork in Tonga suggested that remittances were important to the economy (Figure 4.4) (overleaf), with 90 percent of households in receipt of remittances, constituting an average of 28 percent of household income (Ahlburg

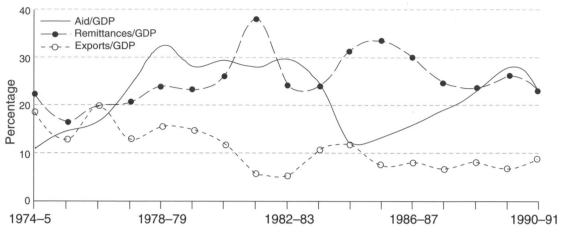

Figure 4.4 Ratios of exports, aid and migrant remittances to gross domestic product in the Kingdom of Tonga (source: Brown and Connell 1993).

1991). A survey of flea-market stall operators in Nuku'alofa, the capital, found that financial and, especially, goods remittances from abroad were supporting a highly successful informal economy; 91.3 percent of vendors were engaged exclusively in the sale of goods sent as remittances from abroad (Brown and Connell 1993). Importantly, 26.1 percent of those surveyed said that the income generated was used exclusively for savings or reinvestment, suggesting that migrant remittances may provide a catalyst for economic growth.

Economic models of internal labour migration in the developed world

In the developed world there has been considerable interest in the (im)mobility of labour. Neoclassical interpretations have often adopted rigorous quantitative models to understand inter-regional flows of people where the apparently simple assumptions of this economic theory can be tested.

Macro-level economic models

Various neoclassical macro-level economic models have been developed from the early *Lowry model* (Lowry 1966), and many of them can be fitted as multiple regression models, as described in Chapter 2 (W. Clark 1982; Greenwood 1975, 1985; Greenwood *et al.* 1991; Mueller 1982). The Lowry model is an extension of the basic gravity model:

$$\ln(\hat{M}_{ij}) = a_0 + a_1 \ln(u_i) + a_2 \ln(u_j) + a_3 \ln(w_i) + a_4 \ln(w_j) + a_5 \ln(L_i) + a_6 \ln(L_j) + b \ln(d_{ij})$$

Migration between i and j is expected to be positively related to the non-agricultural workforces at the origin and destination, L_i and L_j (these are equivalent to the mass terms); the respective percentage of the economically active workforce that are unemployed, u_i and u_j; and the hourly manufacturing wage rates, w_i and w_j. Movement is expected to be from low- to high-wage areas and from high to low-unemployment areas.

Even though this model represented only a marginal improvement over the gravity model, variants of it have been implemented extensively. A notable extension of the approach by Todaro (1969) and Harris and Todaro (1970) was implemented in Nigeria, where variables relating to the *probability* of finding employment at the destination, rather than simply measures of wages and unemployment, were incorporated. The notion of job vacancies was introduced by weighting the relative earnings differentials by the relative unemployment rates, although these models failed to fit well.

Following Sjaastad (1962), Molho (1984) adopted a human capital approach, where the migration decision is regarded as an investment in human capital, with migrants maximising their anticipated future utility (Chapter 3). Accordingly, uncertainty, risk-aversion and adjustment costs are important, and

migration may be viewed as an essentially dynamic process involving lengthy response lags, both in the decision to migrate and the subsequent enaction.

(*ibid.*: 317).

Box 4.3 The Lowry model applied

The Lowry model's results for migration between 1955 and 1960 in the United States were summarised in the table below. If the neoclassical arguments are correct we would expect the parameters a_1 and a_3 to be positive and a_2 and a_4 to be negative but, in the event, the improvement over the simple gravity model was marginal. Of the four additional variables, only the destination unemployment was significant(*) and in the hypothesised direction. Both of the wage variables were in the correct direction but neither was significant, while the origin unemployment variable was insignificant and in the wrong direction – a negative sign means that the higher the origin unemployment, the less likely it is that people will move away.

Lowry's regression model results

Basic gravity model		Economic gravity model	
Variable	Coefficient	Variable	Coefficient
Intercept	-7.91	Intercept (a)	-12.75
		$\log u_i$ (a_1)	-0.133
		$\log u_j$ (a_2)	$-1.294*$
		$\log w_i$ (a_3)	-0.027
		$\log w_j$ (a_4)	0.242
$\log L_i$	$1.019*$	$\log L_i(a_5)$	$1.047*$
$\log L_j$	$1.023*$	$\log L_j(a_6)$	$1.086*$
$\log d_{ij}$	-0.257	$\log d_{ij}$(b)	$-0.493*$

Source: Lowry 1966; Plane and Rogerson 1994.

Most prior studies of this type had been forced to model economic migration cross-sectionally, because of the dearth of time series data, or had relied on annual time-series information. Using quarterly data from the National Health Service Central Register (NHSCR), Molho modelled flows between the ten planning regions of Britain, and the short period between observations revealed some interesting results. He found that, in the short run, unemployment rates significantly affected employment migration but that, over longer periods, the rate of employment growth in a region was more important.

Greenwood and Hunt (1984) also analysed time series data, derived in this case from a sample of individuals employed in Social Security-covered jobs (1957 to 1975) in the United States. They aimed to understand the contribution that employment growth in an area makes to encouraging in-migration, as 'migration to cities has probably been

a self-reinforcing and cumulative phenomenon' (*ibid.*: 957). While employment growth may attract labour, in-migration itself may generate more employment opportunities. They calculated the number of net in-migrants attracted to each of 57 urban centres for each additional job. Through averaging these parameters by broad region, they demonstrated that the south and west of the United States attracted more migrants for each additional job than the north central and north east (Table 4.4). Ten additional jobs in the latter regions attracted approximately one less migrant than the equivalent job growth in the south and west. Furthermore,

because the migrant-attractive power of another job is greater in southern and western areas, migrants are drawn to these areas away from those of the Northeast and North Central regions … the former regions gain a disproportionate share of such employment.

(*ibid.*: 968–9).

Regional migration need not reduce unemployment *significantly*, however (Gabriel, Shack-Marquez and Wascher 1993). A series of models were estimated to explain the migration between the nine census regions in the United States, between 1986 and 1987, using data from the Internal Revenue Service (IRS). A variety of economic variables were included and the hypothesised neoclassical relationships appeared to be upheld (Table 4.5) (overleaf). Migration was positively related to unemployment at the origin and the average destination wage, and negatively related to the destination unemployment rate and the average origin wage. However, even when the results were manipulated to simulate more extreme inter-regional differences in wages and (un)employment,

the direct effects of regional migration flows are insufficient to offset shocks to the regional distribution of unemployment rates, at least in time intervals of less than several

Table 4.4 The impact of employment growth on net in-migration in the United States.

Region	Parameter value (α_1*)
South	0.503
West	0.471
North East	0.438
North Central	0.365

*The parameter α_1 reflects the number of net in-migrants for each additional job in the region.
Source: Greenwood and Hunt 1984.

Table 4.5 Parameter estimates related to inter-regional migration in the United States, 1986–87.

Explanatory variables	Parameter estimates
Constant	−2.17
Origin population aged 20–29 (%)	−9.19
Population with post-secondary education (%)	35.44
Origin population	−0.01
Destination population	0.16
Urban share in origin	14.75
Distance	−0.81
Snowbelt–sunbelt migration	0.46
Origin unemployment rate	0.47
Destination unemployment rate	−0.08
Average origin wage	−2.97
Average destination wage	0.65
Origin house prices	−0.00
Destination house prices	−0.29
R^2	0.69

Source: Gabriel, Shack-Marquez and Wascher 1993.

years, once the migration flows are adjusted to reflect the *labor market status of movers*.

(ibid.: 231, emphasis added).

This discovery reflected the fact that many of the inter-regional migrants did not necessarily join the labour market.

Micro-level economic models

Macro-level models are used to predict and explain the broad patterns of migration between areas. In contrast, micro-level models use individual-level data to explore how migrant characteristics influence migration (for example, Herzog and Schlottmann 1984). For example, search theory models originated from the economics of uncertainty (Clark 1987). They retain a neoclassical theoretical grounding but are used to theorise about the processes and consequences of information collection. It is assumed that the (in)efficiency of the market economy rests on the collection, synthesis and distribution of information, and (in)efficiencies can be related to the ability of market actors to find relevant information.

A cross-national study by van Dijk *et al.* (1989) compared data from the Netherlands and the United States. Incorporating both individual- and regional-level (un)employment characteristics in a logit model, they found that unemployed individuals were more likely to migrate than employed individuals, but with the rate five times higher in the Netherlands

than in the United States. They attributed this to institutional differences between the two labour markets. In the Netherlands, job vacancies are relayed to local employment offices, where unemployed people register in order to gain benefits. These offices are connected throughout the country, allowing the efficient matching of vacancies with unemployed individuals. However, the model also showed that a high regional unemployment rate had a positive impact upon out-migration in the United States but a negative impact in the Netherlands:

Thus, as certain as migration promotes reemployment within the Netherlands (due in large part to the nationwide information system), perhaps it is exactly such a dependence on job certainty (and generous unemployment compensation) that 'discourages' workers in that country from relocating in the face of high local unemployment and low relative wages. Conversely, American workers ... have historically been quite mobile, and as such have assumed significant labour market 'risk' in the process. Perhaps it is exactly this willingness of workers to assume risk within the labour market, many becoming speculative migrants in the process, that promotes macroefficiency in migration?

(ibid.: 79–80).

Pissarides and Wadsworth (1989) addressed a different but related issue using British Labour Force Survey data from 1976 and 1983. They suggested that higher *overall* national unemployment, which occurs during national recessions, will reduce the migration rates of both the employed and the unemployed. On the one hand, the employed are less likely to change jobs when the employment market is frail, because of the old adage 'last in, first out'. On the other hand, the unemployed expect to be jobless for longer and become unwilling to risk using their meagre financial resources in order to relocate. The capital market constraints facing these people will also be greater, as lending institutions downgrade the expected future earnings of the unemployed, thereby making them less creditworthy. Although the unemployed were more likely to migrate than the employed in both years, unemployed individuals in 1976 (unemployment rate 3.2 percent) were twice as likely to move as unemployed individuals in 1983 (unemployment rate 8.6 percent). Thus, in addition to the personal and regional differences identified above, the changing national economic context influences migration propensities.

A recent review of these types of model in the United States (Herzog, Schlottmann and Boehm 1993) suggested three broad conclusions. First, personal unemployment augments migration. Second, this stimulative effect of joblessness on migration tends to decrease with search duration. Third, at a regional level, out-migration is increased by higher area unemployment rates, although the evidence for this effect is not overwhelming.

Refining explanations of labour migration in the developed world

A critique of neoclassical modelling

Despite the successes of the neoclassical economic modelling approach, a series of inadequacies have been identified in the literature. Besides philosophical concerns over its positivism (Chapter 3), there are also some more specific criticisms. First, some academics argue that the underlying assumption that labour mobility will equalise wage and employment rates is flawed. Matsukawa (1991) and van der Laan (1992) have drawn attention to the geography of *structural* changes in local industries and the consequent instability of the economic environment. As growing sectors in a local economy attract labour, the declining sectors suffer redundancies and labour migration increases. Quite simply, the 'increase in employment in some sectors attracts labour from every region, regardless of the level of total employment growth in each region' (Matsukawa 1991: 745). Thus, total demand for labour in a local economy may not be a useful determinant of inter-regional migration.

Second, economic effects make the definition of unemployment itself difficult. The *added worker hypothesis* suggests that when unemployment rises, the loss of jobs by some family members may encourage others to engage in employment. These additional workers mean that the standard unemployment rate overstates the number of jobs needed to reach full employment, because at least some of the additional workers will withdraw from the labour force when overall unemployment levels fall. The *discouraged worker hypothesis* works in the opposite direction, as in periods of high unemployment some of the unemployed become so disillusioned with the prospect of finding employment that they leave or fail to join the unemployment register. Consequently, the presence of discouraged workers means that the official unemployment rate underestimates the number of jobs required to obtain full employment (Fischer and Nijkamp 1987).

As elsewhere, there has been some debate in the Canadian literature (Courchene and Melvin 1986; Polèse 1981; Savoie 1986) about the theoretical validity of neoclassical economic arguments. The fact that inter-regional variations in wage and employment rates have existed for some time appears to question whether labour mobility will ever equalise inter-regional disparities. Some suggest that economic markets do not respond in this way and that inter-regional migration can, in certain circumstances, *increase* these disparities. If the technological efficiency in one region exceeds that in another, both labour and capital may migrate to the high-wage region (Batra and Scully 1972; Drugge 1985). This dual labour and capital mobility undermines the simple neoclassical assumption that increasing numbers of workers in a high-wage area will naturally result in wage decreases. Consequently, spatial variations in technological capabilities prevent the equalisation of inter-regional wage rates, and labour mobility can increase regional income disparities. The question becomes one of why there are regional disparities in technological efficiency, rather than questions concerning the (in)ability of labour to equalise wage and employment rates (Drugge 1987).

Behavioural surveys and the importance of employment in migration

Neoclassical economic models tell us nothing about the precise factors that stimulated individual decisions to move – questionnaire approaches are more helpful for studying these types of issue – and they rarely consider lifestyle decisions in the migration of labour. Behavioural studies of labour migration focus more on these factors, while investigating the assumption that people act rationally in their decision making. Time and again, such surveys have shown the importance of employment in guiding migrant decision making. For example, using data derived from a British national building society survey, Owen and Green (1992) showed that the main factors influencing overall migration are housing and life-course factors. While work-related factors were only third on the list, they were the most important consideration for moves over 25 kilometres (Table 4.6) (overleaf).

The assumption that migrants have equal and complete access to information about employment opportunities and wage rates has also been questioned through behavioural work. Early studies by Speare (1971) in Taiwan and Shaw (1974) in rural

Table 4.6 The major reasons for moving among house purchasers in Britain, 1981, by distance of move.

Reason for move (percent)	Distance moved (miles*)						
	<5	6–10	11–25	26–50	51–100	>100	All
Housing market/accommodation	44.7	34.8	25.1	15.3	8.3	3.0	36.5
Family cycle/social	30.1	33.3	29.6	19.5	12.4	10.7	28.5
Work-related	2.9	10.3	25.3	53.0	70.4	78.9	14.8
Neighbourhood	8.0	9.4	7.6	5.4	3.4	1.9	7.8
Other	14.2	12.2	12.4	6.8	5.4	5.5	12.7
All	100	100	100	100	100	100	100

Note: * 1 mile = approximately 1.6 kilometres.
Source: Owen and Green 1992.

Saskatchewan (Canada) suggested that there were sub-groups of the population who did not perceive costs and returns to (im)mobility. Thus, potential rural–urban migrants in Taiwan:

only have vague concepts of costs and benefits. Only a small percentage of the migrants knew exactly how much they would earn in the city before they moved and most could only distinguish whether they expected an increase, no change, or a decrease in income to result from the move.
(Speare 1971: 129).

We must also recognise that rational economic decisions – whatever they might be – are not the only concern of migrants. Various authors have considered the migration of the rural poor in the United States and the effects that this has on redistributing poverty (Lichter 1994; Nord 1994). Fitchen (1995) examined the village of Riverside, 165 kilometres north of New York City, where the population exchange of the late 1980s had left the village considerably poorer than it had been previously. Questionnaire data showed an influx of one-parent families claiming welfare benefits. For this group, Riverside offered a solution to their housing problems because of the widespread availability of relatively cheap rental property. More detailed examination of these in-migrants, however, showed that only 17 percent of them had moved directly from New York City, and that the majority had moved elsewhere before arriving in Riverside. A relatively neat sequence of events had occurred. Rising levels of unemployment in rural communities resulted in increasing levels of out-migration and a glut of unwanted housing became available for rent. Simultaneously, low-income residents in metropolitan centres were attracted into the rural towns, where rents were lower. 'Pioneer' migrants were followed by relatives and friends but, eventually, the reserve of housing was taken up and rents began to

rise. This stimulated the movement of these poor people into other rural areas where the rents remained low. As Fitchen (1995: 195) explained:

movement of poor people to economically depressed rural towns is counterintuitive (and perplexing to most economists) because individuals and families who have trouble earning a living are moving from dynamic economic centres with adequate or growing employment to stagnant, jobless rural peripheries with high unemployment and serious underemployment.

Employment prosp.

ects for the incomers were minimal, although this was not seen as a deterrent as many of them had limited job skills and anticipated that it would be difficult for them to get or hold a job anyway.

Segmentation theory and labour migration

Dual labour market and internal labour market theories

Neoclassical theory fails to take account of the social and institutional constraints that restrict workers in their job search processes and it ignores the discrimination against women and minorities, which has persisted over time (Peck 1989). Segmentation theories have developed to address this failure. In particular, *dual labour market theory* (Chapter 3) divides the labour market into primary and secondary employment opportunities. The former include firms with structured ILMs and relatively secure jobs that are well paid and offer good career opportunities. The secondary sector includes jobs at the bottom of the occupational hierarchy that are poorly paid, less secure and without good career prospects. Firms in this sector tend to be small, reliant on larger firms for subcontracted work, labour-intensive, and with low levels of unionisation. Mobility between these two sectors is limited; few workers are forced to move down from primary employment, while social and institutional barriers restrict movement upwards from secondary employment (Jarvie 1985). Ignoring these institutional barriers to employment changes results in an unrealistic impression of the freedom of labour migration. While some argue that dual labour market theory is over-simplistic, as it represents two ideal forms of employment at the ends of a continuum (Gordon 1994; McNabb and Ryan 1990), this approach has been widely adopted in migration studies.

Neoclassical migration models also assume that migrants are *speculative*, moving in response to economic considerations, but *contracted* migration, where a job at the destination has been prearranged (Silvers 1977), is actually more common among labour migrants in most developed countries (Flowerdew 1992). McGregor *et al.* (1992) found that only 7 percent of labour migrants making long-distance moves were doing so speculatively. Even in the United States, where speculative migration is assumed to be more established, the rates are relatively low (Lansing and Mueller 1967).

As with skilled international migration, a great deal of intra-national employment migration occurs within the same organisation. In 1981, over half of the inter-regional migrants in the United Kingdom, who were working both at the time of the Labour Force Survey and one year before, did not change employer (Salt 1990). Consequently, Atkinson (1987) estimated that there are about 250,000 ILM

transfers annually. The moves may merely be a part of a traditional transfer of staff as part of their career development – 'spiralism' (Chapter 9) – or organisations may shift their entire activities between sites, with large numbers of personnel moved. These migrants are almost entirely managerial and professional primary sector employees, who are highly valued by companies. In the few cases where manual workers are offered similar relocations, they often refuse (Salt 1987).

The importance of these moves has been established in a number of countries, such as Australia (McKay and Whitelaw 1977), the United States (Sell 1990) and Scotland (Shapiro 1981). Wiltshire (1990) examined this phenomenon in Japan, where unique employment practices reinforce the importance of personnel transfers within organisations for overall migration. The *lifetime employment system* grants tenure of employment within an organisation to elite employees, although the system is in decline. It has resulted in considerable mobility of workers, both occupationally and spatially, since the price that tenured employees pay for their protected status is that the company may move them frequently. Thus, over 62 percent of manufacturing companies with more than 5000 employees had a declared policy of transferring employees to develop their careers (Inagami 1983). Because ILMs prevail at all levels in the management hierarchy, *simultaneous chains* (White 1970) may result in numerous linked moves. In Table 4.7, taken from the Employment Security Bureau of the Japanese Ministry of Labour – which provides public services – ten moves result from the retirement of a section chief.

Table 4.7 A simultaneous chain migration in Hiroshima, Japan, 1987.

Move	Old rank	Section/Office	Location	New rank	Section/Office	Location
1	Section chief	EIS	Hiroshima (HQ)	Retired	–	–
2	Chief manager	ESS	Hiroshima (HQ)	Section chief	EIS	Hiroshima (HQ)
3	Manager	PESO	Mihara	Chief manager	ESS	Hiroshima (HQ)
4	Manager	PESO	Takehara	Manager	PESO	Mihara
5	Deputy section chief	ESS	Hiroshima (HQ)	Manager	PESO	Takehara
6	Inspector	ESS	Hiroshima (HQ)	Deputy section chief	ESS	Hiroshima (HQ)
7	Manager	PESO	Miyoshi	Inspector	ESS	Hiroshima (HQ)
8	Deputy manager	PESO	Hiroshima	Manager	PESO	Miyoshi
9	Manager	PESO	Hiroshima Saijo	Deputy manager	PESO	Hiroshima
10	Chief clerk	ESS	Hiroshima (HQ)	Manager	PESO	Hiroshima Saijo

EIS   Employment insurance section.
ESS   Employment security section.
PESO   Public employment security office.
Source: Wiltshire 1990.

Box 4.4 Social class and labour migration

Within Britain, there has been an steady increase in service class and *petite bourgeoisie* workers (middle class) at the expense of white collar and blue collar workers (working class) in the last two decades (Savage, Dickens and Fielding 1988). The relationship between social and geographical mobility can be explored using data from the Longitudinal Study. Movement between occupations (and classes) has been studied widely within the sociological literature (Glass 1954; Goldthorpe, Llewellyn and Payne 1980; Payne and Abbot 1990), but only recently has the relationship between class changes and geographical movement been considered in depth. The table below illustrates the association between interregional migration and changes in social class in England and Wales between 1971 and 1981. Table values of 100 would suggest that there was no relationship between social and geographical mobility, but higher values for movers between classes, particularly into the service class, suggest higher rates of mobility.

| Social class, 1971 | Social class, 1981 | | | | | |
	Service class	*Petite bourgeoisie*	White collar	Blue collar	Unemployment	Total
Service class	103	117	89	73	115	100
Petite bourgeoisie	185	69	161	100	190	100
White collar	165	163	77	77	157	100
Blue collar	209	168	136	75	133	100
Unemployment	215	104	113	79	70	100

(Service class = professional, technical and managerial workers; *petite bourgeoisie* = small employers and self-employed; white collar = lower-level white collar employees; blue collar = blue collar employees.)

Despite the relative growth of the service class and their higher migration rates, overall mobility rates do not appear to be rising. This can be explained in part by the occupational strategies adopted by the service class (Savage 1987), which have a direct impact on migration. Brown (1982) identified three career strategies: organisational, occupational and entrepreneurial. Although not mutually exclusive, organisational strategies generally involve movement through the ranks of large organisations via the ILM; occupational strategies involve moving to more responsible jobs within the profession; and entrepreneurial strategies involve becoming an employer of labour. Each of these strategies has a different relationship with spatial mobility, with an organisational career causing a great deal of migration and an entrepreneurial career resulting in much less movement, since the process of starting a business often relies on local resources. As the former careers are declining and the latter are increasing, this may explain the apparent decline in migration among those in the higher social classes.

Source: Fielding 1989, 1992c.

Ethnicity, segmentation theory and labour migration

Segmentation theory emerged in the late 1960s as an explanation for the concentration of black workers in the low-wage inner-city labour markets of the United States (Waldinger 1985). Immigrants offered a labour force that was ideally suited to the requirements of secondary employers, as they were more transitory and less concerned with issues of upward career mobility (Piore 1979). For example, Scott (1992) identified the prominence of ethnic minority labour in the Californian electronics assembly industry. These workers faced unstable employment conditions and little scope for career advancement.

However, segmentation theory does not deal well with divisions *within* the primary and secondary sectors (Lever-Tracey and Quinlan 1988). Scott (1992) found ethnic divisions within the unskilled electronics workforce of California, with Asian workers occupying more technical and responsible positions than Hispanics. Waldinger (1985) also pointed out that immigrant labour has increasingly been employed in immigrant-owned firms (Light and Bonacich 1988). Employment in New York in such firms provided substantial opportunities for skill acquisition, which often resulted in workers setting up their own businesses in the future. Hence, Dominican entrepreneurs organised their firms around migration chains from a common site in the Dominican Republic:

In a typical case, relatives and friends comprised the entire workforce of an eighteen-person [clothing] factory owned by three brothers from a small agricultural town in the northern Dominican Republic.

(Waldinger 1985: 219).

Despite these qualifications, there remains strong evidence that ethnic groups often experience poor work opportunities. Campbell, Fincher and Webber (1991) traced a sample of 272 Greek, Yugoslav and Vietnamese immigrants who worked in manufacturing in Melbourne, Australia, and had immigrated between 1952 and 1987. Nearly all of the group entered the Australian labour market at a lower level than their qualifications and experience justified. Throughout the period studied, they became concentrated in lower-skilled manufacturing jobs, despite the declining importance of these jobs in the national economy. Moreover, there was evidence of 'powerful mechanisms of allocation that were not related either to prior experience and skills or to the relative growth or decline in industry sectors' (ibid.: 180). Instead, issues of racism and discrimination were likely to be involved (Chapter 9).

Gender, segmentation theory and labour migration

Increasingly, the importance of gender in understanding the patterns of regional labour migration is being acknowledged (Bielby and Bielby 1992; Halfacree 1995a). In a British context, Boyle and Halfacree (1995) showed that service class (mainly professional and managerial) women were more attracted towards metropolitan areas than men, through observing their relatively high rates of movement into the south east region and London (Figure 4.5) (overleaf). Duncan (1991) suggested that the prevalence of nursery places for children, a more 'feminist' political climate in local government and companies, and the role of public transport (which is well provided for in London) in women's daily lives were important in explaining this attraction.

Merrill Lynch (1986) and Salt (1990) showed that ILM relocation of women is rarer than for males and that those who were relocated within firms were relatively young and unencumbered by family commitments. Gordon (1995) has related the differential geography of male and female labour migration to segmentation theory, arguing that females are stereotyped as being less 'stable'. Distinguishing between *sponsored* (migrants working for the same firm before and after the move, or migrants receiving financial assistance for the move), *unsponsored* (speculative migrants and contracted migrants who were not sponsored by the employer), *port of entry* (those moving from full-time education to primary sector jobs) and *dependent* (economically active females whose husbands moves were job-related) movers, there were considerable differences between the sexes in the 1988–89 British Labour Force Survey data (Table 4.8) (page 99). Job-related movement by females was 40 percent lower than that for males, caused by the low levels of job-related movement for married females and low rates of sponsored female moves, which, even for the non-married, were approximately half of those of men. These results were confirmed in regression analyses, with men being

two and a half times as likely as women with similar characteristics and labour market position to make a sponsored move. The absence of any equivalent effect for unsponsored moves suggests that this gap essentially reflects gender discrimination on the part of employers'.

(ibid.: 146).

The discrimination experienced by females in ethnic minorities is often especially severe. Peck (1992) studied home working in the largely unregulated low-wage Australian clothing industry. Labour turnover was rapid within firms that constantly poached labour from each other. Consequently, the industry was relatively insecure but had managed to deal with this by recruiting from the continual supply of immigrant women, notably those from southern Europe and more recently Vietnam, who fitted neatly into this secondary sector employment.

Housing influences on labour migration

Public housing and long-distance migration

Migration rates in the United Kingdom are lower than those in many other countries, such as the United States. Hughes and McCormick (1989) showed that the difference between the migration rates of manual workers in these countries was especially marked (Table 4.9) (page 99). While inter-state migration rates were higher for manual workers than for non-manual workers in the United States, the equivalent inter-regional rates for manual workers in the United Kingdom were lower than those for non-manual workers. This suggests that

Figure 4.5 Migration efficiencies for female service class workers in Great Britain, 1981 (source: Boyle and Halfacree 1995, reprinted by kind permission of Carfax Publishing Company).

Table 4.8 Job-related moves by gender and marital status in Great Britain, 1988–89.

| | Percentage of economically active | | | | | |
Type of move	Married	Males Non-married	Total	Married	Females Non-married	Total
Job-related	2.5	4.0	3.0	0.8	3.3	1.8
Sponsored	1.8	1.6	1.7	0.4	0.9	0.6
Port of entry	0.0	0.2	0.1	0.0	0.3	0.1
Other unsponsored	0.7	2.2	1.2	0.4	2.1	1.1
Dependent	–	–	–	1.6	–	1.0

Data from the British Labour Force Survey.
Source: Gordon 1995.

Table 4.9 Migration rates for heads of household in the labour force in the United Kingdom and the United States.

| | Migration rates (percent) | | |
| | | Occupation of head of household | |
	All	Non-manual	Manual
*Migration between UK regions**			
All	1.14	1.83	0.62
Job-related	0.45	0.93	0.1
*Migration between US states***			
All	3.09	2.67	3.56
Job-related	1.16	0.59	1.8

* Data from the 1973–74 General Household Survey.
** Data from the 1980 Panel Survey of Income Dynamics.
Source: Hughes and McCormick 1987.

Table 4.10 Tenure characteristics of long- and short-distance movers in Britain, 1991.

Tenure	Short-distance (<50 km)	Long-distance (≥50 km)
Owner occupied (buying)	11,582 (42.1)	2702 (39.0)
Owner occupied (bought)	1421 (5.2)	672 (9.7)
Privately rented	6552 (23.8)	2853 (41.2)
Council housing	6573 (23.9)	505 (7.3)
Housing Association	1386 (5.0)	188 (2.7)

* Percentages in brackets; data from the 1991 British Census Sample of Anonymised Records.
Source: Boyle 1995c.

there may be structural effects in the United Kingdom labour market that severely hinder the migration of manual workers, and this will inevitably affect labour market efficiency.

Hughes and McCormick (1981, 1985, 1987) linked this immobility to the housing market (Black and Stafford 1988; Minford 1983; Minford, Peel and Ashton 1987), arguing that working class people who reside in council (public) housing are constrained from moving because of the administrative rules that govern the allocation of this property. Although council tenants move frequently over short distances, as their changing housing needs are accommodated by the local authorities that administer the property, they are less able to move over long distances (Table 4.10). A number of factors are taken into account when awarding points to those waiting to be housed and considerable weight is given to those who currently live within the local

authority area, compared with those who apply from outside. This is perhaps inevitable, as local authorities feel obliged to house people from the local area. Consequently, Minford, Peel and Ashton (1987) identified council housing as an important factor in explaining the failure of neoclassical theory in the British regional labour market. They argued that the sale of council housing to residents (see below) will inevitably result in greater geographical mobility.

The assumption that council tenants, through their housing tenure, are less likely to move over long distances than people in other tenures may be flawed, however, because it ignores the fact that a disproportionate number of council tenants have working class occupations, which are less likely to involve long-distance migration for many of the reasons cited in this chapter. Consequently,

It is not hard to imagine, however, that people who were in every respect like council tenants except that they owned their houses, would also tend to move only very locally.

(Fielding 1993b: 183).

It has also been argued that it may be the limited *availability* of council housing, rather than council housing *per se*, that restricts migration (Boyle 1995c). Thus, by increasing the availability of council housing it might be possible to relax the administrative restrictions on longer-distance moves.

The view that council housing was restricting labour migration was accepted by the British Conservative government, because it introduced the 1980 Housing Act as one of its first pieces of legislation after gaining office in 1979. This major restructuring of the housing market introduced the 'right to buy', which gave tenants the right to purchase their council house at a considerable discount from the market price. As a result, sales of council houses have been considerable. Nevertheless, detailed studies since that time have not suggested that individuals who bought their council houses have become increasingly mobile. The majority of purchasers were middle-aged or older, with no children (Forrest and Murie 1988). They also appeared to be from the higher social classes (Williams, Sewel and Twine 1986) and few intended to move on in the near future (James, Jordan and Kays 1991). Therefore, 'freeing up' council tenants does not appear to be contributing significantly to inter-regional labour migration and there appear to be other factors that restrict the mobility of the working class (un)employed.

Labour migration and regional variations in house prices

Hamnett (1992) considered the impact of regional house price differentials on migration in Britain. Rapid house price inflation saw a twelve-fold increase in average prices between 1970 and 1990 but there were also significant regional differences. Prices doubled in London and the South East between 1983 and 1987, while northern regions witnessed increases of only 30 percent. This house-price gap fluctuated over time (Hamnett 1988, 1990), but was regarded as one cause of the labour shortages in the south and rising unemployment in the north, as labour could not afford to move southwards (Champion, Green and Owen 1988; Healey 1987). Indeed, economic models of migration have shown inter-regional flows negatively related to relative house prices (Harrigan, Jenkins and McGregor 1986; Mitchell 1988).

Bover, Muellbauer and Murphy (1988) suggested a number of implications from this discovery. First, home owners outside south east England could not afford to move into the region. Second, owners in

the South East were reluctant to move away because they might miss out on further price appreciation and because they might find it difficult to move back later. Third, once the relative prices in the South East began to fall, people were reluctant to move into the area because investment in property appeared unwise. This led Forrest (1987: 1612) to argue that

Contrary to the beliefs and claims of the past, it is owner occupation rather than council housing which is emerging as the major *housing* barrier to labour's mobility.

Economic restructuring and the redistribution of labour

Migration for employment reasons is one of the more persuasive theories that have been postulated to explain counterurbanisation tendencies in the developed world (Chapter 1). Fielding (1990) argued that moves within ILMs, along with moves into new firms by managers and professionals may be a significant feature of rural population growth. This has been stimulated by a move from *regional sectoral specialisation* (RSS), where similar industries clustered spatially (such as the textile industry in Lancashire, England), to a *new spatial division of labour* (NSDL), where the choice of geographical location is more flexible. In this 'post-industrial' society, there has been a shift towards small-scale industries practising flexible production techniques and away from large-scale Fordist mass-production approaches. These small industries have begun locating in more remote rural areas, stimulated by the availability of a 'green' labour force – which is primarily female and unlikely to be unionised – the low land rents in the periphery, and the improving services and communications in these areas, which make them less disadvantaged compared with urban areas (Massey 1984). These firms are unlikely to be able to recruit senior employees in peripheral areas, however, and consequently they are likely to move with the firm into these rural locations. Indeed, the management will have been instrumental in choosing the location in the first place, and its choice will frequently have been influenced by a desire to live in a more rural setting (Chapter 6).

At the same time, it is argued that the increasing globalisation of the economy has resulted in global cities, such as London and New York, where financial services and institutional headquarters cluster (Frost and Spence 1993; Sassen 1988, 1991).

Consequently, these large cities also become attractive destinations for certain types of both international (see above) and internal migrants.

Refining explanations of labour migration in the developing world

Informal labour markets

Neoclassical models have been criticised in a developing world context for being Western-biased and overly concerned with economic maximisation behaviour (Chan 1981). Cultural differences between societies may make these market-based approaches invalid; for example, how do we reconcile the wanderings of young adult males between African urban and rural areas, as part of initiation rites, within a framework of rational economic behaviour? The relative importance of the formal and informal employment sectors also varies between developed and developing societies. Typically, developing countries have a considerable *informal economy*, which is not acknowledged in official employment statistics as it is outside government systems of social security and taxation (Brown and Sanders 1981). Consequently, economic migration models, which utilise official statistics on wages and unemployment, would be mis-specified.

According to Portes (1981), the informal sector provides cheap goods and services to workers in the formal sector, indirectly allowing capitalists to pay their workers low wages. The informal–formal sector relationship is therefore exploitative, favouring those in formal employment. Many rural–urban migrants in the developing world move into the informal economy, as job opportunities in the formal sphere are often limited. A national survey in the Philippines (Koo and Smith 1983) showed that recent in-migrants to cities were especially likely to move into informal employment but that, with time, migrants were gradually absorbed into the formal economy (Table 4.11). This suggests that there are opportunities for migrants to move into formal employment, but this conclusion needs to be treated with caution as more detailed analysis showed considerable differences between male and female in-migrants, as the table shows. Recent and long-term female migrants were far more likely to move into the informal sector; the assumption that urban in-migrants tend to move into informal sector jobs was

Table 4.11 Formal/informal sector distribution by migrant category and sex in Philippine cities, 1968.

Migrant category	Sectors (percent)		
	Formal	Informal	Agriculture
Manilla			
Natives			
Male	68.0	25.9	6.1
Female	53.7	44.5	1.8
All	*62.7*	*32.7*	*4.5*
Long-term migrants			
Male	74.3	22.7	3.0
Female	45.2	54.1	0.7
All	*62.9*	*35.0*	*2.1*
Recent migrants			
Male	71.0	27.7	1.3
Female	23.1	76.5	0.4
All	*48.0*	*51.1*	*0.9*
Secondary cities			
Natives			
Male	39.0	29.5	31.4
Female	32.9	54.0	13.0
All	*36.7*	*38.8*	*24.5*
Long-term migrants			
Male	51.9	32.3	15.9
Female	27.3	71.9	0.8
All	*42.0*	*48.3*	*9.8*
Recent migrants			
Male	72.4	13.8	13.8
Female	4.2	93.8	2.1
All	*29.9*	*63.6*	*6.5*

Source: Koo and Smith 1983.

not upheld for males but it was certainly the case for females, who were economically disadvantaged as a result.

Systems of political control

In South Africa, the former apartheid system impacted strongly upon many people's everyday lives, including their migration behaviour. Soni and Maharaj (1991) have described how black workers were excluded from the residential areas of white South Africans and were concentrated in poor and unsanitary townships:

Black workers are seen and largely used as anonymous 'units of labour', with little attention being paid to their dignity and rights as citizens and human beings.

(Parnwell 1993: 64).

A migrant labour system began in the late nineteenth century, when gold and diamonds were discovered in South Africa. The influx of black workers to the mining areas resulted in demands for segregation from the whites, and the 1923 Native (Urban Areas) Act established the system of repression that has only recently begun to be dismantled. Black workers could only move into urban areas temporarily, they had few political rights, and they were barred from owning land. D. Simon (1989: 141) suggested that the political economy of the region was

predicated on subjugation and dispossession of the indigenous peoples, who were then forcibly concentrated in small, inadequate reserves or Bantustans away from the best agricultural land and usually towards the geographical periphery of the respective territories. Imposition of hut and poll taxes, payable in cash, coupled with the inadequacy of peasant subsistence in the reserves, compelled Africans to seek progressively more wage labour in the colonial money economy.

Labour policies enforced by the state (Chapter 7) prevented families from moving with male migrants who sought work in the mining areas; they remained in the reserves while the male workers were accommodated in towns in prison-like hostels, from which they had to return home annually as a condition of their contracts (Gordon 1977). Thus, despite the disparity between the poor rural subsistence economies and the richer urban centres, it is unrealistic to attribute the migration of South African black workers to neoclassical assumptions about labour markets. The restrictions on household migration and the control over the precise destinations of the male labour migrants resulted in biased and predictable patterns of migration.

More recent state reforms have been instigated, at least in part, because of the impacts that the migrant labour scheme had on capitalist expansion:

The migrant labour system, as practised, meant an unstable labour force of inadequate productivity, affected by the disruption of family life. Job reservation and meagre black education operated together to create a growing shortage of skilled labour without which economic and technological growth were gravely hampered.

(Pomeroy 1986: 123).

The distorted labour market created by the migrant labour system meant that in some cases labour shortages were not being met. Yet the policy of 'orderly urbanisation' introduced in 1986, which replaced the influx control legislation, was designed not to improve the conditions of the black families but to meet the needs of business for a more regulated workforce. The policy retained racial residential segregation but allowed for industrial decentralisation

and the devolution of municipal services; black migration into urban centres actually became more controlled as squatter laws were tightened (Hindson 1987). Only now, with the end of the apartheid era, might we expect labour migration in South Africa to resemble more closely that observed in comparable countries.

Gender and household strategies

Increasingly, work on migration in the developing world has favoured household perspectives, which recognise that household maintenance or reproduction is as important as labour opportunities in explaining migration (Chapter 5). While gender-differentiated migration in the developing world is related to the way that 'sexual divisions of labour are incorporated into spatially uneven processes of economic development' (Chant and Radcliffe 1992: 1), the household strategies approach argues that migrants are influenced by complex intra-household decision making, where individual actions are mediated within the household's requirements, rather than being based purely on rational employment decisions.

As described in Chapter 1, the bias towards female urban-ward movement in many developing countries is related to employment factors. However, a narrow economic interpretation of this movement would assume that women are free to migrate when, in fact, they are constrained by social and cultural constructions of gender and the household. Moreover, the employment factors themselves are mediated by gender relations:

Thus female migration experiences are determined both by intra-household resource and decision-making structures, and by the socially determined, gender-segregated labour markets available to them.

(*ibid.*: 23).

Radcliffe (1990) considered the recruitment of peasant women from the rural village of Kallarayan (Peru) into both urban areas and commercial agricultural developments in jungle areas. Female migration away from the village was more long-term and more urban-biased than male migration. In particular, female out-migrants were almost entirely single; once married their role within the family unit changes and their role in the reproduction of the household assumes central importance. Even when single, the migration of women was organised by parents and older siblings:

Migration by young girls is thus prompted by the resource base of the household and household composition (in terms of age, sex, marital status, numbers), rather than by monetary calculations on the returns to labour.

(*ibid.*: 238).

Household relations, therefore, influence female migration, but additional factors include the various agents of recruitment (other than family members) such as godparents and *conocidos* (acquaintances), figures of authority such as teachers, employers, and *enganchadores* (recruiting agents).

Similarly, a study of migration in Guanacaste (Costa Rica) showed that household decision making influenced female migration into cities. Structural factors, such as the changing nature of production in rural areas, with the move to mechanisation and the subsequent loss of jobs for women, and the increase in service sector employment opportunities in urban areas, also contributed (Chant 1992a). Additionally, the breakdown of traditional societal constraints has freed women to a degree, as temporary migration of males away from the family reduces male control over wives and daughters. Increasing access to education has also led to more independence for women. Thus, we should not ignore the changing role of cultural and religious values in determining society-specific female migration patterns (Chant 1992b).

Economic restructuring and labour migration

Shifts towards export-oriented economies in the developing world have had significant impacts on internal migration, as the tendency has been a decentralisation of industrial growth away from the major urban centres (Gilbert 1993). In Mexico, this has occurred on the outskirts of cities and along the

---

Box 4.5 *Purdah*, patriarchy and migration in Bangladesh

Most female migration that has occurred in Bangladesh has traditionally been associated with marriage or as part of a family group. The patriarchal family system maintained this situation, such that few women could be described as labour migrants, and this is reflected in the high ratio of males to females in urban areas (126:100). The system of *purdah* (female seclusion) imposes many constraints on female life; women are segregated in the public sphere, suffer from patrilineal inheritance rights, and are bound by strict rules related to marriage and family negotiations.

Although *purdah* continues to be a powerful cultural norm within Bangladesh, it is becoming increasingly untenable as increased poverty undermines the 'patriarchal bargain' (Kabeer 1988) and the number of female-headed households rises. The latter has been caused by increasing levels of male desertion and the inability of parental households to take in abandoned women. Additionally, traditional forms of 'secluded' women's employment in rural areas are being removed, because of mechanisation, and they are forced to seek work elsewhere; thus women are being seen in new and visible work outside the culturally prescribed areas they have participated in in the past. This has inevitably involved the migration of women, as labour migrants, towards urban centres, where job opportunities exist and female employment is more culturally accepted.

An example is the case of Momena, whose arranged marriage into a wealthy landowning family failed after she was harshly treated and regularly beaten by her husband. A second arranged marriage to a widower ended when he was killed at work and the husband's brothers took charge of the land he owned. Momena was not given enough food to live with her three children through the winter and she finally migrated to Khulna (south west Bangladesh) to live temporarily with her brother's family. Their poor existence meant that she had to find work and alternative accommodation. She achieved this but, after a series of poorly paid jobs for herself and her teenage son, she found herself in serious debt four years after moving into the urban centre. Her household income was 59 percent of the slum poverty line and she was in debt up to three times the household's monthly earnings; the only person who was being fed properly in the family was her teenage son, who needed sustenance to continue working as a rickshaw puller. Women in Bangladesh increasingly find themselves in this kind of situation, but the lack of reasonably paid work for women in the patriarchal Bangladeshi society means that their existence is often one of constantly striving for survival:

While the social consequences of independent geographical mobility may ultimately be positive for women in that they allow them an opportunity for slightly greater control over their own lives, the material consequences, as we have seen, are often extremely negative and usually involve continued reliance on men from their kin groups.

(Pryer 1992: 152).

Source: Pryer 1992.

US border, funded primarily by American invest-ment. These *maquila* industries have been criticised for their failure to reinvest in the local areas:

> By their very nature, the maquila and its territory are totally disarticulated from the local productive structure and from the local markets supplying the raw materials and inputs, and they fail to generate any local multiplier effect. Their only impact is to multiply the banking and service sectors. Not only do they not integrate with the local econ-omy but they disintegrate it, linking parts of it to those of imperialist countries which are the origin and destination of both the capital and the products.
>
> (Pradilla 1990: 99).

While they have helped break down the traditional monocentric urban growth of the Mexican economy, this industrial relocation does not appear to be at-tracting industry into the poorer regions, and devel-opments on city outskirts have been regarded by some as an intensification of industrial concentration (Gwynne 1990).

Conclusion

This chapter has detailed a wide range of literature concerned with the links between employment and migration at a range of spatial scales. Numerous studies, of both international and internal migration, have concentrated on neoclassical economic in-terpretations of migration and it is certainly true that these have contributed considerably to our under-standing of the characteristics of labour migrants and the regional variables that stimulate and dis-courage migration. Many labour migrants are motiv-ated by improving their economic circumstances, and wages and employment rates will figure in this calculation to some extent. However, the migration process is far more complicated than a simple reac-tion to economic disequilibrium. Increasingly, nar-row interpretations of labour migration have been questioned. Behavioural work highlights serious misgivings about the rationality of migrant behav-iour, while more structural perspectives have shown that individuals' apparent freedom to migrate is bound up with 'hidden' forces that manipulate mi-gration patterns. Finally, despite the considerable at-tention that employment-related migration has received in the migration literature in the past few decades, various other factors influence both inter-national and internal movement independently and also mediate migration linked to economic consider-ations. One important set relates to demographic factors, notably the life-stage of the person con-cerned.

CHAPTER 5

Migration and the life-course

Introduction

The view that individuals and families move through a series of changes from birth to death is by no means a new one. Shakespeare, for example, wrote of the 'seven ages of man', and described how males began as babies and then subsequently became schoolboys, lovers, soldiers and justices, before retiring and then entering a second childhood. The idea of a life-cycle had even entered the social science literature by the early years of this century, when Rowntree (1902), writing on urban poverty, commented that 'the life of a labourer is marked by five alternating periods of want and comparative plenty'. While Rowntree and others, such as Sorokin, Zimmerman and Galpin (1930), were not suggesting that such patterns were inevitable for all, they established a perspective that relied upon cross-sectional analysis and the belief that subsequent cohorts would experience the same cycles and stages. They also made a direct link between stages in the family life-cycle and consumption levels, on the assumption that the former drove the latter.

While the idea of *life-cycle* stages had thus been introduced to the literature before the Second World War, Glick is often credited as the academic who brought the formal model of the family life-cycle into the mainstream. Glick worked for the United States Bureau of Census, and in 1947 he used census data to define seven stages in the development of the 'average' American family (Table 5.1) (overleaf). Glick subsequently extended his model.

Up to the mid-1970s, the family life-cycle was widely used both as a framework for analysing other phenomena and as an object of study in its own right. The model was refined by building in non-demographic variables, such as when children first went to, or left, school (Wells and Gubar 1966), but it also began to relate the stage of family development to patterns of housing consumption or employment (Duvall 1977).

The family life-cycle concept and migration

Rossi's legacy

A great deal of work has linked migration behaviour to stage in the family life-cycle. Rossi (1955), in his seminal work *Why Families Move,* was one of the first to make such an explicit link. He studied four different census tracts in Philadelphia, stratified by socio-economic status and mobility levels. In each tract he undertook about 200 interviews designed to collect information on the frequency of residential moves and the motivation for both leaving the previous home and selecting a new one. Rossi argued that mobility arose for five reasons, namely the creation of new households, the circulation of existing households, mortality, household dissolution, and moves related to work (Chapter 5). The first four of these reasons are clearly related to stage in the life-cycle. Rossi argued that moves stimulated by work, divorce and death of a household member were less important than those generated by existing families searching out different housing. His own survey suggested that only about one-quarter of all residential mobility was 'forced' (driven by eviction, demolition, work, marriage, divorce or downward social mobility), with the remainder being voluntary moves in which households engaged in a continuous

Table 5.1 The classic family life-cycle.

Phase	Beginning	End
I. Formation	marriage	birth of first child
II. Extension	birth of first child	birth of last child
III. Completed extension	birth of last child	first child leaves home
IV. Contraction	first child leaves home	last child leaves home
V. Completed contraction	last child leaves home	first spouse dies
VI. Dissolution	first spouse dies	death of survivor

Source: Hohn 1987, after Glick 1947, reprinted by permission of Oxford University Press.

process of matching their accommodation to their changing housing needs. In summary,

the major function of mobility [is found] to be the process by which families adjust their housing to the housing needs that are generated by the shifts in family composition that accompany life-cycle changes.

(*ibid.*: 9).

Rossi focused on this majority group in his research and argued that the amount of space available in existing accommodation was the issue that encouraged households to move so frequently, with over half the respondents listing this as their main reason for moving. Space was also the most frequently cited characteristic determining the choice of a new property. Nonetheless, Rossi was careful to note that space was not the only factor determining residential mobility, the social status of the neighbourhood being an important additional motive for leaving an area. Even here, though, status was seen not only as a visible symbol of the social mobility of parents but also as a way of ensuring that children were socialised in a preferable social milieu and were interacting with a suitable pool of potential marital partners. The arrival and needs of children were, therefore, powerful determinants of mobility, although Rossi (p. 178) concluded that 'the substantive findings stress space requirements as the most important of the needs generated by life-cycle changes'.

Since Rossi published his ideas they have become central to much of the research on residential mobility. Two strands of research that have developed from Rossi's model are the attempts by researchers to demonstrate that there is a clear empirical link between stage in the life-cycle and mobility levels, and efforts to prove that voluntary moves driven by the need for greater residential space do indeed outweigh mobility arising from changes in employment, marital status, or the housing market. For example,

McCarthy (1976) studied residential mobility in Brown County, Wisconsin, in the United States between 1973 and 1974. He demonstrated that

housing choices are powerfully conditioned by the demographic configuration of the household, as measured jointly by the marital status and ages of the household heads, the presence of children in the household, and the age of the youngest child.

(*ibid.*: 55).

Criticisms of the family life-cycle model

Criticisms of the general model

In the second edition of *Why Families Move*, Rossi noted that the family life-cycle model had not initially been enthusiastically received and, although by the mid-1970s there was a huge literature built upon the model (Young 1977), there was a growing body of criticism. Some writers objected to the basic analogy drawn by the model. Bryman *et al.* (1987: 2) asked 'precisely what is it that is returning to its original position?' and denounced the 'deterministic implication that life is irreversibly leading something back to where it came from.' Harris (1987: 22) amplified this point:

A living species inhabits cyclical time in which the same process, characterised by a regular sequence of stages, recurs until the species becomes extinct or mutates. ... Our lives are not constituted by the endless repetition of orderly sequences. In so far as our lives constitute a sequence we move through this sequence once and then stop at death. Personal time, like historical time, is linear not cyclical.

For the same reason, different terminology has been proposed, designed to embody the beliefs that individuals can influence not only structures but also their own destinies, and that the human species is not simply rooted to the spot. Frankenberg (1987), for example, considered the terms life-cycle, life tra-

Box 5.1 Residential mobility in Brown County, Wisconsin, United States

The results of a survey of 3700 households in Brown County in 1973 and 1974 generated an eight-stage model of life-cycle stages:

1 Young single head without children.
2 Young couple without children.
3 Young couple (less than 46 years old) with young children (less than 6 years old).
4 Young couple (less than 46 years old) with older children (6–18 years old).
5 Older couple (head more than 45 years old) with older children (one child less than 18 years old).
6 Older couple, no children.
7 Older single head (more than 45 years old), without children.
8 Single head (less than 60 years old) with one child (less than 18 years old). Includes divorced, separated, widowed and never married.

The economic characteristics of households differed in each of these stages:

Stage	Percentage of sample	Average age of working head	Percentage with working wife	Median income ($)
1	8.6	25.4		7,564
2	7.3	26.4	67	13,433
3	26.0	31.5	31	12,656
4	10.2	38.9	49	14,593
5	11.8	51.8	34	17,549
6	18.0	62.8	27	10,965
7	13.0	67.1		4,697
8	5.1	37.2		5,704

Note how household income grows progressively with age until retirement, but also how this steady growth is broken once a couple decide to have a family and the wife either chooses or is forced to give up work. Note also how income is drastically reduced in retirement with the death of one partner.

Data were then provided on the housing characteristics of households:

Stage	Average number of rooms		Percentage owners	Average number of persons per room	
	Owners	Renters		Owners	Renters
1	5.14	3.69	<10	0.25	0.46
2	5.65	3.99	40	0.37	0.54
3	6.10	4.66	75	0.80	0.83
4	6.52	5.39	>90	0.82	0.98
5	6.61	5.81	95	0.84	0.96
6	5.57	4.42	90	0.43	0.52
7	5.52	3.81	60	0.24	0.32
8	5.79	4.77	50	0.70	0.68

The data show how the size of house inhabited increases with family formation and expansion, which is tied closely to the desire for owner occupation, but then decreases as children leave home. Note how the decrease later in life is smaller than the initial increase; how renters are less able to afford larger properties; and how later movement into communal or shared accommodation has the effect of increasing the density of occupation.

Finally, information was obtained on the percentage of households that had moved in the preceding year and the most cited reason for having moved:

Stage	Percentage mobile in preceding 12 months	Main reason for moving
1	68.5	Change in family circumstances
2	58.6	Change in family circumstances
3	20.8	Change in tenure or property type
4	10.4	Change in size of property
5	1.3	More convenient location
6	3.7	More convenient location/space
7	9.4	Family circumstances/had to leave
8	29.3	Change in size of property

Stages 1 and 2 thus represent leaving home and getting married; stages 3 and 4 are a period of familism, as parents strive to provide their children with stability, space and a suitable physical environment in which to enjoy their early years; in stage 5 parents begin to think about their own needs as their children leave home and they look to match their address to their lifetime career progress, although inertia stifles mobility; and stages 6 and 7 represent adjustments to retirement, loss of income or loss of a partner.

On the basis of this evidence, McCarthy (1976: 76) concluded that the life-cycle model had been broadly validated and exemplified. He commented:

Young single individuals typically set up their households in small rental units in large multiunit buildings. As households progress to the middle of the life cycle they adjust their consumption accordingly, moving first to larger rental units (often single family homes), then buying a home. After peak household size is reached in the middle of the life cycle, households begin to reduce their housing consumption by moving to smaller single family homes and rental units.

Source: McCarthy 1976.

jectory and *pilgrimage*, and preferred the latter. He suggested:

Lifedeath Pilgrimage fruitfully combines individual choice with awareness of the necessity of structure, the instrumental with the expressive, the passage of time with passage through space, and the symbolic with the merely signifying.

(*ibid.*: 137).

Yet other writers have objected to the assumptions made by the model. Hohn (1987) commented that the model defines a normative course of events when, in reality, normativity is always socially defined. Uhlenberg (1974) voiced similar sentiments and described how, to have followed the 'preferred' life-cycle, a woman should, by the age of fifty, have married, remained within this marriage, and borne at least one child. Trost (1977) went further, suggesting that while the concept of the family life-cycle was harmless *per se*, its use as a research and educational tool might actually stultify social change. Equally, some of the terms and vocabulary associated with the life-cycle model are value-laden. The term 'empty nest', often used to describe that stage when a married couple have successfully launched their children into independent households, does not do justice to the full, vibrant and satisfying lives enjoyed by many couples in their fifties and sixties (Laslett 1989).

Other researchers have criticised the life-cycle model for being time- and place-specific. Anderson (1985) argued that, while for many people in contemporary advanced societies both their expectations and their experiences closely follow that predicted by the life-cycle model, the same could definitely not be said about earlier historical epochs. A different set of critics pointed to the cultural specificity of the model. Collver (1963) noted some of the key differences between circumstances in the United States and in India, especially the absence of birth control and effective health care in India and the early age at marriage there. His study of twenty villages near Benares led him to conclude that:

the cycle is a continuous flow, with little to mark off the transition from one stage to another. Each stage of life, whether of the individual or the family, overlaps the previous one to some extent ... the concept of stages is inadequate to describe this situation in which early marriage, prolonged childbearing and early death combine to run the stages together, or to obscure entirely the final stages.

(*ibid*: 87–8).

Collver also commented that, in developing countries, the extended family defies attempts to impose stages upon it, since it is, in effect, 'immortal'. As cohorts are lost through mortality, others are added through fertility. Thus, while the position of individuals within the extended household changes over time, the actual shape and structure of the household does not. Consequently, Murphy (1987: 36) concluded that

the family life cycle model as it is usually elaborated, refers to the family circumstances of white urban middle-class Americans in the 1950s and 1960s – a particularly child-orientated period.

This period was, of course, when Rossi was formulating his ideas.

The life-cycle model also mixes scales of analysis by trying to incorporate households and the individuals who make them up. Two families that outwardly have similar life-cycles may well contain individuals who had very different life experiences. To group them together for purposes of classification and further analysis might well be mistaken. Such inflexibility is reflexive of the model's general inability to accommodate 'deviance' or social change. In the 1990s, even within US society, the 'traditional' life-cycle is followed by a decreasing proportion of individuals and households. Divorce, cohabitation, remarriage, study away from home, increasing life expectancies, and the decision by many couples to postpone or even reject childbearing have all created new forms of life pilgrimage. The supposed 'golden age of the family' (Hall 1995) of the 1950s and 1960s, in which people married, did not divorce, and reserved sex for marriage, is no more.

Criticisms of the life-cycle model in relation to mobility behaviour

Critics have claimed that there are two main specific problems with applying the life-cycle model to migration behaviour. First, there are problems with the data that have been used to test the model. Courgeau (1985) has argued that because of basic flaws in the research design of much of the subsequent empirical research, the assumptions that underlie the model remain relatively untested. He points out that most of the data that are readily available on mobility are cross-sectional in nature, indicating only a person's present place of residence and that at some previous point in time. Consequently, as described in Chapter 2, we have no way of knowing how many moves have taken place between these arbitrary dates; nor can we ascertain their duration. People might also have moved away but then returned to the same location, being falsely recorded as non-migrants, a scenario particularly common for young adults. Much of the other data that analysts link with that on mobility are also cross-sectional, so that, for example, we might know only that an individual was unmarried at one census but married at the next, not the exact date of the marriage. We might know that migration and birth of a child have both taken place in the inter-census period, but we would have no way of knowing whether family creation spurred migration or was made possible by it. Many studies of mobility and stage in the life-cycle are thus attempting to correlate two very imprecisely measured and potentially incomplete variables in the search for causal relationships. Courgeau (ibid.), therefore, made a strong plea for research that was more longitudinal in nature and that relied upon life-histories and event histories (Chapter 2).

A second methodological problem with using the life-cycle model is that other characteristics of individuals or households also change over time in parallel with those that underpin the model. Consequently, it is essential to separate life-cycle stage from other time-related variables when seeking explanations for changing behaviour. Income, for example, tends to change through the working life of an individual, and it may be the growth in disposable income that actually allows the realisation of mobility encouraged by the birth of a child. Spanier, Sauer and Larzelere (1979) found that while the socio-economic status of wives was best explained by life-cycle stage, the same was not true of husbands, for whom age was more important. On the basis of this and other evidence, Murphy (1987: 49) concluded that 'empirically it does not prove to be a particularly powerful classification for discriminating among different families'.

Even the most ardent supporter of the life-cycle model would now have to admit that, with the benefit of hindsight, it suffered from overapplication and overenthusiasm. The classical model was certainly a creature of its time and place. Thus:

with hindsight the formulation seems plausible for middle class white Americans in a dominantly private sector housing market but singularly inappropriate for black Americans or low income Europeans dependent upon public or state-subsidised housing. ... the earlier representations of the life cycle and of the most common reasons for moving have become outdated, partly because of changes

in the housing market but principally because of altered social forms.

(Warnes 1992a: 186).

It is possible to take this criticism much further. By concentrating upon the internal demographic and familial characteristics of a household and seeing these as a dominant cause of migration behaviour, the life-cycle model deliberately downplays other determinants. Structural determinants, such as shifts in the geography of opportunity or state intervention in migration (Chapters 4, 7 and 8), are largely ignored. Warnes (ibid.) concluded that geographers had simply adopted the life-cycle model as if it were timeless and had unthinkingly reiterated it since the 1960s. However, to condemn the model for the overenthusiasm of its supporters would be a mistake. Unlike the inventors of many other models used in geographical research, Glick, its key proposer, did attempt to modify it to keep pace with social change, and he also argued for its more parsimonious application. As late as 1977, Glick suggested that the model should be applied only to stable first marriages and noted that many other life-paths did exist that were outside the scope of his proposed chronology. Murphy (1987), although critical of the model, defends its empirical usefulness by noting that, despite over forty years of social change, the model is still applicable for around half the British population. He also observed that social change should not allow the real merits of the family life-cycle concept to be obscured, namely that it emphasises those dynamic aspects of family life that can be overlooked in cross-sectional analysis. Even simple life-cycle models had made a substantial contribution to demographic analysis.

Broadening the concept: life-courses

From life-cycle to life-course

While Glick and his supporters struggled to limit the application of the classical life-cycle model to fewer and fewer circumstances, thereby circumventing complexity, others sought to develop concepts that could accommodate such diversity. Family historians such as Elder (1978) were some of the first to recognise diversity in life pilgrimages and responded to this by pioneering the concept of individual life-courses based upon personal biographies and event histories.

The essence of the life-course approach is that the unit of analysis becomes the individual sited in geographical, social, historical and political space, and that the study of the household or family becomes the study of conjoined life-courses. By studying the individual, life-course analysis allows an infinite number and form of life pilgrimages and does not attempt to impose a 'normal' or ideal life-path on every person. In addition, what is central to the concept of the life-course is not the notion of *stage* but that of *transition*. As Harris (1987: 25) put it:

the study of the life course involves an examination of what transitions the members of different social categories within a given *cohort* typically experience and puts the question as to whether those transitions are of such a *nature* and so *timed* as to constitute life transitions: as to constitute major changes in the whole way of life of the person concerned.

Furthermore, the life-course approach emphasises that individuals negotiate their own paths to a much greater extent than was allowed for in mechanical and deterministic life-cycle representations, and also that early transitions within the life-course have implications for later ones.

Concern has been expressed about the extent to which, by focusing upon the individual, knowledge about how the family or household functions may be lost. Family historians have responded to such disquiet. Hareven (1982: 6) noted how, although 'concerned with the movement of individuals over their own lives and through historical time', the life-course approach is also concerned with 'the relationship of family members to each other as they travel through personal and historical time'. Harris (1987) concurred, arguing that individual transitions did not just occur in 'personal time' and 'historical time' but also in 'family time'. Elder (1978) also recognised the importance of recognising interdependence.

While the shift to life-courses rather than life-cycles leaves analysts ideally placed to respond to diversity, it does not absolve them of the responsibility to generalise. The classical life-cycle model clearly went too far in trying to generalise the varied experiences of individuals in the 1960s into only one pattern, but an infinite number of patterns would be no more helpful. Authors have therefore tried to chart out alternative multiple life-courses for individuals within specific demographic and social groups. Uhlenberg (1969) provides an early example, with her typology of women's life-courses during the period 1830–1920. She suggested that the

Table 5.2 A typology of life-courses.

- Both adults marrying for the first time, remain married
  - without children
  - with one child
  - with two children
  - with three or more children
- Both adults remarrying, or one remarrying and one marrying for the first time, no children from the previous marriage
  - without children in the current marriage
  - with children in the current marriage
- Both adults remarrying, or one remarrying and one marrying for the first time, children from a previous marriage
  - without children in the current marriage
  - with children in the current marriage
- Marriages that end without remarriage
  - without children
  - with children
- Never married
  - no children
  - children

Source: Hohn 1987, reprinted by permission of Oxford University Press.

experience of individual women would tend to follow one of six alternative and parallel life-courses. More generally, Hohn (1987) provided a twelve-fold typology of life-courses, which she suggests is the minimum number that could reasonably capture the experience of the majority of the population (Table 5.2).

Age migration schedules

While most geographers now agree that there is no *one* life-cycle into which all individuals or families can be fitted, research does still demonstrate regularities, both in the propensity of individuals to migrate at different ages and in the way migration is often associated with other demographic events (Courgeau 1985). A sizeable body of evidence now exists on *age migration schedules* for inter-region/inter-district moves within affluent mixed-economy countries. Rogers and Castro (1981, 1986), for example, have brought together 500 such schedules, while Rogers (1988), Veergoosen and Warnes (1989) and Warnes (1983a, b) have added others from different parts of the world. On the basis of their own data, Rogers and Watkins (1987) have proposed a model migration schedule (Figure 5.1) which has a constant migration component aug-

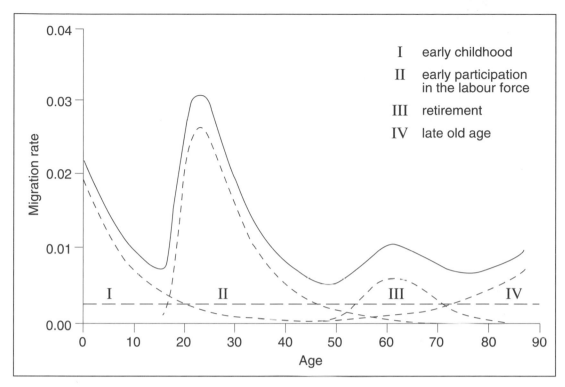

Figure 5.1 A model age migration schedule (source: Rogers 1992b: 240, after Rogers and Watkins 1987).

mented by four peaks of migration during different life transitions (early childhood, early participation in the labour force, retirement and late old age). Figure 5.2 (page 112) provides us with empirical data to compare against the model. Although these data relate to different countries (England and Wales, the United States, Japan, the Netherlands, and Australia) and are taken from different periods (1950s through to late 1980s), what is striking is their similarity. The migration rates follow a pattern, with relatively high rates of mobility shortly after birth, a trough of mobility between ages 3 and 13 followed by a steadily rising trend towards a mobility peak between 17 and 30. Thereafter, mobility falls steadily with age, although there is an interruption to this trend between 57 and 67 years of age.

Warnes (1992a) has sought to associate this pattern of mobility with particular life transitions, adding information on the likely frequency of moves and the distance over which they might take place (Table 5.3) (page 113). His schema is an improvement on that of Rossi since it not only allows for parallel life-courses – note how Warnes has different pilgrimages for 'good' and 'low'-income groups – but

it also responds to migrations motivated by reasons other than family expansion and the need for domestic space. He thus denotes mobility associated with career progression, illness, divorce and remarriage as well as familism.

Warnes (1992b) has also sought to investigate whether age migration schedules have changed significantly over time. His findings are summarised in Figure 5.3 (page 114) and clearly owe a debt to Zelinsky's (1971) mobility transition (Chapter 3). In pre-industrial societies (Figure 5.3a), migration peaks when young people enter the labour force, since this step often involves migration to take up employment as farm workers or domestic servants. Migration declines steadily with age, with no distinctive 'retirement' phase, since the absence of a welfare state means that people must work until death. With the onset of industrialisation (Figure 5.3b), migration increases sharply, as people leave the land for towns, either permanently or as temporary migrants. Age of entry to the workforce is delayed, as full-time education for children takes hold and people live longer. In late industrial societies (Figure 5.3c), compulsory education to a later age is the

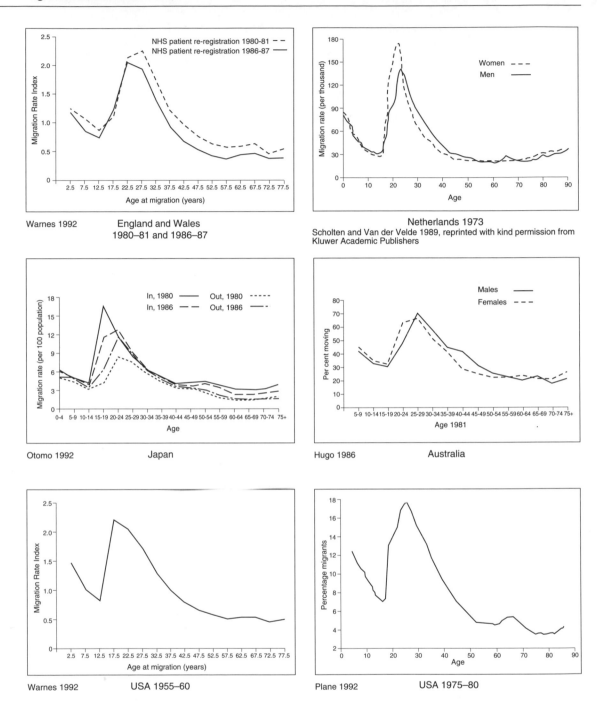

Figure 5.2 Selected age migration schedules.

norm, retirement is possible because of state pensions and other welfare provision, and increasing longevity encourages those reaching retirement to have a positive view of their future. Finally, in post-industrial societies (Figure 5.3d), full-time education is further extended and therefore entry into the workplace is delayed again, with the search for appropriate education prompting migration. Pre-

Table 5.3 Life course transitions associated with migration.

Life course transition	Housing needs and aspirations	Distance and frequency of moves per year	Age (years)
Leaving parental home	low cost, central city, temporary shared	short and long 1+ moves	16–22
Sexual union	low–medium cost, short tenancy	short 0.3 moves	20–25
Career position	low mortgage, flat or house	long 0.5 moves	23–30
First child			
• good income	medium mortgage, 2+ bedroom house	short	23–30
• low income	social housing	very short	21–28
Career promotion	higher mortgage, larger house	long 0.1 moves	30–55
Divorce	low cost, short tenancy	short	27–50
Cohabitation and second marriage	medium cost rental or low mortgage	short and long 0.1 moves	27–50
Retirement	buy house outright, medium–low cost	long	55–68
Bereavement or income collapse	low-cost rental, shared	short or return	70+
Frailty or chronic ill health	low-cost rental, shared or institutional	short 0.3 moves	75+

Source: Warnes 1992a.

retirement migration is more important, as people prepare for the end of their working lives and the beginning of the 'Third Age'. Retirement migration also increases, as more people can afford to make amenity or recreation moves, and as the very elderly seek refuge either with children or in homes.

Life transitions and migration

Leaving the parental home

Many have studied the migration patterns produced by the decision of young adults to leave the parental home. Work indicates that three main reasons prompt this decision: first, the need to move to institutions of further and higher education in countries where the policy is not to have local catchment areas; second, migration in search of a first job, whether this be speculative migration to look for work or contracted migration to take up pre-arranged employment (Chapter 4); and third, migration to form a new sexual union. Clearly, each of these different types of migration would have different geographical patterns.

Both national and international migration are occasioned by the transition to further or higher education. Li *et al.* (1996) have recently investigated the international migration of young adults for higher education. They make a strong case for studying this group, arguing that they are not only numerically significant, as likewise shown by Hugo (1994b) in Australia and Sellek (1994) in Japan, but also that while their migrations are often temporary, they are no more temporary than those of many skilled international migrants moving within internal labour markets. They also note that a significant proportion of students remain overseas after completing their education, often overstaying the duration of their temporary visas. Student migration can thus be one of the first links in a developing chain migration and can also be an important element of illegal migration. Their own work on students looked at overseas migrants from Hong Kong to the United Kingdom. Hong Kong is the sixth largest source of students studying overseas, with over 34,000 abroad at any one point in time, and the United Kingdom is a major destination for foreign-born students, with about 104,000 such migrants in 1992/1993. In the case of the sample of engineering undergraduates in the United Kingdom studied by Li *et al.* (1996), the

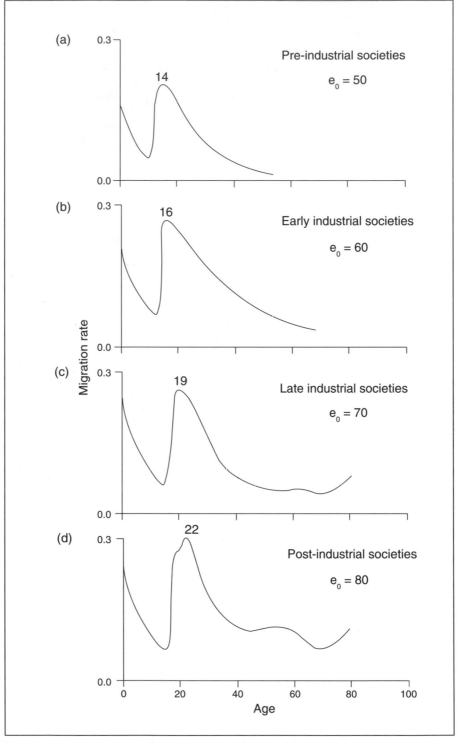

Figure 5.3 Changing age migration schedules over time (e_0 is life expectancy) (source: Warnes 1992b).

prime motivation for their long-distance migration was the international value of qualifications gained in the United Kingdom and the perceived quality of the tuition available there. Most did intend to return once they had completed their studies, but evidence also suggested that residence abroad had changed the perceptions that students had of the desirability of permanent settlement overseas.

Seyfrit and Hamilton (1992) have looked at the opposite end of the migration to take up educational opportunities, namely its impact upon source areas. Their postal questionnaire of youths in the Scottish Shetland Islands probed intentions to leave the island. They found that around one-third of all respondents did think they would leave the islands, and that – unsurprisingly – this was particularly true of those aspiring to higher education and professional employment. Children of long-term residents were less likely to express such intentions because of the strength of their local social ties. Interestingly, the influx of new employment opportunities provided by the exploitation of North Sea oil seemed to have had little effect on migration intentions within this age cohort. Seyfrit and Hamilton (p. 272) reiterated a common conclusion found in work on migration from rural areas, that 'the Shetland and Orkney Islands, like many rural areas, are drained by out-migration of bright and ambitious youth'.

Developing their insights elsewhere, Hamilton and Seyfrit (1993) also surveyed young Canadian Inuit about their plans for migration from Alaska. The work highlights not only the importance of education as a motivator for migration but also employment. Their survey of students in Bristol Bay and in the north west Arctic region revealed many similarities to their Scottish findings. They cited Condon's (1987: 171) comment upon the aspirations of the Inuit in earlier eras:

Before the impact of Eurocanadian culture ... [a] young man would aspire to be a good hunter with a hardworking wife and healthy children. He would seek to be respected as a wise and knowledgeable member of the community. ... Beyond this ... no other opportunities were possible.

Now, however, 63 percent of young adults said they would leave their region, with those from towns being more likely to express this view and being more oriented to leaving Alaska altogether than those from villages. Concern about the quality of local education was one motive for wanting to leave but paucity of local employment opportunities was also widely mentioned. Limited social life was also seen as a major disadvantage of rural living, with respondents describing their teenage lives as 'boring' and bemoaning the lack of shops and facilities. Young women provided a separate set of reasons for seeking to leave. They were highly critical of the quality of 'bush' life for women and argued that they, too, now wished to get jobs and have careers. Because of the gendered nature of available work in the local mineral extraction and oil industry, this meant that most women seeking work had to leave their villages and head for the regional centres or large towns. This gender bias to out-migration has had a profound impact upon sex ratios. Whereas in the north west Arctic region as a whole there are 113 15–39-year-old men per 100 women of the same age, in villages the ratio is often 132:100 and in Anchorage (the largest city) it is 83:100. The authors conclude that migration is having a central role in facilitating gender differences in Inuit acculturation:

through migration, education, employment and intermarriage, it appears that many northern women are following their own acculturation path, independent and deeper than that of men.

(Hamilton and Seyfrit 1993: 263).

In England and Wales the migration of young people has a major influence on national migration patterns, with 20 percent of all movements between United Kingdom Family Practitioner Committee Areas in the 1980s made by 20–24-year-olds (Stillwell, Rees and Boden 1992a). Moves made by such young people were both concentrated into specific streams and less affected by distance than those of older age groups. Almost all parts of England and Wales were experiencing a drain of young people to the South East, a trend that intensified as the 1980s progressed, focusing especially on Greater London. Coombes and Charlton (1992) concur, also noting how the major northern cities all contributed more young migrants to this stream than would be expected. They also suggested that *step-wise* migration was taking place, with young adults moving from small towns or villages to their local city, and then from there to London. This finding ties in well with Fielding's (1989) concept of the South East as an 'escalator region' (Chapter 9), although Coombes and Charlton (1992) also note how many young adults are attracted to London only to enter the urban underclass.

Staying with the British example, Grundy (1992) has explored the changing nature of the process of leaving home, whether it be motivated by work, education or marriage. She argues that, in previous epochs, the timing of departure from the parental home was often determined by the needs of the parents themselves. It was they who would decide which children would remain in the family home, eventually to inherit it, and it was they who would decide when the other children were ready to seek employment or were released from domestic work or labouring on the family farm. Previously, too, departure from the parental home did not necessarily coincide with the creation of a new household. Young adults would lodge for a number of years, with either landlords, employers or relatives, but they would rarely return to their parent's house having once left it. Married couples would also often live with their parents because of their inability to find suitably priced accommodation. In the 1990s, young adults are increasingly determining the timing of their own departure, either to take up work or to attend educational institutions. Increasingly, too, leaving home is associated with the creation of new sexual unions, with 50 percent of women and 40 percent of men in Britain leaving their childhood home to marry or cohabit. Furthermore, 'leaving home' is becoming more of a phase than a final event, with young adults leaving for education but then returning, and leaving to enter a relationship but returning when this fails. Finally, Grundy also notes some short-run changes, with young adults leaving home at an older age (22 years for men, 20 for women) in the 1980s than in the 1970s, due to youth unemployment, the contraction of the rented sector and the trend towards later marriages.

Kiernan (1986) added a broader European dimension to our knowledge by drawing comparisons between the living arrangements of young adults in Denmark, France, Ireland, the Netherlands, (West) Germany and the United Kingdom, based upon a survey of 15–24-year-olds. While most 16- to 17-year-olds still lived with their parents, the speed with which young adults left the parental home after this age varied considerably. Although by age 18 to 19 over half of Danish women had left home, this was true of only 7 percent of Irish women. Setting aside what she terms the anomalous case of Denmark, it is clear that age 20–21 for men and 18–19 for women are major turning points for this particular life transition. Of those young people who have left home, a significant proportion were living with a partner: 25–30 percent of men and 55–60 percent of women. The nature of this union varied a good deal between countries. In the United Kingdom and Ireland,

unions were mostly to be legalised through marriage, whereas in Denmark, France and Germany cohabitation was more common. Sharing accommodation accounted for only 5–10 percent of the sample, although living alone was not unusual in countries where cohabitation was frequent.

Finally, young adults seeking work form an important component of undocumented migration. Sellek (1994) describes how young Taiwanese, Thai and Philippine women who previously worked in the sex industry in their home countries have had to migrate to Japan to access their clients, since both their own and the Japanese government have tried to curtail sex tourism. Because of the very tight restrictions on legal entry of foreigners to Japan, the 'entertainment' industry has had to smuggle the women into the country, using forged passports and visas and illegal entry through quiet ports. Once there, they have no legal rights and are often exploited by their employers, who later abandon them when their earnings decline with age. Sellek notes that, in 1979, 80 percent of those exposed as working in unauthorised occupations in Japan were young Taiwanese women in the sex industry, while Loiskandl (1995) has shown that 88 percent of illegals discovered in 1985 were women. Sellek estimates that there are now over half a million such women in Japan.

Marriage and migration

A substantial literature has developed around the link between marriage and mobility, both on an international and on a local scale. Jackson and Flores (1992) have undertaken one of the most thorough studies of international migration for marriage and its demographic consequences. They focused on the migration of Filipinas to Australia, noting that perhaps 20,000 Filipina brides had moved to Australia since the late 1960s.

British census data often make it impossible to link specific acts, such as a change of address with a change of marital status. However, 80 percent of 16–29-year-old women who had married in the twelve months prior to the 1971 census had also changed their address, with 10 percent of movers crossing a regional boundary (Grundy 1992). Put another way, one quarter of all intra-county moves and one third of all inter-regional moves were made by 16–29-year-old women who had recently married. Coleman and Haskey (1986) have attempted to indicate indirectly the migration field that might be associated with marriage. They discovered that

Box 5.2 Marriage and migration: Filipina migration to Australia

One of the most thorough studies of international migration for marriage focused upon the migration of brides from the Philippines to Australia, noting that *circa* 20,000 Filipinas had engaged in such migration. The migration began in the late 1960s, when a handful of hostesses had returned to Australia to marry Australian men they had met through the sex industry in Manila. Through the 1970s, this migration increased significantly in scale and became more organised. Hostesses continued to find partners but so too did well-educated and professional Filipinas, who met Australian men through introduction agencies or mutual friends. Filipina brides are now widely distributed across the whole of Australia, unlike the Filipino labour migrants, who are largely concentrated in Sydney and Melbourne. They also sponsor more relatives after arrival than labour migrants with, for example, every 1 000 Filipinas in 1987 sponsoring 77.1 siblings. Despite this, the Filipino population as a whole in Australia now has a highly skewed sex ratio with only 44.5 men for every 1 000 women.

What is perhaps most interesting about this work is its attempt to explain why Filipinas seek Australian husbands. In the Philippines, unmarried women are 'objects of pity, ridicule and hostility' (Jackson and Flores 1992: 23) and are marginalised.

Filipinas thus aim to be married before their midtwenties but for some women this becomes a difficult task, for three reasons. First, women have traditionally been more mobile in the Philippines than men and there are always more women in cities seeking marriage than there are men. Urban women are thus less likely to find a partner who is also a Filipino. Second, men in the Philippines still expect their brides to be virgins, and prostitutes would, therefore, be ineligible for marriage to local men. Third, highly educated women outnumber their male counterparts and, because men tend to marry spouses who are less well educated than themselves, educated women have a lower likelihood of ever marrying. Overall, therefore, both urban female sex workers and highly educated women have particular problems that often lead to the attenuation of links with their families, their social marginalisation and marriages with Australian partners. Marriage itself helps with their reintegration into the family, marriage to a (wealthy) Australian adds additional prestige, and the opportunity for the woman to send remittances or sponsor relatives to migrate to Australia completes their rehabilitation.

Source: Jackson and Flores 1992.

while one-third of a sample of newly-weds gave the same address on their marriage certificates, and had therefore been cohabiting, two-thirds did not. Of the latter, the median distance separating the premarital addresses of the couple was 4.7 kilometres, although the distribution was bimodal, with over-representation amongst short distances (33 percent within 2.5 kilometres) and relatively long distances (25 percent beyond 14 kilometres). While the median distances differed considerably by social classes (75 percent of unskilled grooms lived within 5 kilometres of their bride, compared with 30 percent of professional and managerial grooms), the results show that most couples marry locally and, therefore, presumably make a local first move. Kendig (1984) considered some of the housing implications of such mobility, using a data set from Adelaide, Australia. He noted how the combination of two incomes produced by marriage is the most effective means of gaining entry to the owner-occupied sector, so geographical shifts may well be paralleled by tenure shifts.

The association of marriage with both partners leaving their previous homes and establishing a new shared home is much less true in the developing world. Particularly in societies that are highly patriarchal in nature, marriage may require the woman to leave her parental home and move in with her husband's extended family. If this tradition is also associated with village exogamy (men marrying women from outside their village), then a considerable proportion of young women will be involved in short-distance migration in their teenage years. Pryer (1992), for example, cited three studies of villages in Bangladesh, where teenage female migrants outnumbered their male counterparts. In one of these studies, fully 67 percent of out-migrants were female, with most being aged 10–20 years old. Pryer's empirical work in a *bustee* of Khulna City in Bangladesh showed how, despite female migration outnumbering male migration in the teenage years, beyond this age the situation sharply reversed. She was able to find few independent female rural-to-urban migrants, other than the destitute, and argued that women have their freedom of movement severely curtailed by the cultural practice of *purdah* and the patriarchal control of resources (including female labour). She also argued that such constraints are weakening and that, in conjunction with the deepening of rural poverty and the increased availability of industrial employment in the cities, the number of female-headed households and female migrants is rising.

Landale (1994) has taken a different perspective on the link between union formation and migration. Rather than seeing the union as the cause of migration, she sees migration as a powerful modifier of the way in which women form unions. Puerto Rican women in the United States have begun to decouple childbearing and marriage and, as a result, an increasing proportion of Latino households are now headed by unmarried women. These trends were shown to be counter to those at work in Puerto Rico itself. The different behaviours in the United States and in Puerto Rico were not the result of changes in behaviour brought on by the act of migration or exposure to different social norms in the United States. Rather, they resulted from the selective nature of migration, with migrant Puerto Rican women tending to be less advantaged, more sexually active, lower achievers at school and the daughters of women who had themselves been very young when they first gave birth. This selectivity 'contributes to a set of migrant traits that increase the rate of exit from the single state and encourage the formation of informal first unions' (*ibid.*: 151). This example, and those cited earlier, show how the relationship between the formation of human partnerships and migration behaviour is continually in a state of flux, with experiences and patterns of movement becoming much more differentiated over time.

Divorce and migration

Before looking at the direct link between divorce and migration, it is important, briefly, to describe how divorce rates have changed over time. Divorce rates in many Western European countries began to increase in the late 1950s, with particularly rapid increases until the early 1970s, slower rates of increase in the later 1970s and very low rates of further growth in the 1980s and 1990s (Haskey 1996). The impact of this has been to increase sharply the likelihood that a marriage will end in divorce. Haskey (1989) has estimated that if 1993/1994 divorce rates were to persist, around 41 percent of all marriages in the United Kingdom would fail, up substantially from earlier estimates. If this prognosis were to be correct, 11 percent of all marriages would end before their fifth anniversary, 22 percent before their ninth (compared with 0.5 percent in 1926) and 27 percent

before their twelfth. Hall (1995) notes that such estimates would be in line with expectations elsewhere in Europe, where divorce rates are high in Sweden, Denmark, Norway and the United Kingdom, but lower in southern Europe and Ireland. While 'the instability of marital relationships is now regarded as the norm' (*ibid.*: 40), marriages still last as long as they did in the nineteenth century, when they were often ended prematurely by death.

Hayes, Al-Hamad and Geddes (1995) argue that the life transition of divorce can have both a positive and a negative influence on mobility. Although the very act of separation usually forces one partner from the marital home, divorce can also break ties to a locality and facilitate hitherto latent migration intentions. Hayes, Al-Hamad and Geddes used the United Kingdom Sample of Anonymised Records from the 1991 census (Chapter 2) to calculate that the divorced were almost twice as mobile as their married counterparts (a 12.5 percent migration rate, as against 6.8 percent). Such mobility peaks at various times, being most frequent after *de facto* divorce, when 54 percent of wives move out of the marital home (Grundy 1992), but also being commonplace in the following twelve months, as separated individuals either move in with other partners or make moves to secure better accommodation; 25 percent of separated women move at least once in the twelve months following their departure from the marital home. Another mobility peak occurs as cohabiting couples remarry and move to a new home (58 percent of men and 40 percent of women will have remarried within ten years of divorce) (Coleman and Salt 1992). Even those women who remain in the marital home after separation frequently move later because they are unable to make mortgage repayments. Most of this mobility is short-distance, although inter-county rates for divorcees are still higher than for those who remain married. Longer term, Grundy (1985) also notes how the remarried retain higher rates of migration than those who have married only once, reflecting the high rates of fission of remarriages (Coleman and Salt 1992).

Divorce is now one of the main causes of new household formation, contributing 20 percent of 1991's annual growth in England and Wales. This process is associated with housing tenure shifts to bedsits, rented rooms and social housing. Sullivan's (1986) analysis of longitudinal data from the Family Formation Survey showed that 50 percent of all women experiencing marital breakdown went on to live in local authority housing at some point, whereas only 27 percent went on to own an unshared house. There was also a substantial temporary return of young women to their parental home after separation. The tenure shift of divorcees into privately rented accommodation and council housing has clear geographical consequences:

> Unless there is an upturn in the proportion of dwellings available for renting ... it would seem inevitable ... that the clustering of the divorced in inner-city areas, where the bulk of privately rented and small unit council accommodation is to be found, will become more concentrated.
>
> (*ibid.*: 47).

In extreme conditions, marital breakdown can be a cause of homelessness. Dail (1993), for example, has calculated that 38 percent of America's homeless are now female-headed single parent households, and the size of this group is growing at 34 percent per annum. Such people are apt to migrate considerably in an effort to secure a new home.

Childbirth and migration

Much empirical work has confirmed Rossi's (1955) assumption that households move in order to adjust the space at their disposal to the size of the family. Evidence from the United States (Morris 1977), the Netherlands (Clark, Duerloo and Dieleman 1984) and France (Courgeau 1984) all confirms that the birth of a child is still one of the most important causes of local migration. In some cases, such migration occurs after household expansion, with those in shared or overcrowded accommodation being most likely to move (Grundy 1986), but in others a residential move is made in anticipation of childbirth (Grundy and Fox 1984).

Less well researched is the way in which migration can forestall childbirth. In an empirical study using panel data from the British census, Grundy (1986) showed that women who had undertaken long-distance migration between the 1971 and 1981 censuses were likely to have postponed having a first or second child, although she was not able to control for the effect of social class. Grundy also drew attention to the way in which unfavourable circumstances for mobility could have a depressing effect on fertility, as parents wait to create the right conditions for childbirth. She cites two examples of this relationship. First, because of the Depression in the 1930s, many young couples could not afford to move out of overcrowded housing, and therefore delayed childbirth, hence depressing fertility levels. Second,

house price inflation in parts of the United Kingdom during the 1980s may have deterred couples from having children, as they could not afford to move to larger housing or to give up a second income.

Retirement and migration

Several factors have radically changed the transition from work into retirement. As Anderson (1985) notes, in earlier historical periods retirement as such did not really exist, with workers shifting from demanding to less demanding work as they grew near to the end of their lives. Few could afford ever to give up work voluntarily. In many countries, the advent of state and personal pensions has changed the situation and has allowed men and women to cease working without fear of destitution. Indeed, for those with occupational pensions, retirement can now coincide with the receipt of lump sum payments and worthwhile regular income, both of which allow today's elderly to be adventurous and consumerist. This would, of course, be of little value if people were not living long enough to enjoy their post-work period, but increasing longevity is meaning that more people can look forward to lengthier periods of retirement. In the United Kingdom, for example, average life expectancies for men have risen from 48 years in 1900 to 72 years in 1985–87, with women living an average of five and a half years longer. In addition, people are now healthier in old age and therefore more active later in life.

The growing number, proportion and longevity of elderly people has encouraged a sizeable research effort designed to explore this group's mobility patterns and processes. Law and Warnes (1982: 53) succinctly summarise the rationale for such work:

> Retirement has become a clear and distinctive stage in life to which many look forward, partly because of the opportunities to do things impossible during working life. One of these is the opportunity to live in another area, for accessibility to employment ceases to be a locational constraint. People can move to be nearer their relatives or friends, to return to their native or childhood areas, or to residentially, climatically, or socially more attractive areas.

Much of the subsequent research effort has been directed at seeing whether these possibilities have been realised in practice, whether elderly people are mobile, and whether they seek out residence in particular areas or particular types of areas.

Both the United Kingdom and the United States display characteristic profiles for rates of migration

Box 5.3 Cultural factors in retirement migration

In the main text we have demonstrated how, in countries such as the United Kingdom and the United States, there is a minor peak of migration associated with the legal retirement age for men. We will also show how the desire to improve the quality of life is the driving force behind much long-distance retirement migration, with couples seeking out more environmentally conducive areas in which to live. This has increasingly taken them to coastal areas and more recently to inland rural areas, often of a remote nature (Chapter 6). However, it is important to note that both the age of retirement and the factors underpinning the scale and direction of retirement migration are specific to different countries and cultures. A contrasting example demonstrates this clearly.

From an analysis of the pattern of inter-provincial migration in China in the 1980s, the graph below demonstrates how the propensity for Chinese people to migrate (relatively) long distances increases greatly after the age of 55 and only falls again beyond the age of 70. Even 75-year-olds have a migration propensity twice that of 50-year-olds and three times that of 25-year-olds. This pattern of age-specific migration rates is very different to those found in the West. The pattern can be explained by reference to Confucianism. Followers of this philosophy have a moral responsibility to attend to the graves of their ancestors, and those who left their home region in early life are expected to return to it later in life. Consequently,

return migration is a very strong motivating force in explaining Chinese population movement. This supports the

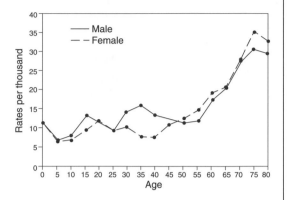

hypothesis that Confucian belief in *Ye lo hui gen* (falling leaves return to their roots) plays a very important role in return migration in China. Migrants make great efforts to return home, especially when they are old.

(Li and Li 1995: 143).

Thus, analysing Chinese inter-regional migration using a Western perspective based upon migration for monetary gain, for example, would be unsuccessful. Rather, migration in this instance needs to be seen as being driven by a certain type of 'psychic gain'.

Source: Li and Li 1995, figure is from Special Characteristics of China's Interprovincial Migration by Wen Lang Li and Yuhui Li, *Geographical Analysis*, **27** (2), reprinted by permission. © 1995 by the Ohio State University Press. All rights reserved.

through later life (Warnes 1983a). The likelihood of migration declines steadily after the age of 50 but, for men, this is interrupted by a small peak of migration at 65–66 years in the United Kingdom or 66–69 years in the United States. This peak tends to reflect long-distance migration, although in the United Kingdom there is also a much smaller peak at this age of short-distance migration. For women the pattern is quite different, with no noticeable peak of movement in the mid- to late sixties but, instead, a much less pronounced bunching around the 59–63-year age point, presumably because of the earlier retirement age for women and the fact that men tend to marry women younger than themselves.

Law and Warnes (1982) found that the decision to migrate near retirement age was often relatively unplanned and came to fruition very quickly. Over a quarter of their sample said that less than a year had elapsed between the first suggestion of a move and

that move taking place. The short period of search also meant that most potential migrants considered very few alternative destinations; 40 percent considered only one area, although married couples tended to look more widely than widows or divorcees. In addition, the search was often confined to coastal areas, although single people and those who retired early again tended to search further inland. As Law and Warnes (p. 74) put it:

a least effort principle was strongly operating in most areas. ... We found little support for the view that the selection of a migration destination area follows an elaborate comparative evaluation of place utilities.

Rather, most people were guided by their past holiday experiences and by the present location of children, other family and friends (Chapter 6).

Comparable work on a sample of elderly migrants who had moved to western North Carolina, in the

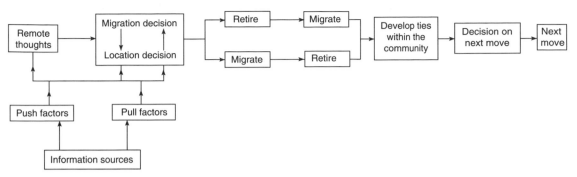

Figure 5.4 A model of the retirement migration process (source: Haas and Serow 1993, *The Gerontologist*, **33**: 212–20. © The Gerontological Society of America. Reprinted by permission).

United States, found that respondents had also considered relatively few possible destinations (an average of 1.9), but that they had been actively considering retirement migration for an average of six years (Haas and Serow 1993). A significant minority had also chosen to migrate and then retire at a later date, usually after four or more years. A model of the process of retirement migration (Figure 5.4) not only looked at the initial retirement migration but also at any subsequent migrations motivated by an inability to settle into a new community.

Central to the decision making of many elderly people is the desire to live close to kin, especially children and siblings. Motivations vary from seeking companionship, to gaining proximity to someone who might care for them in the future (Meyer and Speare 1985) or immediately if they have just been widowed or disabled (Longino 1990). Clark and Wolf (1992) used empirical data to demonstrate that the location of children is more important for the 'old old' than for the 'young old', as it is for those with fewer qualifications, those who have been widowed or those who have a number of children.

Many authors have researched the distinctive geographical patterns produced by elderly migration, with much recent work attempting to test the validity of the *elderly mobility transition* (EMT) proposed by Rogers (1989), which attempts to model the geographical distribution of the elderly population over time and to ascribe different countries to three temporal phases. Rogers (1992b) argues that the United States is just moving from the middle to the late phase of the EMT, while Law and Warnes (1982) and Rees (1992) show the United Kingdom to be well into the late phase. Rogers (1988) and Bonaguidi and Terra Abrami (1992) find evidence that Italy is still in the middle phase. Rogers (1992a)

suggests that this is because Italy still has a relatively weakly developed pension system, with the elderly therefore having fewer resources to move, and being more tied to family for assistance. Finally, Otomo (1992) considers Japan, which is considered still to be in the first phase. There, many people are excluded from the pension system, social services for the elderly are very limited, and family ties are still very strong. Not surprisingly, therefore, many elderly Japanese move only to be with their families.

Law and Warnes (1976) addressed the spatial dimensions of the EMT in England and Wales. They demonstrated how, in the 1960s, the elderly migrated between specific areas (Figure 5.5) (page 123) and how flows were rarely balanced by return or counter flows. Distinctive 'migration areas' appeared (Figure 5.6) (page 124). They also noted the significance of long-distance migration for the elderly, and how retired or retiring people were moving over longer distances than previously. They commented, for example, on the strong distance decay effect that existed in patterns of elderly long-distance migration, with those leaving the southern and western suburbs of London rarely looking north of the Severn–Wash line and preferring instead to head for nearby seaside resorts on the south coast. On the other hand, those previously resident in the north and east of the capital were more likely to look to nearby East Anglia, while those leaving Manchester and Liverpool often headed for the Lancashire or North Wales coast. Allon-Smith (1982) added to this analysis by considering temporal changes in the pattern of retirement migration. The South East region was the first to be targeted as a retirement area, followed by centres in Wales and the North West region. Places in the South West and East Anglia have been the most recent to take on a retirement function. The drift of

---

Box 5.4 The elderly mobility transition

The elderly mobility transition (EMT) was originally proposed in 1989 and represents an elaboration of Zelinsky's mobility transition (Chapter 3). The EMT is a model of how the number and geographical (re)distribution of the elderly population in an advanced society changes over time. The model can be formalised into a series of propositions:

Population ageing
- The final stage of the demographic transition sees advanced societies experiencing a greying or ageing of their populations. Societal ageing will be accelerated as 'baby boom' generations reach old age.
- Increasing longevity results in a larger elderly population.

Migration of the elderly
- As societies become more advanced, more elderly people migrate.
- The destinations to which elderly people migrate change over time:
  - Early – if the cohort reaching old age were rural–urban migrants, they may well choose to return to their rural origins on retirement. This may well be to take advantage of family support in the absence of state provision.
  - Middle – as state provision for old age improves and people acquire greater personal mobility, migration will become concentrated into flows directed towards areas offering a good climate (sun) or perceived quality of life (coast). The concentration of the elderly into these areas/towns will lead them to become 'retirement resorts'.
  - Late – as elderly people acquire a greater share of societal wealth, their migration flows will become more disparate, with some people seeking out remote or inland areas in order to avoid traditional retirement resorts, and others (especially those in late age) moving to be near their children, who are likely to be located into rural or semi-rural areas.

Migration of the non-elderly and its impact upon the elderly
- The migration of the young also affects the proportion that elderly people form of an area's population:
  - Early – during urbanisation it is likely to be the young who move to the city, leaving behind an 'accumulation' of residual population of older people.
  - Middle – as elderly people target specific retirement resorts, young people might choose to leave these areas in search of other young people and the services they support. This form of concentration is termed 'recomposition'. Equally, during the later stages of urbanisation, the young frequently opt to move away from the metropolitan core into the suburbs, thereby leaving behind a residual elderly (accumulation) population.
  - Late – since elderly people are migrating to the same areas as young people (either to be with them or to gain access to the same natural amenities), regions may acquire concentrations of elderly in parallel with concentrations of other age groups; concentration through 'congregation'. Within urban areas, as the young flee the suburbs for ex-urban living, so the elderly are left concentrated in the former (again, accumulation).

Source: Rees 1992; Rogers 1989, 1992a.

---

the elderly population from the metropolitan areas, especially those in the north, towards non-metropolitan destinations in the south of England continued into the 1980s (Rees 1992).

Patterns of elderly migration in the United States have been far more stable than in the United Kingdom. Frey (1992) argued that, in the 1970s, the patterns of population redistribution for the elderly and the non-elderly converged, with both groups fleeing the large metropolitan areas of the north in favour of smaller, low-density locations in the sunbelt. He suggested that the same quality of life factors underpinned both migrations. However, in the 1980s, this synergy broke down, with elderly and non-elderly populations pursuing different trajec-

tories. The non-elderly were again forced to seek residence in (coastal) metropolitan areas as economic restructuring undermined uncompetitive industries in smaller urban areas. In contrast, the elderly continued to migrate in channels to particular states – most notably Florida, Arizona, Nevada, Arkansas, and New Mexico (Rogers 1992b) – and, within these, to rural and urban retirement destinations. Alternatively, they aged *in situ*, especially within the suburbs, so that by 1977 the number of elderly living in suburbs exceeded that in the inner city (Golant 1990, 1992).

At a broader geographical scale, Salva Tomas (1992) has considered the international flows of the retired, and has shown how elderly migrants are now

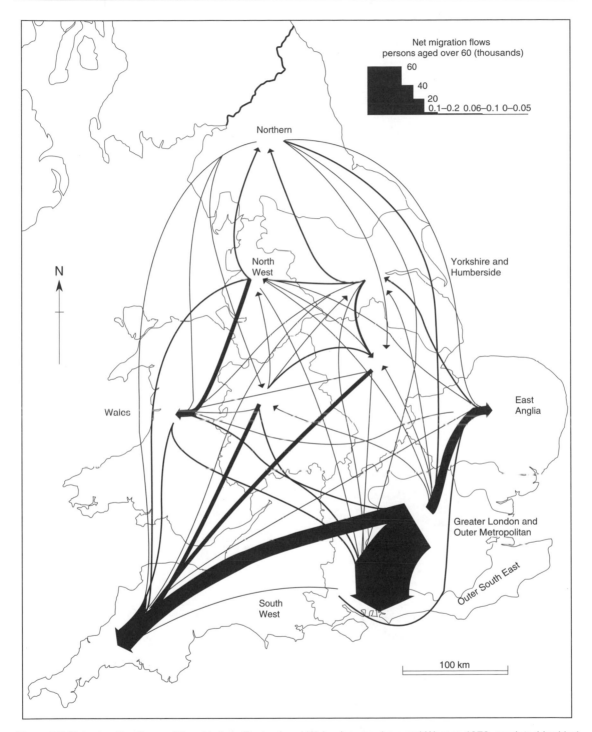

Figure 5.5 Net migration flows of the elderly in England and Wales (source: Law and Warnes 1976, reprinted by kind permission of the Royal Geographical Society from *Transactions of the Institute of British Geographers*).

Figure 5.6 Retirement areas in England and Wales (source: Law and Warnes 1976, reprinted by kind permission of the Royal Geographical Society from *Transactions of the Institute of British Geographers*).

an important element within the social demography of parts of Spain. The number of foreign residents in the Mediterranean and Balearic zone had risen from about 100,000 in 1981 to 229,000 by 1990. Most of this growth had been concentrated into the Balearics themselves, Malaga, Alicante and Barcelona. Around one-third of these foreigners were aged over 55, and many came as couples from the United Kingdom or Germany, having migrated to Mallorca and Menorca to retire to a place in the sun. Likewise, Buller and Hoggart (1994a) have noted the presence of British retirement migration to rural France. However, Warnes (1991) bemoans the general lack of accurate data and research on such migrants, suggesting that there now may be as many as 12,000 elderly Britons in Spain and that this number will rise sharply as the cohort of people who have taken holidays regularly in Spain ages into retirement.

Demographic change and migration

Concentrating upon the potential life transitions through which an individual might move during his/her life tends to ignore the effects and consequences that can be generated by a large number of individuals passing through the same life transition at the same point in time or space. Consequently, there is considerable value in taking a cohort approach as an alternative. This demonstrates how, when aggregated, individual life transitions can have an impact upon regional or even national trends.

One of the most convincing demonstrations of the value of a cohort approach has been the analysis of the experience and effect of America's *baby boom* generation maturing into adulthood. Plane and Rogerson (1991) noted how 72 million Americans were born in the baby boom of 1946–64 and how the survivors have then moved through different life transitions as a large cohort. Figure 5.7 (overleaf), which is calculated from US census data, shows how the 'boomers' first began to experience the life transition of entering the workforce in the late 1960s. Plane (1992) showed how, because of previous patterns of urbanisation, the boomers were over-represented in the north east and parts of the Mid-west, and argued that as this cohort tried to enter the labour force there they experienced considerable difficulties of two kinds. First, those areas relied upon stagnant industries that were not creating sufficient new jobs to absorb the increased number of new entrants to the workforce. Second, with the increased number of women seeking work, those jobs

that were available were more heavily oversubscribed. As labour markets in traditional core areas choked under the weight of the boomers, so short-distance intra-regional moves dwindled but the smaller number of long-distance moves increased. Inter-regional migration became a more viable proposition, with the south and west being especially attractive destinations because of their combination of growth industries and better climate. Many boomers sought to escape from the congested labour markets of the north and move instead to the south and west. This group was one of the main driving forces behind the dramatic population redistribution of the 1970s. Net inter-regional outflow from the north east and Mid-west thus doubled between 1965–70 and 1975–80, while the south gained 1.3 million people over the same period. Thus, the reaching of a particular life transition by an expanded cohort had a major part to play in explaining national population redistribution. Had the cohort been smaller, and labour markets in the north east and Mid-west therefore been better able to cope, the flood from rustbelt to sunbelt might have been considerably less dramatic and been made up largely of those retiring.

Plane (1993) went on to generalise what had been discovered about the baby boomers in the United States into a more widely applicable model of the relationship between the size of a cohort and the mobility rate of young adults. Figure 5.8 (overleaf) shows how, as the lead components of a boom cohort enter the workforce, mobility rates decline (point A). These reach their lowest rates (point B) with the last of the boom cohort, who have to experience the most congested labour markets and the most depressed economic conditions (brought about by suppressed demand). In contrast, as the first of the following 'baby bust' cohort seek work (point C), there are fewer new workers about, and the previous cohort of boomers have reached the stage in their lives where they are engaged in peak consumption. Hence, baby busters will be particularly mobile. Towards the end of that cohort, though, a further change occurs, with demand falling, as the large number of baby boom consumers are replaced by the much smaller number of baby bust consumers.

Conclusion

We have seen how the idealised and uniform family life-cycle has, over time, been replaced by the much

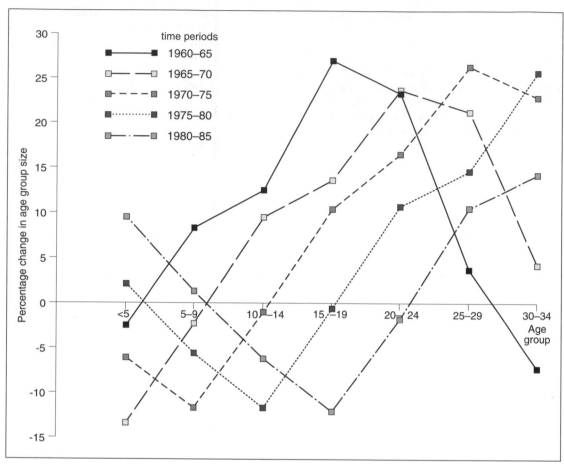

Figure 5.7 Percentage change in the size of different age groups in the United States, 1960–65 to 1980–85 (source: Plane 1992).

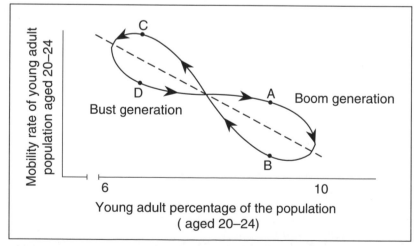

Figure 5.8 The relationship between cohort size and the mobility rates of young adults (source: Plane, D. 1993, Demographic influences on migration, *Regional Studies*, **27**, (4), pp. 375–83. Reprinted by kind permission of Carfax Publishing Company, PO Box 25, Abingdon, Oxfordshire, OX14 3UE, UK).

more flexible and diversified concept of life-course, with its emphasis upon the individual and upon life transitions, rather than stages. In addition, much of the material demonstrates how social conventions and norms are now changing so rapidly and frequently that almost every cohort faces a different range of choices at successive life transitions than those offered its predecessor. However, despite this continual change, and the growing complexity and variety of modern life-courses, there do still appear to be observable regularities in the relationship of age with rates and types of migration. Finally, it must be reiterated that all migrations take place within a structural context, and that voluntary life-course moves depend heavily upon structural factors such as the availability of employment, the housing market, cultural norms and the application of state policy.

CHAPTER 6

Migration and the quality of life

Introduction

'Quality of life' concerns an individual's or a group's state of well-being, defined as the 'degree to which the needs and wants of a population are being met' (Johnston 1994: 568). From the outset, this focus on wants as well as needs indicates how any one person's understanding of what constitutes the quality of life will be both individually and socially defined through being specific to a given society or subdivision of that society. As an idea that is not, therefore, some kind of objective absolute, defining the key features of a high quality of life can be subject to considerable debate and dispute. For example, the issues defining quality of life that feature most prominently in this chapter may well be quite different from those that would be picked up by authors from a different social and/or cultural tradition.

Assessment of quality of life can be made either through the use of objective indicators or, as is the case with most migration research concentrating on this issue, more subjectively by those concerned. Quality of life is implicated in almost all forms of migration. For example, migration to improve job prospects represents a clear attempt to improve quality of life, since economic security is essential for well-being. Likewise, life-course moves, such as those linked to marriage or retirement, all involve attempts by the individual or household to improve their everyday living situation in tune with their changing wants and needs. These moves for employment and life-course reasons have, however, been discussed in previous chapters. They must also be seen in terms of demands placed on the migrant through economic pressures and household norms. This chapter concentrates on the extent to which migration can be more completely evaluated in terms of *individual* personal, household or other group attempts to improve the quality of life, notably in migration representing an active effort to obtain a preferred living environment. In other words, the chapter concentrates on migrations driven by environmental concerns in the broadest sense, whether these concerns are linked to the physical or to the social environment. People here are, in the memorable phrase, 'voting with their feet' much more explicitly than was apparent in the types of migration discussed previously.

The role of quality of life issues in migration decision making has been apparent for as long as migration has not been reduced to a narrow income-maximising strategy (Chapter 3). The change from 'the invisible hand to visible feet' (Kearney 1986) lets the broader quality of life perspective in. For example, Wolpert's (1966) description of migration as an 'adjustment to environmental stress', or Cebula's and Vedder's (1973) account of migration as an 'investment decision', give clear openings for quality of life concerns. By the mid-1970s, Svart (1976) had concluded that there was enough evidence that environmental preferences could determine migration destinations, a conclusion amply reflected in subsequent studies, even if the precise role of 'values and attributes' in guiding migration was unclear (Ritchey 1976). Recently, a review by Rudzitis noted the growing recognition of the importance of non-economic factors for driving migration, suggesting that the strictly economic dimension should be seen within an overall lifestyle context:

Wages ... are what people trade off for a particular lifestyle, stress level, or happiness that they perceive they can find in particular locations. ... migration should be considered within the substance or context in which people want to live their lives.

(1989: 403).

While Rudzitis's observation may have much greater relevance to his study area of the north west United States than to many other places, notably the developing countries, his tone reflects the broadening of the migration debate to include quality of life factors.

This chapter concentrates on migration linked to quality of life concerns that are underpinned by *images of places*. Geographers and others have given considerable attention to the roles played within everyday life of images of place (for example, Pocock and Hudson 1978; Rodwin and Hollister 1984; Shields 1991), and studies of human migration are no exception. In this chapter, a focus on migration to places perceived to be of a higher 'quality' is concerned almost exclusively with internal migration, although, as demonstrated most clearly in Chapter 9, international migration can also be influenced by such issues. More generally, we must reiterate how almost all types of migration involve some quality of life elements when explored in all their complexity.

The first part of the chapter examines different aspects of the appeal of urban living or the 'lure of the city'. After describing some contrasting images of the city, it examines links between these images and urban-focused migration, intra-urban migration, gentrification and suburbanisation. This is followed by a section on the 'lure of the countryside', which again details this appeal before exploring its association with rural-focused and intra-rural migration. A third section takes a more place-specific focus and examines migration to sites where key amenity values are expected to be found, such as a warmer climate. Throughout, emphasis is given to the interlinkage between quality of life and other dimensions of migration, as outlined elsewhere in this book.

The 'lure of the city' and urbanisation

Contrasting images of the city

With the advent of the Industrial Revolution in what are now the most highly developed countries of the world, 'the city' became the focus for daily life, experience and reality as never before (Chapter 1). Consequently, it was around this time that often highly contrasting images of the city and the country developed more fully amongst the general population, as opposed to the artistic community among which they had largely been confined previously (Williams 1973). Polarisation occurred between those who saw the city as a 'wild and wicked place' and those who saw it as 'an earthly paradise' (Short 1991: 81). The city and its fate could be seen as a metaphor for the future, as an embodiment of progress or chaos. As Table 6.1 illustrates, the city represented either 'Babylon' or 'Jerusalem' (Girouard 1985).

Quality of life migration spurred on by the lure of the city relies upon the migrant regarding the city in Jerusalemic terms. Such a perspective regards the city as a source of potential liberation, especially for the individual, who can choose the lifestyle he or she wishes (Raban 1974). Migrants can lose themselves in the myriad spaces of the city and express themselves there in ways that are not possible in the countryside, where life is regarded as being 'hemmed in by deference to authority and tradition, a suffocating network of ritual obligations' (Short 1991: 37).

The *cultural construction* of these contrasting images of place must be recognised explicitly. Thus, the image of the city as a Jerusalem can be seen as being highly Judaeo-Christian, although such an image emerges in a similar form within other cultures. Findlay (1994a: 166) noted that a key image of urbanism is 'an Islamic image derived from the long traditions of orderly urban living, with the use of public and private spaces being dictated by interpretation of the *Sharia* [Islamic law]'. Likewise, in other religions and cultures worldwide, the ubiquity of urbanism attests to its status as a mark of civilisation. This is clearly expressed in the cosmological sacred and profane spaces that underpinned the elaborate structure of the classical Asian city (Duncan 1990; Wheatley 1971). Furthermore, even

Table 6.1 Contrasting images of urban and rural.

| City as Babylon | | City as Jerusalem | |
Urban	Rural	Urban	Rural
Mob	Community	Community	Society
Disorder	Order	Freedom	Repression
Work	Retreat	Progress	Regress

Source: Short 1991, reprinted by kind permission of Routledge.

Box 6.1 The 'community studies' tradition and the 'idiocy of rural life'

With the advent of the Industrial Revolution and the drastic changes that it wrought on everyday life came a growing intellectual concern to be able to classify different types of settlement or 'community'. In the late nineteenth and early twentieth centuries a number of classifications arose based on simple dichotomies between urban and non-urban settlement types. A summary selection of these 'theories of contrast' are given in the table below. Although these classifications were rooted in theories of environmental determinism that have now largely been discredited, whereby society in all its aspects was seen as very much the product of the underlying environment, they proved highly influential in both academic and popular thinking.

Author	Urban category	Non-urban category
Becker	Secular	Sacred
Durkheim	Organic solidarity	Mechanical solidarity
Maine	Contract	Status
Redfield	Urban	Folk
Spence	Industrial	Military
Tönnies	*Gesellschaft*	*Gemeinschaft*
Weber	Rational	Traditional

Source: Reissman 1964.

The non-urban category usually referred to the rural world, seen largely as an endangered relic of the past, as it was slowly transformed through contact with urban society. While this transformation was typically mourned in highly nostalgic terms, there was also the feeling that such change was inevitable. It was in the context of theories and beliefs such as these that Marx and Engels were to make their famous statement in *The Communist Manifesto* on the backwardness of the countryside:

The bourgeoisie has subjected the country to the rule of the towns. It has created enormous cities, has greatly increased the urban population as compared with the rural, and has thus rescued a considerable part of the population from *the idiocy of rural life*.

(quoted in Tucker 1972: 339).

Source: Phillips and Williams 1984.

within the Judaeo-Christian world, perceptions of the city as Jerusalem vary. For example, it is much weaker in English culture (Williams 1973) than in southern Europe, where urban life is contrasted with a countryside strongly associated with the peasantry (Weber 1977; White 1984).

Empirical support concerning the association between settlement size and perceived quality of life in the developed world usually shows a bias in favour of more rural or non-metropolitan destinations (Blackwood and Carpenter 1978; Ilvento and Luloff 1982). Work includes studies of people's *mental maps* (Gould and White 1974; Svart 1976), which tend to rate rural areas as being most desired for residence. However, the city does come out favourably in many other aspects that might be expected to influence migration. For example, Fuguitt and Zuiches (1975) showed that American respondents with 'big city' preferences tended to emphasise social, services and employment quality of life factors, as well as the benefits of personal freedom, individualism and non-conformity and, to a lesser extent, the greater racial and sexual equality of the metropolitan destination (Christenson 1979). Robson (1988: 56) argues that cities should be recognised as presenting a 'dense mesh of opportunity and of stimulus' (Elkin, McLaren and Hillman 1991; Mulgan 1989) and, in contrast to what he sees as the overemphasis on anti-urban sentiments,

It is as convincing to argue that there is a positive liking for urban life: for the range of choice which it offers, for the sheer symbolism of large massed buildings, for the greater opportunities it offers.

(Robson 1987a: 11).

For Robson (1987b), adopting the words of former US President John F. Kennedy, to neglect the city is to neglect the nation.

Migration to the city

While for some the city may be imagined in a positive light compared with the countryside, it takes another step to link this assessment with actual migration. However, such evidence is available, with Heaton *et al.* (1979) providing the context with their discovery from the United States that people who preferred a different size place or a different location to where they were living were five times as likely to move as those who were less bothered. Such probabilities could be translated into actual moves to urban areas. For example, in a study of white male migration in the United States between 1970 and 1980, Clark and Hunter (1992) combined economic, life-cycle and quality of life considerations into an overall model. They concluded that quality of life, as reflected in demands for amenities, followed a life-cycle pattern, being especially important for older migrants (Chapter 5), but also that younger adults showed a clear association with migration to the inner city, even when controlling for economic considerations.

Quality of life features among the variety of push and pull factors (Chapter 3) generating migration from many rural areas in the developed world, leading to continued net population losses in many rural areas, such as much of southern Europe and the prairies of North America (G. Robinson 1990). Experience of urban life, such as that obtained through the media and direct visits, generates a desire, particularly among the young, to leave the rural environment. Brandes (1975) observed this in respect to Spain in the 1970s, where migration from the village of Becedas could not be understood solely in terms of economic factors. He noted how

Moving to the city also means the acquisition of modern conveniences: stoves, refrigerators, gas heaters, bathtubs, and toilets, not to mention luxuries such as radios, telephones, televisions and automobiles. Further, there is easy access to a wide array of bars, movies, theaters, parks, and other centers of diversion. The city has more *ambiente*, 'action', than the *pueblo* and, almost for this reason alone, is to be preferred over the restricted and comparatively uneventful life of the village.

(*ibid.*: 57).

Since the 1970s, the 'luxuries' cited by Brandes may have become more ubiquitous throughout rural Spain, but the *ambiente* of the city remains. Elsewhere, in a study of secondary school children in the small town of Temuka on the South Island of New Zealand (G. Robinson 1986), those wishing to leave emphasised the importance of education, training, and a desire to broaden experience and travel, as well as employment. The bad points of Temuka mentioned by these potential migrants included, as well as employment concerns, the lack of entertainment, privacy, shops and services, and the over-small and quiet character of the town. Perceptions of such negative features only become more acute as the notorious *cycle of decline* (Figure 6.1) (overleaf) grinds away to destroy the quality of life in many rural areas.

The importance of quality of life factors in rural-to-urban migration in the developing world tends to be overshadowed by economic issues, with most research emphasising economic motives (Chapman and Baker 1992a, 1992b; Dickenson *et al.* 1983; Findlay and Findlay 1987; Simpson 1987). This focus

is understandable since, as Gilbert and Gugler (1992: 67) bluntly put it, 'poor people who ignore their material circumstances are rapidly threatened in their very survival.' Describing migration as being driven by quality of life considerations is, for many people, at best rather naïve and at worst morally irresponsible. Nonetheless, the city does act as more than just a potential source of daily livelihood, providing overall 'better prospects' (Simpson 1987: 157) than the countryside. Gilbert (1994: 44–5) phrased the situation carefully:

If lack of land, starvation, or poverty were the principal factors behind out-migration, then the figures should show a relatively higher proportion of poor migrants in the total flow. The fact that they do not suggests that, however difficult rural conditions, there is an important component of migration flows that can only be explained in terms of choice. The people who move are those who under current conditions can best adapt to the city.

In Gilbert's Latin American context, those who can best adapt tend to be young adults, the skilled and the educated, and, increasingly, women. A social selectivity in access to quality of life is clear.

As regards the broader attractions of city life, Gilbert and Gugler (1992: 65) drew attention to the *collective consumption* role of the city. This refers to the city's provision of public facilities, such as clean water and electricity, the availability of which appears more restricted in rural areas. The idea of collective consumption indicates how the state can construct quality of life. The concept can be extended to include educational facilities; an extremely important draw even if privately supplied, since education is widely seen as a means of escaping poverty. Likewise, the city tends to be the primary location of training opportunities. Better health care and housing provision are key aspects of the city's higher quality of life in the developed world, and these associations are also apparent in the poorer countries. Finally, the city can provide greater political freedom and scope for individuals both to obtain privacy and to make more of their own lives (Dickenson *et al.* 1983).

Illustrating these general points in greater detail, Altsimadja (1992) noted from Central Africa how the colonial authorities were partly responsible for making the cities attractive to migrants by disproportionately developing the health and education infrastructure within them rather than within the countries as a whole. Consequently, living in the city, with proximity to services such as hospitals, provides

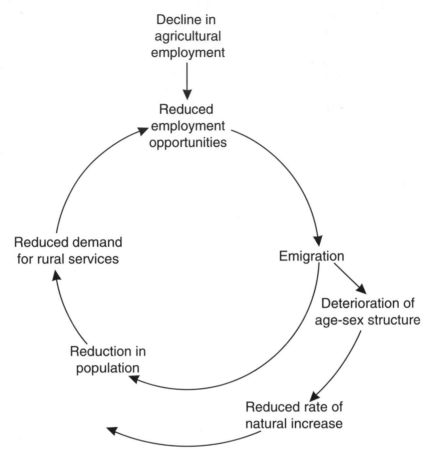

Figure 6.1 The cyclic nature of rural decline (source: Gilg 1983: 95).

greater levels of personal security. The range of ways in which the African city is associated with improved quality of life was also noted by O'Connor (1983), with the city providing higher incomes, better health facilities and schooling, and the possibility of escaping family pressures and experiencing the 'bright lights'. Within India, even in the Delhi slums, which contain most rural-to-urban migrants, the majority 'find less discrimination, and an enhancement of their opportunities in life' (Chapman 1992: 22). In Thailand, Rigg and Stott (1992) described the hundreds of thousands of migrants who migrate from the country to the cities each year in their search for work and 'a better life'. This new life is focused on the capital city, Bangkok, whose primacy in everything from health services to social quality of life is considerable (Table 6.2).

The cultural aspect of urban-focused migration draws attention to gender differences in the mi-gration of developing world peoples to their cities. This migration stream has historically been most associated with men, largely due to the greater economic opportunities presented to them and their greater involvement in formal education, with the migration of women being linked to family concerns. For example, in the 1981 Indian census, which asked for migrants' reasons for moving, 'employment' was cited by 55.5 percent of male migrants but 80 percent of female moves fell into the 'marriage' and 'consequent on family movement' categories (Skeldon 1986). However, recent years have seen this situation changing, as greater urban economic and educational opportunities open for women (Hallos 1991), promoting higher levels of autonomous migration (Baker 1992; Brydon 1992; Hugo 1992). With the presence of these greater opportunities, any further incentive for women to migrate has a greater chance of being translated into actual moves.

Table 6.2 Indicators of primacy in Bangkok, Thailand.

	Bangkok	Whole country	Bangkok relative to the whole country
Population (1988)	5,670,692	54,465,056	10.4%
Economy			
Gross regional/domestic product (millions baht)	495,310	1,098,366	45.1%
Average household monthly income (baht, 1986)	7427	3710	2 times
Domestic telex services (Jan.–June 1987)	1,726,758	2,456,940	70.3%
Telephone line capacity	861,392	1,251,102	68.9%
Health			
Number of hospital beds	16,461	84,438	19.5%
Number of physicians	4142	9464	43.8%
Number of dentists	883	1395	63.3%
Number of pharmacies	2783	3356	82.9%
Maternal death-rate (per 1000 population)	0.04	0.35	–
Education			
Number of state institutes of higher education	8	14	57.1%
Number of graduates (1985)	37,858	53,492	70.8%
Transport			
Number of passenger cars registered (1986)	593,505	753,326	78.8%
Social			
Number of divorce licenses (1987)	8773	31,068	28.2%
Divorce rate (per 1000)	1.77	0.69	2.5 times
Number of cinema seats	99,932	399,818	25.0%
Colour television sets per 100 private households	56	21	2.5 times

Source: Rigg and Stott 1992, reprinted by kind permission of Routledge.

This will especially be the case if this migration can be incorporated into a general family strategy, as demonstrated by Trager (1984) in the Philippines (Chapter 5) and Datta (1996) in Botswana.

One clear incentive for women relates to the greater likelihood of being able to avoid in the city the onerous traditional restrictions and norms placed upon women in the more confined rural settings (Momsen 1991; Obbo 1980; Wilkinson 1983). As Radcliffe (1986) observed in her study of Cuzco, Peru, independent migration by women was rare but it did occur, being driven in particular by the experience of domestic violence, the search for employment, or as a challenge to their subordinate position within Peruvian rural society. As regards the more commonplace moves of women in male-headed households, Chant's (1991a, 1992a) study of low-income households in Guanacasteco, Costa Rica, found that while there was no clear 'employment' lure of the towns, there was an

interesting emphasis on factors associated with ... 'reproductive' (e.g. welfare, kinship) aspects of household survival, as opposed to 'productive' (e.g. work, income) imperatives.
(1992a: 59).

Once in the towns, the women were apt to remain there – as Chant (1991b) also found in Mexican cities – even when their partners were regularly doing waged labour elsewhere. In the towns, the women were *relatively* freer than their rural counterparts, reflecting this frequent absence of the men and their less isolated lives (Chant 1992a).

Overall, rural-to-urban migration appears to provide satisfaction for those involved (Gilbert and Gugler 1992), in spite of the hardships of urban life. For example, the wage–labour and urban life in South Africa experienced by many Basotho miners resulted in many remaining in the capital city, Maseru, on returning to Lesotho, clearly illustrating a linkage between work and lifestyle (Wilkinson 1983). While there is evidence that migrants can be misled by false images of the city's bounty (Dickenson *et al.* 1983; Gilbert and Gugler 1992), received from television and the general media, urban employers and even fellow former villagers – who may present an unduly positive picture of urban life from a feeling that they have failed to 'make it' in the city – most new migrants have a good idea of what to expect. This is facilitated by the connections

---

Box 6.2 The non-economic determinants of rural-to-urban migration among Hausa women in Nigeria

Katsina, in the far north of Nigeria close to the border with Niger, is one of the ancient walled cities of Hausaland and provided the setting for a study of rural-to-urban migration among Hausa women. The Hausa culture is notable for its seclusion (*kulle* or 'locking up') of women, justified by Islamic injunction but uncommon elsewhere in Islamic West Africa. During the day, young women are rarely seen on the streets of Katsina. Yet, autonomous as well as associational migration is to be found among the Hausa women. The latter is almost always linked to 'virilocal' marriage, with the wife moving to her new husband's area of residence. Although this suggests little room for broader quality of life issues to influence the migration process, this was too simple an understanding of associational migration, with Hausa women actively seeking an urban residence:

for Hausa women at least, an additional positive factor in marriage choice is the fact of their prospective husband's urban residence. And whereas rural women will cheerfully marry urban men, urban women will rarely marry into the rural area. ... an urban–rural marriage is generally seen as a misfortune for the bride. ... Thus, while rural–urban migration is desired, urban–rural migration is avoided if at all possible. While the opportunity to move to the city is not likely to be refused by the rural wife, the elements of choice, and interest, remain, and should be recognized as factors in the wives' migration.

(Pittin 1984: 1300).

Additionally, in spite of the seclusionary practices of Hausa culture, there is also a substantial amount of autonomous migration by women. In particular, this is associated with young and unattached women becoming courtesans (*karuwai*), reflecting the fact that autonomous migration in itself is 'often seen as tantamount to prostitution' (*ibid.*: 1312). This type of migration is generally precipitated by marriage-related pressures, such as plans for an unwanted arranged marriage or pressures to return to a marriage that had broken down. Hunger and poverty can also be important in driving women to the city. Nonetheless, perceptions of the benefits of city life in general also encourage women to migrate. These benefits include economic advantages but also the relative social and cultural freedoms that played a role in generating the associational migration patterns.

In conclusion, the study showed the benefits and necessity of adopting a gender-sensitive perspective in understanding and explaining migration patterns in northern Nigeria. Such a perspective allows us to see how general quality of life factors mesh with more economic and life-course concerns to bring about the migration patterns observed.

Source: Pittin 1984.

---

between urban and rural, which tend to be maintained after migration, linking village and city in a migration network with its attendant cumulative causation fuelling further quality of life migration (Chapter 3). The value of connections is symbolised by the importance of circulation migration in Africa and the Pacific islands (Gilbert and Gugler 1992; Prothero and Chapman 1985), where 'urban dwellers [remain] loyal to a rural home' (H. Jones 1990: 224).

Migration within the urban environment

Urban environments are not, of course, homogenous and may conform to the demands and expectations placed upon them by their inhabitants to radically different degrees. This gives rise to quality of life migration between different urban milieux or even within the same milieu, even when this migration is not also associated with other factors, such as employment change. Consequently, the United States has a *Places Rated Almanac* (Boyer and Savageau 1989), which ranks cities on the basis of nine quality

of life factors and is promoted as 'a guide to finding the best places to live.' Work exploring the quality of life in different British cities (Grayson and Young 1994) by the Glasgow Quality of life Group found that people rated cities according to typical quality of life dimensions, such as crime levels, access to health care, pollution levels and availability of different facilities (Rogerson *et al.* 1989). The detailed quality of life factors are listed in Table 6.3, showing clearly how they varied between young adults and more elderly respondents, plus the intimate association between quality of life and housing, employment and life-cycle considerations. Moreover, in their investigation of migration between and within cities, the Quality of life Group found that 68.5 percent of inter-urban and 75.9 percent of intra-urban migrants stressed that choice of 'living environment' was important in destination selection. At the aggregate scale there was a positive link between observed migration patterns and the quality of life of the cities studied (Findlay and Rogerson 1993).

As regards intra-urban variations in perceived quality, which again may be linked to migration pat-

Table 6.3 Quality of life dimensions stressed by British urban residents.

Dimension	Percentage indicating dimension as important or very important
Aged 65 and over	
Health service provision	93.3
Violent crime levels	91.7
Non-violent crime levels	91.2
Cost of living	90.8
Pollution levels	86.8
Shopping facilities	81.6
Access to areas of scenic beauty	75.8
Climate	68.5
Owner-occupied housing costs	58.5 ·
Quality of council housing	55.5
Aged 25–34	
Education facilities	93.9
Health service provision	93.2
Violent crime levels	90.7
Pollution levels	88.8
Employment prospects	88.6
Wage levels	88.1
Non-violent crime levels	87.9
Cost of living	87.2
Shopping facilities	82.2
Owner-occupied housing costs	81.6

Reprinted with permission from Findlay and Rogerson 1993: 40, in T. Champion (ed.) *Population Matters* (2nd edition), © 1993 Paul Chapman Publishing Ltd, London.

terns, Lewis (1982), in a study of Leicester in England, found respondents favoured areas that were local, high status and of *social proximity*, the latter referring in part to concerns over the ethnic composition of the area. Further evidence of the importance of intra-urban quality of life perceptions comes from studies of residential mobility, where there has been a recognition that economic reasons for moving what are often quite short distances should not be overstated. For example, Rossi's (1955) classic study of Philadelphia in the United States noted the significance of neighbourhood quality in the choice of a specific area to live in, even if his primary concern was with life-cycle factors (Chapter 5).

Lee, Oropesa and Kanan (1994) combined objective and subjective measures in a model that sought to describe how properties of urban neighbourhoods influence whether or not their inhabitants remain within them or migrate elsewhere. The authors noted that individuals are aware, to a greater or lesser extent, of the qualities present in different neighbourhoods and, within constraints, adjust their residential location accordingly. The resulting model (Figure 6.2) suggested that an individual person's or household's position combines with their subjective perceptions of different neighbourhoods to generate 'thoughts about moving'. These subjective perceptions have both substantive and temporal aspects, the former being concerned with the social and physical character of the areas and the latter with the extent to which they are changing or remaining the same over time. While some perceptions are rooted in objective 'facts', others are of a more subjective nature. The model was applied in Nashville in the

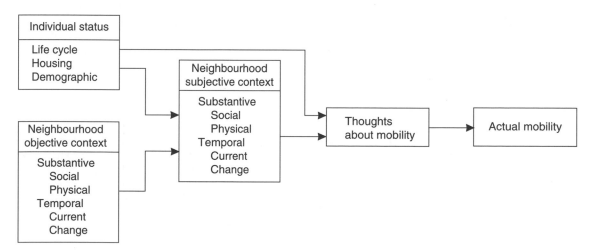

Figure 6.2 A revised decision-making model of residential mobility incorporating differing 'neighbourhood contexts' (source: Lee, Oropesa and Kanan 1994: 254).

United States and suggested that the link between neighbourhood perception and migration was indirect; perceptions influenced thoughts about migration rather than directly correlated with actual decisions to migrate. Linkage between thoughts and actual mobility must take further note of the individual statuses mentioned in the figure.

Gentrification

Recent years have seen an urban-focused migration trend of a highly selective nature back to certain sections of the inner city. This trend forms part of the gentrification of the inner city (Chapter 1), which can significantly alter its social geography (Hartshorn 1992; White 1984). Numerous reasons have been proposed to explain the trend (Illeris 1991), including economic restructuring and the urban orientation of increasing numbers of immigrants. However, much attention has been given to the role of quality of life and residential choice, with Champion (1992: 475) arguing that re-urbanisation reflects 'the ageing of the 1960s baby boom into the city-loving young adults of the 1980s'. Consequently, given the demographic and socio-economic selectivity of these urban-focused migrants in favour of young and affluent single people or couples without children, Lever (1993: 271) claimed that the best of both urban and rural worlds is attainable, since a 're-vival in preference for an urban life-style for some high-income households can be combined with an enjoyment of rural environments through the purchase of a second home'.

The revival of urban centres can be illustrated by the example of Glasgow in Scotland, where economic and housing changes have transformed the city, spurred on by urban redevelopment schemes – also apparent elsewhere in Britain (Cameron 1992) – and the favourable promotion resulting from the city hosting a Garden Festival in 1988, being the European City of Culture in 1990 and developing an annual Mayfest arts festival; Lever (1993: 278) observed that the 'overall social mix of the central city has diversified in a way which would have not been anticipated in 1980'. 'Brownfield' sites have been developed by private developers and housing associations in partnership, there was dockside redevelopment after the Garden Festival and there has been a transformation of the 'merchant city' area of the central business district.

Efforts within Glasgow to bring people and work back into the city – making it 'Miles Better' in the words of its promotional slogan – dated back to the early 1970s, when the image of the city was seen as a serious handicap to its prosperity:

Glasgow was seen as the City of mean streets and mean people, razor gangs, the Gorbals slums, of smoke, grime, and fog, of drunks, impenetrable accents and communists.
(Taylor 1990: 2).

For Glasgow, most of the lure of the city had long been lost. Thus began a sustained campaign to improve the city's image, to promote Glasgow as a 'new dynamic and sophisticated European capital for the development of culture' (Boyle and Hughes 1991: 220). Its future was to be linked with service employment, notably in tourism, and with creating a truly post-industrial city. While this strategy was criticised by socialist groups, such as Workers' City, for its neglect of working class history and experience and its general lack of attention to the everyday concerns of ordinary Glaswegians, it proved successful in re-presenting Glasgow both to the rest of Britain and to the world at large. Consequently, it might be expected to have encouraged further migration to the city.

Gentrification does not, however, rest solely upon images 'constructed from the drawing board' (Boyle and Hughes 1991: 221) or through the active promotion of local authorities. There are rival theories concerning the ultimate causes of the phenomenon, notably approaches linked to structuralist theoretical perspectives and those linked to more humanistic concerns (Hamnett 1984, 1991). The former, such as N. Smith's (1979, 1982, 1987) *rent gap theory*, argue that gentrification is brought about by the producers of housing and is driven by capitalist, profit-maximising concerns. In contrast, Ley (1980, 1986) stressed how gentrification was a consumption-led phenomenon, the product of the choice of key middle class individuals to live in the city centre. This is not the place to evaluate these competing explanations, or attempts at integration (Figure 6.3). Nonetheless, almost all rely on the inner city being perceived as a place of opportunity, whether this opportunity is marketed and sold by the producers of housing or more independently perceived and reacted to by its consumers.

The inner city as a place of opportunity is clear in the spread of gentrification since Glass (1964) introduced the term for the movement of the middle class into working class areas of London in the mid-1960s. Gentrification is found throughout the developed world, in cities of all sizes and in a wide range of

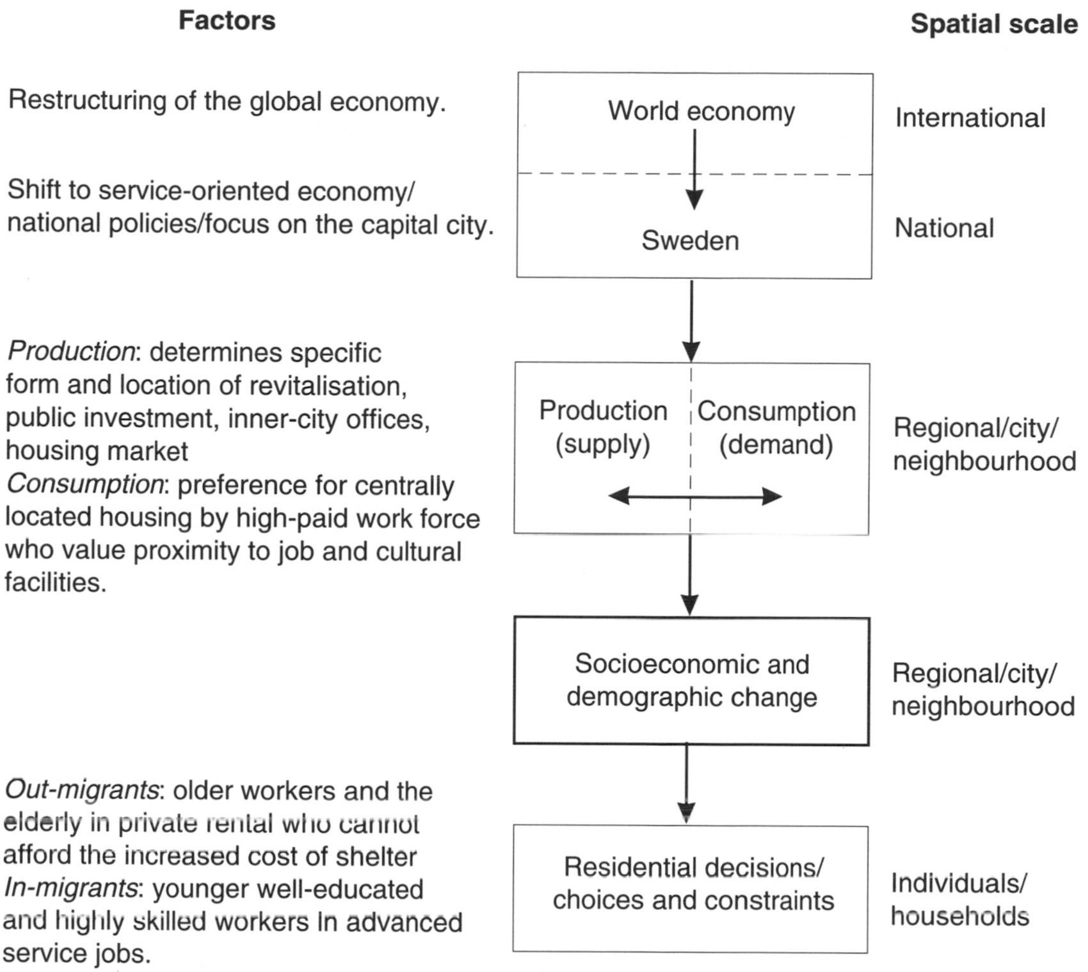

Factors   **Spatial scale**

Restructuring of the global economy.

| World economy | International |

Shift to service-oriented economy/
national policies/focus on the capital city.

| Sweden | National |

Production: determines specific
form and location of revitalisation,
public investment, inner-city offices,
housing market

| Production (supply) | Consumption (demand) | Regional/city/ neighbourhood |

Consumption: preference for centrally
located housing by high-paid work force
who value proximity to job and cultural
facilities.

| Socioeconomic and demographic change | Regional/city/ neighbourhood |

Out-migrants: older workers and the
elderly in private rental who cannot
afford the increased cost of shelter
In-migrants: younger well-educated
and highly skilled workers in advanced
service jobs.

| Residential decisions/ choices and constraints | Individuals/ households |

Figure 6.3 Model of urban restructuring and inner city revitalisation (reprinted from *Tijdschrift voor Economische en Sociale Geografie*, **84**, Borgegård and Murdie, Socio-demographics, impacts of economic restructuring on Stockholm's inner city, pp 269–80, © 1993, by kind permission of the Royal Dutch Geographical Society).

neighbourhoods (Smith and Williams 1986; Van Weesep 1994; Van Weesep and Musterd 1991; Zukin 1987). It generally involves renovation of existing dwellings, although the term can be expanded to large-scale new developments, such as in London's Docklands (Short 1989).

In order to understand the attractions of urban living that provide the demand for gentrification, we must note at least two key social trends that have occurred over recent decades. First, there was the restructuring of the economy and the growth of professional, managerial and other higher-level white collar occupations at the expense of blue collar jobs. In particular, there was the growth of a new middle class or *service class* (Chapter 9). Second, there was

an increase in paid employment by women, a trend that often intersects with the growth of the service class, as many of these new jobs require high levels of qualifications. Consequently, we also saw the growth of service class households that remained childless, in part so as not to jeopardise either partner's career. Gentrification has been linked strongly to these trends (Williams 1986). The typical profile of gentrifiers shows them to be young adults, typically dual-career, living in small households. From evidence in the United States, they also tend to be white and from the greater city area rather than from further afield (Le Gates and Hartman 1986; Smith and Williams 1986; Zukin 1987). Attractions of gentrification are, first, economic, in that the

properties purchased are typically a good investment (Beauregard 1986; Short 1989). This investment is underpinned by the cultural value of a
gentrified neighbourhood, since gentrification typically does much to satisfy the consumption needs of
service class and/or dual-income households.
Gentrifiers are often attracted by the aesthetic appeal of places such as Victorian Melbourne (Jager
1986) and bohemian Soho in New York (Zukin
1989), a process that becomes self-reinforcing as
formerly depressed areas such as Park Slope in New
York City, Barnsbury in London and the Marais in
Paris become fashionable addresses (Carpenter and
Lees 1995; Lees 1994).

Gentrification is especially well suited to households where a dual career structure necessitates the
renegotiation of domestic responsibilities (England
1991). Indeed, gentrification can be seen as a middle
class female strategy to reduce the time–space constraints of their dual spheres of waged labour and
domestic responsibilities (Bondi 1991; Warde 1991).
For Rose (1989), changing gender relations were the
key to determining the character of the service class
and their association with gentrification. For
example, the Mont Royal area of Montréal:

may be particularly attractive to women professionals in
the public, non-profit, and cultural sectors due to its proximity to downtown, its well-developed social and information networks which, for example, might help
contractual workers to obtain employment, its modest-
priced housing stock, and extensive provision of a diverse
range of community services.

(*ibid*.: 126).

These advantages extend to unconventional households (England 1991), such as female single parents
(Rose and Le Bourdais 1986) and the gay community, which Knopp (1990) argued has been crucial
to the success of gentrification. The inner city milieux

enable a diversification of ways of carrying out reproductive work; they offer a concentration of supportive services;
and they often have a 'tolerant' ambience.

(Rose 1989: 131).

Although identity through style and distinction is
of crucial significance to the service class (Chapter
9), some more mundane attractions of the inner city
complete the picture. An inner city location provides
access to the 'pulse of city life' (Borgegård and
Murdie 1993: 274) – leisure, services and entertainment. This access can be especially important for
single people and childless couples, often highly
dedicated to their employment, since it provides

extra opportunities for meeting others socially
(Beauregard 1986). From the example of Vancouver, the inner city has its parks, beaches, marinas,
and culture and retail activities, plus the lifestyle,
ethnic and architectural diversity of the gentrifying
neighbourhoods (Ley 1981). Ley's (1981: 129) assertion that 'these desiderata of the culture of consumption should not be underestimated in
interpreting the revitalization of the inner city' was
supported by the high degree of residential satisfaction, especially for amenity reasons, shown by residents of the city's Kitsilano district.

Gentrification has a number of important consequences for people living in the area prior to gentrification, consequences that expose some of the costs
of producing spaces where other people's high quality of life can be attained. The 'exclusionary landscape' (Carpenter and Lees 1995) of privatised
security measures symbolises gentrification's displacement of the existing population, who can
neither afford the increased rents of the gentrified
neighbourhood – they are often not owner-occupiers
– nor live with the area's new cultural demands.
Gentrification is thus associated with out-migration
as well as in-migration, with out-migrants in the US
context being predominantly white people of all
ages; low- to middle-income blue and lower white
collar employees; and from a range of household
types (LeGates and Hartman 1986). Bearing in mind
these negative social experiences (Short 1989), we
must assess critically summaries of the effects of
gentrification, such as the following:

In the space of two decades, gentrification has lifted residential capital formation and ground rents; boosted spending power and reinvigorated local shopping (store refits,
specialist arcades, galleria, village markets); given rise to
movements for adequate social services, urban conservation, better street design, and high-quality public spaces; in-
jected a new vitality into the cultural and artistic life of the
city; and spread an awareness among suburban Australians
of the virtues of inner-city living.

(Badcock 1995: 84).

Although predominantly a feature of the developed world, gentrification is also found in poorer
countries. First-hand evidence comes from South
Africa, a transitional case between rich and poor
worlds. The community of Woodstock, Cape Town,
successfully resisted official efforts to make the area
'coloureds only' in the apartheid era through the
Open Woodstock campaign of the 1980s. Since then,
however, gentrification has occurred with the in-migration of middle class professionals. These migrants

Box 6.3 Migration and population change in Tel Aviv inner city, Israel

The Tel Aviv–Jaffa municipality in Israel forms the heart of a highly urbanised region. It contained around 323,000 people in 1990, 30 percent of whom lived in the inner city districts of Old North, Sheinkin, Lev Tel Aviv and Kikar Hamedinah. Tel Aviv is the financial capital of Israel and strongly reflected the decline in manufacturing employment relative to services that took place in the country throughout the 1970s and 1980s. These economic changes were accompanied by changing migration experiences for the inner city. Throughout the 1960s, 1970s and 1980s the inner city areas had low gross migration rates compared with the rest of the city. However, a back-to-the-city migration trend affected the area from the 1970s, maturing in the 1980s. Rates of both in- and out-migration increased but the former increasingly outpaced the latter. Spatially, the trend spread from Sheinkin, eventually to reach Kikar Hamedinah in the late 1980s. In-migrants were

predominantly young (59 percent under 30); single people or childless couples; and career-oriented, with professional/managerial status. They came mostly from the Tel Aviv metropolitan area (68 percent), although net migration from outside Tel Aviv enabled the inner city to grow in population as there was a net loss of population from inner Tel Aviv to the rest of the city.

Quality of life factors were very much in evidence when 411 migrant households were surveyed regarding their principal motives for living in the inner city. The area was strongly regarded as a centre for cultural, shopping, social and leisure activity, a nearby residence providing excellent access to these facilities. Characterising the migrants further, the table below describes seven clusters of migrant according to their motivations. It illustrates clearly the place of quality of life next to factors linked to life-course and employment.

Cluster	Group	Percent	Major motivations
Urban lifestyle (1)	Young urbanites	32	Close to leisure and cultural activities.
Urban lifestyle (2)	Yuppies	20	Close to shopping and job opportunities. Increase in income and investment opportunities.
Life-cycle (1)	Young, mobile households	14	End of tenancy, change of job, close to university.
Life-cycle (2)	Before or after children	9	Marriage or decrease in family size.
Family lifestyle (1)	Upper-class families	16	Close to high-quality services, suitable environment for children, upper class social *milieu* and improved self-owned apartments.
Family lifestyle (2)	Elderly households	6	Well-maintained, upper class neighbourhood. Nostalgia.
Family lifestyle (3)	Religious families	3	Religious neighbourhood, small apartments.

Do the changes occurring in Tel-Aviv comprise gentrification? Schnell and Graicer argue that they do not, because in-migration resulted in little change to the socio-economic status of the inner city. The four districts were all built initially for middle and upper middle class households and had not experienced significant deterioration or decline in housing values prior to in-migration increases. Instead,

the key change was the rejuvenation of Tel Aviv, with younger new middle class households supplanting the more elderly old middle class residents. Hence, we must broaden our appreciation of the varied forms that inner city revitalisation can take.

Source: Schnell and Graicer 1993, 1994.

are attracted both to the proximity of the central business district and to

Woodstock's Victorian architecture, its close proximity to Table Mountain, and hotpotch mixture of residential, retail and warehousing activities which was markedly different to the bland uniformity of much of suburban Cape Town.
(Garside 1993: 33).

More generally, however, migration linked to quality of life factors features less prominently in de-

veloping world gentrification. In St. John's, Antigua, there has been gentrification of the waterfront area, but this has been mainly for the benefit of tourists and not for residential purposes (Thomas 1991). Elsewhere, in Puebla, Mexico, Jones and Varley (forthcoming) consider that gentrification of the city centre represents a reassertion of Spanish values and an elimination of unacceptable land uses and people (notably the *ambulantes* (street traders)) by the local middle classes. Although residential colonisation of gentrified properties through migration is increasing, it is restoration for museums, cultural centres, hotels and offices that predominates. Even among the better-off sections of developing world societies, migration back to the city centres for quality of life reasons does not appear to have taken hold with any strength. This, of course, reflects the fact that such countries are still very much engaged in the initial urbanisation of their societies discussed near the start of this chapter.

Suburbanisation

A key migration-related process that forms a bridge between the urban and rural environments is suburbanisation (Chapter 1). Throughout their history suburbs have been strongly associated with quality of life considerations (Bunce 1994). Suburban development began originally through efforts to provide residential environments that would enable the upper classes to escape the pollution and misery of the burgeoning industrial cities. However, as we saw in Chapter 1, suburbanisation soon became a mass phenomenon, increasingly encompassing broader sections of society, aided and abetted by transportation developments. While there is not the scope here to review the theories of suburbanisation, quality of life issues feature prominently. For example, Walker's (1981) structuralist interpretation described the emergence of suburban growth as a solution to the *crisis of overaccumulation*, diverting surplus capital into housing and other development. Nonetheless, he also emphasised the *ideological* role played by the suburbs. This involves demands generated by the repulsion of the city (due to pollution, crime, noise, misery) being stemmed by the escapism of a semi-rural existence or 'the Arcadian ideal of ruralized living at the edge of the city' (Walker 1981: 396). This is ideological to Walker in the sense that suburban escapism diverts attention from the real cause of the demand, namely the alienating experience of the capitalist system (Harvey 1978). Suburbanisation is also seen as ideological in the way in which it is strongly associated with and serves to naturalise key institutions of capitalist society, notably

the so-called nuclear family, the single-family home, home-ownership, the neighborhood school, and a certain limited type of 'community', conjoined with a localized political jurisdiction.

(Walker 1981: 392).

Again, the socially constructed character of quality of life is apparent.

Quality of life concerns with regards to suburbanisation centre around the demands, whether ideological or otherwise, for the various features identified by Walker. Indeed, the early suburbs captured the public's utopian imagination, becoming associated with decongestion, space and low residential densities, a separation of the home from the workplace and, not least, the prospect of home ownership (Baldassare 1992). Moreover, it was asserted, often through the use of ethnographic evidence, that the suburban experience generated a certain type of lifestyle, namely a greater sense of community and a decline in alienation, anonymity and deviant behaviour (Tittle and Stafford 1992).

The suburban dream soon subsided, as it became apparent that suburban living was often little different from mainstream urban living, in that it was also highly problematic. For example, much attention has been given to the selectivity of the suburban experience. The suburbs of the early post-1945 period were predominantly peopled by white, middle class families. Since that time the ethnic, class and demographic composition of the suburbs has broadened somewhat but, for example, the 'black suburbanization' noted by Rose (1976) is still relatively rare and distinctive (Baldassare 1992). Moreover, the association between suburban living and consumption has been heavily criticised by feminist scholars, who emphasise the suburb as a place of women's productive and reproductive work, and also of their effective entrapment (England 1991). Other aspects of what Baldassare (1986, 1992) regarded as a suburban crisis are outlined in Table 6.4.

Whilst suburbs are no longer seen today in such utopian terms by migrants, they often retain quality of life attractions, notably in terms of space and housing. Moreover, once obtained, a place in the suburbs is often guarded vigorously, through preventing other development taking place there (Ambrose 1992; Evans 1984; Short, Witt and Fleming 1987). These political struggles to maintain a certain quality of life are characteristic of the activities of preservation groups and can form a key component of local politics. They also take us away from a focus on the urban environment to address the quality of life associated with rural living and the 'lure of the countryside'.

Table 6.4 Dimensions of the 'suburban crisis'.

Issue	Example
• Political fragmentation:	There is often a lack of a coordinated local political response to problems such as pollution, traffic congestion and affordable housing. Suburban residents also often display a resistance to local taxation and welfare.
• Growth revolt:	Many residents are concerned with and attempt to limit new housing and other developments due to NIMBYism (Not In My Back Yard) and/or the perceived decline of suburban quality of life. From another perspective, many governments have increasingly stressed conservation and rehabilitation policies, which impact most heavily on the urban centres.
• Community quality:	Although residential satisfaction is usually high, there is concern with issues of crime, congestion, social and ethnic conflict, political tension, roads, pollution and quality of neighbours amongst many suburban residents.
• Affordability:	A key attraction of suburban living, affordable housing, has been found to be unattainable for many residents, leading to problems in finding suitable accommodation. This is reflected in the high cost of land and mortgages. The potential of rising energy costs may also deter people from commuting long distances.
• Changing family structures:	The stereotypical nuclear family with its 'dream house' that is 'designed around the needs of a breadwinning male and a full-time housewife who would provide her prince with a haven from the cold world outside' (Jackson 1985: 300) is increasingly rare within many developed world societies.

Source: Baldassare 1992: 483–8; Jackson 1985.

The 'lure of the countryside' and counterurbanisation

The 'lure of the countryside'

Returning to Table 6.1 (page 129), the city can be seen as Babylon rather than as Jerusalem. In this image, the city is the arena of the mob and is associated with disorder and the realm of (waged) labour. In contrast, the countryside is the space of the order of community and provides a retreat from urban tensions. In other words, the countryside attains a pastoral status of a *rural idyll* and, as a critique of the hectic pace of modern (urban) life (Bunce 1994):

The countryside as contemporary myth is pictured as a less-hurried lifestyle where people follow the seasons rather than the stock market, where they have more time for one another and exist in a more organic community where people have a place and an authentic role. The countryside has become the refuge from modernity.

(Short 1991: 34).

More generally, this anti-urbanism or 'ruralism' (Halfacree 1996a) has a much longer historical pedigree, dating from the emergence of cities, and can be found in books such as the Bible and in the writings of prominent Muslim thinkers (Hadden and Barton 1973).

Ruralism is present in the cultures of much of the developed world, often becoming the embodiment of that nation (Short 1991). For example, an imagined ruralness is central to the dominant ideas of Englishness (Samuel 1989; Wiener 1981; Wright 1985), captured expressively in the following quotation from a wartime radio broadcast by the naturalist Peter Scott in 1943:

Friday was St. George's Day. St. George for England. I suppose the 'England' means something slightly different to each of us. ... But probably for most of us it brings a picture of a certain kind of countryside, the English countryside. If you spend much time at sea, that particular combination of fields and hedges and woods that is so essentially England seems to have a new meaning.

(quoted in Wright 1985: 83).

In the United States, invigorated by Jeffersonian visions of agrarianism but extending back to the first European settlement, the pastoral idea(l) has been central to the definition of the nation (Hadden and Barton 1973). Here, it refers more to wilderness than to the domesticated countryside of the British rural idyll. Pastoralism exists both in high culture and in the popular culture that feeds the 'flight from the city':

Box 6.4 Building the quality of life in village England

The new settlement of Watermead, on the edge of the town of Aylesbury in the county of Buckinghamshire in south east England, is an excellent example of a development – supported by planners, local politicians and the private sector – that has promoted itself to incoming residents strongly in terms of a particular idea of a high quality of life. This is a quality of life, targeted at young, higher socio-economic group adults, that has been explicitly and self-consciously constructed within the development.

In the mid-1980s, Aylesbury Vale District Council found itself unable to identify sufficient land within its area of jurisdiction to fulfil its requirements for new housing. Therefore, after considerable negotiations, permission was given to a private developer to build what was eventually intended to be 800 houses on a largely self-contained site just outside Aylesbury town. The development was to be characterised by substantial amounts of prestigious executive housing, which, the planners hoped, would ease demands for such housing in surrounding villages. In addition, the development was also to feature two lakes designed for water sports (jet-skiing on one, more sedate pursuits such as punting on the other) and an artificial ski slope. The settlement itself was meant to represent a 'traditional village'. Overall, this whole 'lifestyle development incorporating health and glamour' (Murdoch and Marsden 1994: 78) was highly stylised and designed explicitly to resemble a film set.

The quality of life aspects of the rustically-named 'Watermead' came across most strongly in publicity and marketing material for the development. This spoke of:

- 'the perfect location for a way of life you previously only thought available when you are on holiday';
- 'entering another world';
- 'the warmth and charm of a traditional Edwardian village';
- 'timeless designs';
- 'distinctive village square, with its pink and cream-painted pub, restaurant and shopping mall, set around an attractive piazza.'

Such publicity was accompanied by photographs of windsurfers, (Alpine!) skiers, joggers and exotic birds.

The strategy of the producers of Watermead paid off handsomely at first, with a £5 million investment creating a site worth around £40 million. The development received awards and keen initial uptake of the properties. Unfortunately for the developers, this early success was subsequently dampened by a recession in the property market and other problems with overall site management. Nonetheless, when residents were surveyed by Murdoch and Marsden, there remained clear statistical evidence that the intended exclusivity of Watermead had been achieved: 72 percent of residents were from socio-economic groups 1–3, and 70 percent were aged 25–44. Lastly, looking at migration to the settlement: 26 percent of migrants had come from Aylesbury town and 31 percent from elsewhere in Buckinghamshire, displaying a clear local bias; and 'housing type' was cited by 29 percent of residents as the principal reason for moving there, compared with 22 percent citing 'work', 21 percent citing 'environment' and 21 percent citing 'marriage'. Overall, a quality of life to which certain people have responded to through migration has been constructed in this corner of village England.

Source: Murdoch and Marsden 1994: 75–83.

The soft veil of nostalgia that hangs over our urbanized landscape is largely a vestige of the once dominant image of an undefiled, green republic, a quiet land of forests, villages, and farms dedicated to the pursuit of happiness.

(Marx 1964: 6).

This long pedigree of ruralism in the United States, especially amongst intellectual thinkers, is illustrated in White and White (1962), *The Intellectual versus the City*.

Besides its Anglo-American form, a version of the rural idyll can also be seen in eastern Europe, expressed eloquently in Milan Kundera's novel, *The Unbearable Lightness of Being*:

Tereza looked into the farm worker's weather-beaten face. For the first time in ages she had found someone kind! An image of life in the country arose before her eyes: a village with a belfry, fields, woods, a rabbit scampering along a furrow, a hunter with a green cap. She had never lived in the country. Her image of it came entirely from what she had heard. Or read. Or received unconsciously from distant ancestors. And yet it lived within her, as plain and clear as the daguerrotype of her great-grandmother in the family album.

(1985: 168).

As Kundera's passage suggests, the rural idyll may be an urban perspective on the countryside, refracted through various media and not based on direct experience, but it nevertheless can be a strong force guiding migration. Moreover, the rural appeal that this idyll expresses need not be seen as being essentially backward and reactionary, since it reflects people's very real concerns with coping with the complexities of the modern world, and its link to migration should be judged likewise (Halfacree 1997).

Evidence for popular levels of ruralism with respect to residence are numerous. For example, from the United States, besides evidence presented below, studies building on the surveys of researchers such as Fuguitt and Zuiches (1975) and Blackwood and Carpenter (1978) have detailed popular associations between the countryside and various high quality of life indicators: environmental quality, lack of crime and criminality, sense of community, cheap living, suitable environment for children's upbringing, and so on. Work has also illustrated how Jeffersonian ideas are reflected in desires to 'escape industrial civilisation' through migrating westwards (Meinig 1991; Stegner 1992), a contemporary version of Turner's (1894) 'frontier thesis'.

Developing world images of the city relative to the countryside are less negative and ruralism is much less expressed than in the developed world. While the legacy of Biblical and Qur'anic condemnations of the city and urban life may be reflected in popular attitudes, the central question of everyday material survival and its link to urban opportunities tends to position the countryside as the sphere of poverty, backwardness and oppression. In contrast, as we have seen already in this chapter, the city represents not only economic potential but also more general opportunities for people to obtain a better quality of life. Nevertheless, there are exceptions to this representation, such as the dislike of the city as a symbol of sedentary society expressed by nomadic peoples (Chapter 9).

Migration to the countryside

Chapter 1 demonstrated the importance of counterurbanisation in driving dominant contemporary migration trends in much of the developed world. There are strong economic reasons to explain this trend, with particular attention given to the emergence of a new spatial division of labour (NSDL) (Chapter 4). Therefore, we might argue that counterurbanisation should be seen, in Moseley's (1984) terminology, as a job-led phenomenon (Fielding 1982). However, this is not the whole story. Counterurbanisation has also been characterised as being people-led, where a widespread preference for rural living is actively played out, assisted by an improved transport infrastructure and level of personal

mobility. A lure of the countryside appears to drive much counterurbanisation. It must be stressed, however, that this migration option is highly dependent upon a secure and relatively well-paid job or some other substantial level of regular income. Even within the developed world, counterurbanisation is a selective migration experience.

Arguing that some form of the lure of the countryside underpinned migration from urban to rural areas was stated boldly in one of the pioneering studies of counterurbanisation in the United States. Berry (1976b: 24) argued that counterurbanisation represented a 'reassertion of fundamental predispositions of the American culture', including a love of newness, a desire to be near nature, the frontier spirit, a freedom to move and the assertion of individuality. While one can criticise Berry's assertion in several ways, including its arguably narrow, overgeneral, highly selective and idealised view of 'American culture' and his neglect of the economic underpinnings of counterurbanisation, he did recognise how elements of choice can be incorporated into residential decision making and his work set the context for many subsequent studies.

Research suggests that while counterurbanisation moves over long distances tend to be underpinned by employment considerations (Chapter 4) and those over short distances by housing concerns, there is plenty of evidence to support the suggestion that broader quality of life issues are also implicated in this migration if the question of scale is addressed carefully (Vartiainen 1989a). Indeed, instead of talking about national versus local moves, Gordon (1991) suggests a three-fold division, with intermediate-scale regional moves being explained primarily by environmental factors, national moves by employment factors and local moves by housing factors. Thus, while a recent international collection of research on counterurbanisation (Champion 1989) emphasised its economic dimension, several chapters made reference to quality of life considerations: Hugo (1989) suggested a role for residential preferences in urban-to-rural migration in Australia; Kontuly and Vogelsang (1989) saw a filtering down and intensification of counterurbanisation in Germany to younger adults with children, reflecting employment factors plus the 'preference for residences located in areas with abundant natural amenities' (p. 157); and Tsuya and Kuroda (1989) noted the pollution-related push factor behind Japan's muted counterurbanisation and the general increasing importance of environmental factors.

Counterurbanisation has reached some of the more remote countries of the developed world, and here too quality of life factors play a part. From Finland, Vartiainen (1989a, 1989b) described a socio-cultural counter-trend to the urban-oriented Great Move economic restructuring-linked migration trend of the 1960s. Although not a mass phenomenon, migration away from the larger towns, fuelled by incomes earned during the Great Move, reflected a desire among households for private housing on their own land and the last romantic gasps of a peasant spirit. The ability of Finns to move out from the cities was facilitated by tax and housing policies, which favoured owner occupancy, and the weak building controls in rural areas. An earlier case study of Joensuu found obtaining a home of their own to be an important migration motivation for young Finnish families coming into the town, while urban-to-rural migration among older residents was more likely to reflect a desire for country life (Paasi and Vartiainen 1981).

Further details as to why and how residential preferences are reflected in counterurbanisation moves are provided by survey-based work (Chapter 2). These studies show that perceptions of higher environmental quality in the countryside act as important pull factors. Table 6.5 illustrates the varied recognition given in the British literature to the importance of 'environmental' reasons for moving. Crucially, when researchers allow *secondary reasons* to be given for the move, the importance of the environmental dimension is enhanced considerably. This insight returns us to Chapter 3, where it was argued that embedding migration within people's everyday biographies draws attention to the multiple reasons and issues that are likely to mould specific moves.

Looking at one of the studies cited in Table 6.5 in greater detail, the environmental reasons given by the migrants in Halfacree's (1994) research tie in well with the lure of the countryside outlined above. Table 6.6 describes the 'physical' and 'social' features of the (rural) destination emphasised by the migrants. These are almost all features of the (English) rural idyll and it is thus unsurprising that the migrants also tended to subscribe to such an

Table 6.5 Environmental reasons for moving cited in the British literature.

Study	'Environmental reasons'	Percentage citing given reason
Primary		
Radford 1970[1]	Countryside, health, evacuation	5–11
Ambrose 1974	Wanted village amenities/community spirit	6
Connell 1978[1]	Country area	28
	Character of local area	8
Hedger 1981	Liked area/to get away from ...	<25
Sherwood 1984	Live in village	17
Jones *et al.* 1986	To live in a nicer area	
	– physical	50
	– social, community	7
Perry, Dean and Brown 1986	Preferred environment	42
	Escape urban rat race	39
	Enjoyed previous holidays	38
	Better for children	21
	Better for retirement	10
	Better for health	8
Lewis 1989	Rural environment	22–31
Halfacree 1994[2]	Physical quality of the environment	31
	Social quality of the environment	15
Secondary		
Ambrose 1974	Wanted village amenities/community spirit	17
Sherwood 1984	Live in village	40
Halfacree 1994[2]	Physical quality of the environment	51
	Social quality of the environment	27

Notes:
[1] Studies that did not allow more than one reason to be given.
[2] Urban-to-rural migrants only.

Table 6.6 Key 'physical' and 'social' features of the destination for British urban-to-rural migrants.

Physical features
- The area was more *open* and less crowded; one no longer felt hemmed in by houses. There was a more human scale to things.
- It was a *quieter* and more tranquil area, with reduced traffic noise and less hustle and bustle.
- The area was *cleaner*, with fresh air and an absence of traffic pollution and smog.
- The *aesthetic quality* of the area was higher – views, green fields, aspect, beauty. There was stimulating, spiritual scenery.
- The surroundings were more *natural*, with an abundance of flora and fauna.

Social features
- The area allowed one to *escape* from the rat race and society in general. This was underpinned by a degree of utopianism.
- There was a *slower* pace of life in the area, with more time for people. There was a feeling of being less pressurised, trapped and crowded, and of being able to breathe.
- The area had more *community* and identity, a sense of togetherness and less impersonality. The general idea of 'small is beautiful' came across here.
- It was an area of *less crime*, fewer social problems and less vandalism. There was a feeling of being safer at night.
- The area's environment was better for *children's upbringing*.
- There were far *fewer non-white* people in the area.
- The area was characterised by *social quietude* and propriety, with less nightlife and fewer 'sporty' types.

Source: Halfacree 1994: 180.

idyll, albeit critically at times (Halfacree 1995b). Similar sympathies have been described in other British studies (for example, Cloke, Goodwin and Milbourne 1997; Cloke, Phillips and Thrift 1997; Jones 1995).

Survey-based studies elsewhere in the developed world also emphasise the importance given by migrants to the environment (Winchester 1989; Zuiches 1980). For example, in their study of migration in the United States Mid-west, Williams and Sofranko (1979) gave a central role to the environment for non-metropolitan destined migrants (Williams and McMillen 1980, 1983). In terms of reasons for leaving their metropolitan origins, environmental push factors came out as the top category (26 percent of migrants) and environmental pull factors the fourth category (14 percent), while, for choice of destination, environmental pull factors came second (28 percent). Similarly, Roseman and

Williams (1980) found environmental factors second only to 'ties' in choice of destination. The likely link between these factors and subsequent residential satisfaction is illustrated in Figure 6.4 (overleaf), where residents of Josephine and Jackson counties, Oregon, in the United States, tended to state that quality of life factors considered 'much better' in their areas were also important in affecting their likelihood of moving (Stevens 1980).

A substantial amount of the survey work in the United States was undertaken in the 1970s and, as Rudzitis (1991) observes, much has not really been followed up since. However, while the counter-urbanisation patterns of the 1970s and early 1980s have often been tempered considerably in recent years in many developed countries (Chapter 1), the importance of quality of life considerations remains. In his own work on the Pacific and interior north West of the United States, Rudzitis (1993) stressed the non-economic motivations behind much migration to non-metropolitan areas, concluding that 'it is the landscape [not jobs] that attracts and holds people in the interior West' (p. 576). One aspect of this preference is the presence of federally designated wilderness areas in this region, a feature that a geographically broader survey found was cited as being important in encouraging migration to that area by 60 percent of migrants (Rudzitis and Johansen 1991). Supporting Rudzitis, in their study of migrants to the Gallatin Valley, Montana, Williams and Jobes (1990) stressed the importance of quality of life reasons. They emphasised the combined importance of quality of life and economic motivations for the in-migrants of higher socio-economic status.

As Williams and Jobes suggested, environmental and quality of life migration to rural areas tends to be highly selective in involvement (Serow 1991). For example, in rural South Australia, Smailes and Hugo (1985; Hugo 1989) noted a bias in favour of Australian-born young adults with children. In socio-economic terms, the migration was also associated with unemployment and service sector employment. Other key groups noted by Hugo (1989) as being engaged in Australian quality of life migration were hobby farmers, long-distance commuters, people seeking alternative lifestyles and retirees. Better-off people, notably young adults, often with children, plus retired people, tend to form key counterurbanisation streams elsewhere, including the United States (Frey 1989), Germany (although here there appears to be a filtering down to younger age

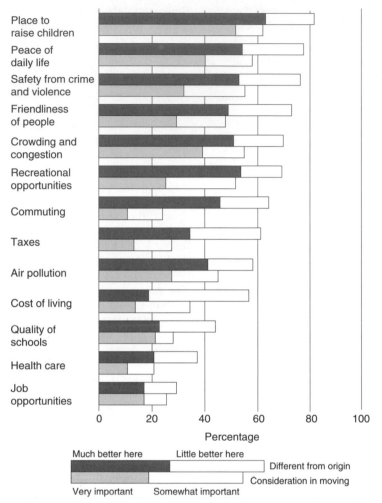

Figure 6.4 The relationship between quality of life factors and likelihood of migration (source: Stevens 1980: 122 from D. Brown and J. Wordell (eds), *New directions in urban-rural migration*. © Academic Press. Reprinted by permission. All rights reserved).

groups) (Kontuly and Vogelsang 1988, 1989) and France (Winchester 1989; Winchester and Ogden 1989). Counterurbanisation is also very much a white phenomenon. This can partly be explained by class-profile differences between the white and black populations of developed countries, possibly enhanced by cultural preferences. However, it also reflects the role played by racial intolerance in 'pushing' whites out of the urban areas (Fliegel and Sofranko 1984) and the racism experienced by many black people in the countryside (Agyeman 1990; Kinsman 1995). Nonetheless, overall, we need to be careful in not caricaturing counterurbanisers. In their work on north Devon in England, Bolton and

Chalkley (1989) and Grafton and Bolton (1987) searched for examples of the counterurbanisation stereotypes given in Table 6.7 but found none of these groups to be predominant. Instead, they concluded that 'counterurbanisation is a phenomenon of the masses' (Bolton and Chalkley 1989: 250).

To date, counterurbanisation has been largely confined to the developed world, as is predicted by models such as Zelinsky's mobility transition model (Chapter 3). Nonetheless, within the developing world, Gilbert (1993) noted how migration patterns are starting to change around some cities, with improved transport systems facilitating rural-to-urban commuting and home-based production. Mabogunje

Box 6.5 Environmental implications of environmentally driven migration

The influx of migrants motivated by the physical environmental quality of an area can have significant and often detrimental environmental consequences for that area. This is illustrated by migration to the Pacific and inland north west of the United States. Just as tourist-based economic and other development can have negative environmental consequences (Frederick 1993), such as the degradation of both the landscape and the social structure of a place, so too can residential in-migration in general. While a desire for a strong 'sense of place', plus a concern to be near wilderness areas, drives much quality of life migration to this part of the United States, it also runs the danger of undermining that very sense of place if it is insensitive to the local physical and social milieu. Thus, instead of a cumulative causation tendency, with quality of life migration begetting further quality of life migration, such movement of people can prove highly unsustainable over time and across space.

On a more positive note, in-migrants to the north west have shown concern with the area's environmental quality and to the threats that it faces. They recognise that

The physical landscape provides the region's real economic base while supporting people both physically and spiritually in a way difficult to imagine occurring in urbanized settings.

(Rudzitis 1993: 576).

This realisation by the migrants has led to conflict with more established residents over the present and future economic development of the region. Protection of the landscape becomes contested as a 'commodity production landscape has become an environmental consumption landscape' (*ibid.*: 577). This conflict is reflected in issues of water use, pollution, mineral and energy development, and forestry. In the case of forestry, a dispute has developed between private timber interests and conservationists, the latter typically represented by relatively recent migrants to the area. Concern has been expressed that the timber harvesters, in both public and private forestry areas, have been working in an unsustainable, environmentally destructive manner, in order to maximise profit and resist takeover. However, many local people are grateful for the jobs provided, given the lack of other employment options in the area. Moreover, potential alternative sources of employment, such as tourism development, can themselves have strongly negative environmental consequences. Indeed, Rudzitis thinks that tourism development in the north west is something of a blind alley and we must instead maintain the environmental quality of the environment for all of its inhabitants in order to attract alternative forms of economic development that make less physical impression upon the landscape.

Source: Rudzitis 1991, 1993.

Table 6.7 Stereotypes of counterurbanisers.

- The elderly spending their twilight years in a quiet rural setting.
- Long-distance commuters combining a rural home with an urban workplace.
- People returning to where they were brought up.
- Refugees from the inner city, escaping its grime and crime.
- Unconventional, anti-materialist commune members.
- Information technology wizards, running high-technology businesses from remote homes.
- Company managers brought in to run businesses relocated to areas of cheap and non-militant labour.
- Urban unemployed who would rather live in the countryside than in the city.

Source: Bolton and Chalkley 1989: 249.

(1990) observed a reversal of migration streams back in favour of rural Africa, a trend that was even more apparent in Latin America (Gilbert 1993). While economic considerations undoubtedly underpin this counterurbanisation trend – if the term counterurbanisation is considered acceptable in this very different socio-economic context – quality of life considerations may have some role to play, especially given the very low environmental quality of many developing world cities.

Finally, where counterurbanisation linked to quality of life becomes established it tends to demonstrate a clear tendency towards self-perpetuation or even the positive feedback of cumulative causation (Chapter 3), at least in the short term. A range of information sources pick up on the phenomenon – academic, press and other media, popular culture – often outlining and publicising its pervasiveness beyond its actual extent. Such a process helps to accord quality of life migration a normative status within that society, even it remains unattainable for most people. Therefore, there is some scope to study such quality of life moves in terms of an act of *collective behaviour* (Boyle and Halfacree 1998; Campbell and Garkovich 1984). This perspective emphasises the 'unnaturalness' of counterurbanisation, since it points to a crisis or at least a severe set of problems within contemporary society that individual acts of migration are serving to overcome in the manner of a 'craze'.

Migration within the rural environment

As with the urban environment, the countryside and its towns and villages do not always live up to the expectations or ideals of their inhabitants as regards quality of life. This can lead to subsequent migration within the rural environment, especially when strategies such as commuting enable a greater flexibility in the spatial relationship between the (waged) workplace and the home. For example, migrants to the countryside often do not wish to become too spatially isolated but to maintain their urban links, whether for work, services or more personal reasons. This is reflected in the relative lack of counterurbanisation in Norway, where a weak trend towards net urban-to-rural migration in the 1970s did not last, possibly because the country's remote peripheral communities were too poorly resourced to retain their indigenous young people, let alone attract inmigrants (Hansen 1989). Norway's experience was in sharp contrast to less sparsely populated Scandinavian countries, such as Denmark (Court 1989) or even Sweden (Borgegård and Murdie 1993), where counterurbanisation, or the 'Green Wave' as it was termed in Sweden, has been more noteworthy.

The study by Halfacree (1994), discussed above, also considered migration within the rural environ-ment. It found 'physical quality of the environment' reasons cited by 12 percent of rural-to-rural migrants at the primary reason level and by a further 24 percent at the secondary reason level. In addition, 'social quality of the environment' reasons were cited by 5 percent of movers at the primary level and by a further 13 percent at the secondary level. The destination features stressed by these migrants are shown in Table 6.8. They show similarities to those listed in Table 6.6 but suggest that, having obtained a rural residence, a more refined and sophisticated perception of ruralness is employed with subsequent migration within the rural milieu. On the social side, a moderating of the 'extreme rural' was apparent, while aesthetic concerns emerged as paramount in the physical dimension.

Migration and location-specific amenities

Modelling migration and location-specific amenities

Most of the work discussed so far in this chapter adopted a behavioural emphasis, with particular attention paid to surveys of migrants. However, there have been quantitative attempts to model migration related to quality of life factors, notably where the search for *location-specific amenities* is involved. These amenities comprise *non-traded goods* (Graves and Linneman 1979) that can only be obtained,

---

Table 6.8 Key 'social' and 'physical' features of the destination for British rural-to-rural migrants.

Physical features
- The area had more varied and/or *attractive scenery*.
- The area had more *attractive houses* and the overall village was more pleasing to look at.
- The area was near or next to *the sea*.
- It was a *more open* and/or remoter area.

Social features
- The area was more *socially active*, with plenty going on. It was not a dormitory or a retirement area but a village community, with friendly, welcoming people and a community spirit.
- The residents were *less parochial* and backward, and had a broader outlook on life. There were more professional people around.
- The area had fewer 'yuppies', company cars and 'company representative' people around. There was a *wider social mix*.
- It was an area of *higher status* and had a better reputation than the origin, with fewer working class people and less council housing.
- There was *less crime* and general trouble in the area.
- There was a *slower pace* generally, it was quieter and the general quality of life was better.

Source: Halfacree 1994: 183–84.

when needs or demands arise, through migration to a place where they are present. Research on the importance of location-specific amenities in driving migration comes from the *equilibrium* school of migration modelling (Greenwood 1985; Hunt 1993). Neoclassical and human capital models are typically of the *disequilibrium* kind, with migration regarded as a function of geographical variations in 'economic opportunity' (Chapter 3). In contrast, equilibrium models assume that amenities can compensate for variations in wages and other economic factors, with *overall* quality of life – expressed in terms of economic opportunity plus amenity factors – in relative equilibrium over space. From the equilibrium perspective, migration is seen not just as a response to income differentials but also as a result of changing demands for amenities within the household, linked to life-course patterns (Chapter 5).

Much work on the place of amenities in equilibrium models has been undertaken in the United States by Graves, although this work is not without its critics (Harrigan and McGregor 1993; Hunt 1993). Graves added climate and amenity variables to an economic model of net migration rates and improved its explanatory power considerably (Graves 1979a), before disaggregating this model by age and race, showing the importance of amenities to be greatest for whites and the elderly (Graves 1979b). Furthermore, he argued that obtaining high amenity values can cost a migrant, in terms of either low wages or high ground rents. The retired, therefore, having left the labour market, are likely to move to locations where amenities are priced mostly in terms of low wages rather than high rents, since wages are no longer of concern to them (Graves and Waldman 1991).

While the precise importance of amenity variables in Graves's models is unclear and somewhat related to the exact model specification (Graves 1980, 1983; Graves and Regulska 1982), other modellers have also demonstrated the importance of amenity. For example, Porell (1982) added a variety of such variables to a modified gravity model, discovering that they were important pull factors. Following Rosen (1979), he suggested that amenities could be financially quantified and traded off against wages. Specifically, he determined that, amongst a wide range of amenity variables, a $1 increase in the weekly wage was equivalent to an extra 3.8 inches of snow per year, a decline of 79.3 public tennis courts per 100,000 population or an increase of 7.87 micrograms of sulphur dioxide per cubic metre of air!

Similar conclusions were drawn in a study by Clark and Cosgrove (1991), who highlighted the importance of climate for snowbelt-to-sunbelt migration in the United States. Schachter and Althaus (1989) also quantified amenities in economic terms, suggesting, for example, that an extra heating degree day compensates for a decline of about 20 cents in annual earnings.

Other insights into migration and location-specific amenities

Away from modelling attempts to quantify the role of amenity in driving migration, demonstrating the role of climate, for example, also exposes the selectivity of much migration driven by demands for location-specific factors. While many aspects of quality of life appeared to be most important for the migration of the elderly (Chapter 5), climate-related migration is particularly associated with the better-off elderly. Indeed, this is also the case for much of the other location-specific migration, such as that linked to recreational facilities (Cebula 1979). An interesting example of this latter type of migration and its selectivity in this respect is seasonal migration in the winter to the hotter areas of the United States, which combines climatic and leisure concerns: the flight of 'Snowbirds to the sunbelt' (Mullins and Tucker 1988). Such migration is particularly associated with higher-income elderly couples in good health.

Survey work supports the significance given to climatic factors in influencing migration patterns. Again from the United States, Long (1988) found one-fifth of inter-state migrants leaving the north east in 1979–81 including climate-related issues at the primary or secondary reason level. Surveys also reveal other location-specific factors linked to migration. For example, ties in the form of the presence of kin and/or friends at a destination can be important. In a study of the US Mid-west, Williams and McMillen (1980, 1983) found that such ties provide an important stimulus to both return and non-return migration. Prior residence in a potential destination is important as it furnishes the migrant with *location-specific capital*, enabling him or her to discover more specific information about the place. While much of the importance of ties concerns minimising risk with respect to employment concerns (Massey *et al.* 1993; Stark and Levhari 1982; Stark and Bloom 1985), broader quality of life concerns may also prove important. Attention can be drawn to the importance of networks for guiding inter-

Box 6.6 Seasonal migration to sunbelt locations of the United States using recreational vehicles

Location-specific factors in the form of favourable climatic and recreational facilities are responsible for a major form of cyclical migration among many North Americans. Every year, hundreds of thousands of Americans and Canadians drive south in recreational vehicles (RVs) to spend the winter journeying around a network of camping sites, resorts and parks. States such as Florida, California, Arizona and Texas are the preferred destinations of these 'RVers'. Phoenix, Arizona, is a good example of a city proving popular for these seasonal migrations and was the subject of a large questionnaire-based study of managers, activity directors and residents in 1988. An estimated 74,000 RVers reside in Phoenix at the height of the season, being heavily concentrated (85 percent) along Highway 60, the Apache Trail, illustrated in the map below. This area is relatively cheap in terms of land rent and provides good access to surrounding recreational areas, such as Superstition Mountains, Tonto National Forest and the Salt River.

The survey found that Phoenix's RVers were overwhelmingly white, retired and married couples, although with some, often recently widowed, 'loners on wheels'. They were also mostly middle- to upper-middle-income households, with higher levels of education than the general population. For 90 percent of the RVers, circulatory nomadism was not all year round, as they maintained permanent housing. Most of these houses were well north of Phoenix, with 16 percent of the RVers coming from Canada, 11.5 percent from Washington, 8.3 percent from Minnesota and 7.3 percent from Iowa. The highest rate of RV migration to Phoenix was from North Dakota, with 28.66 RVers per 1000 persons aged 55–79 years in that state.

As places, the RV resorts of Phoenix are typically quite luxurious and again demonstrate how quality of life can be very deliberately constructed. The resorts exhibit three distinct characteristics that boost their appeal. First, they reflect a strongly recreational lifestyle, with specialist staff supporting swimming pools, jacuzzis, saunas, tennis and shuffleboard courts, bowling, horseshoe pits, golf, bicycle paths, plus a plethora of indoor activities. There are also team competitions taking place outside the resorts and numerous recreational trips further afield. Second, the resorts have a strongly suburban appearance. They are clean and tidy, with paved streets and an orderly layout; security is tight. Third, socially the resorts have a distinctive small town atmosphere. Residents return year after year, to be greeted with 'Welcome Home' signs and the community centre forming a hub of social activity. All in all, the presence of valued location-specific factors in cities such as Phoenix has generated seasonal migration patterns that are stamping a distinctive mark on the residential landscape.

Source: McHugh and Mings 1991; Mings 1984.
Figure reprinted with permission from *Urban Geography*, Vol. 12, No. 1, pp. 1–18. © V. H. Winston & Sons, Inc., 360 South Ocean Boulevard, Palm Beach, FL33 4SO. All rights reserved.

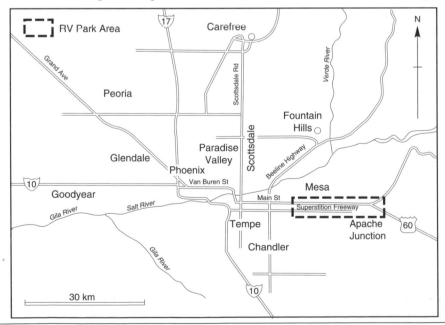

Box 6.7 Migration to Australia's east coast

Australia is one of the world's most urbanised countries, with just 14 percent of its population resident in rural areas by 1971. Nevertheless, in the 1970s there was a reversal of the post-colonisation pattern of increased population concentration into the major cities and a rise in counterurbanisation. In particular, net migration favoured the edges of the major metropolitan areas and the east and south east coasts, the migrants involved being predominantly Australian-born rather than new arrivals in the country. Through the 1980s, counterurbanisation became more selective and increasingly spatially concentrated. A growing emphasis on leisure and climate concerns became apparent, with the east coast of Queensland and New South Wales being favoured, but also the snowfields and the Murray River resort areas. As elsewhere, this counterurbanisation stream was highly selective, being dominated by child-rearing young couples and retired couples.

Looking at the example of New South Wales in greater detail, the expansion of coastal settlements, such as Byron Bay, Port Macquarie and Batemans Bay, is clear from the accompanying map. New coastal settlements were also emerging, such as Lemon Tree Passage between Port Macquarie and Sydney. In contrast, very few inland settlements witnessed expansion in the early 1980s, most continuing to decline. The growth of coastal counterurbanisation in New South Wales was underpinned by a growth in tourism and recreational activities, facilitated by the availability of land and a permissive planning regime. Overall, however, it demonstrates the demands of migrants for a more rural and coastal residential environment, even if by the 1990s many of the coastal settlements were themselves quite urbanised.

The importance of quality of life was apparent from a detailed study of 'sunbelt migration' to Port Macquarie. Of 146 retirement-age migrants surveyed, quality of life considerations underpinned almost all

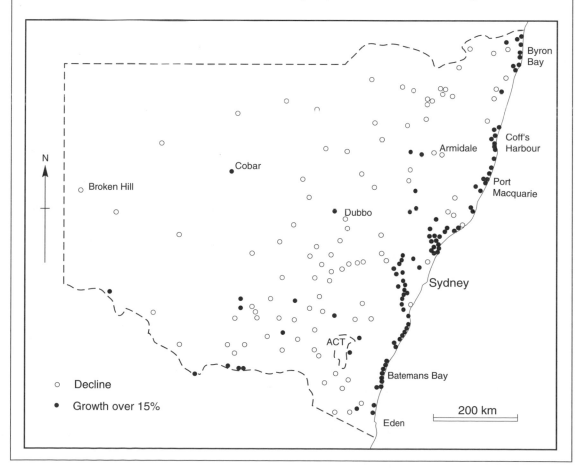

Box 6.7 (cont.)

moves, with migrants giving reasons such as 'peace and quiet' (64 percent), 'improved climate' (55 percent), 'better recreation facilities' (55 percent) and 'healthier lifestyle' (16 percent). Choice of Port Macquarie was related strongly to previous positive experiences of the area, with holiday encounters mentioned by 44 percent of migrants. Migrants generally felt satisfied that they had obtained these characteristics by moving. The distinctiveness of the migrants' own characteristics was also apparent. They predominantly originated locally, with half coming from Sydney and nearly 80 percent from New South Wales; they were especially concentrated amongst people in the higher status/income groups, and one-third had previously been self-employed; and they tended to be married couples rather than single people.

The pattern and development of counterurbanisation in Australia, as evidenced clearly in New South Wales, can be aligned with Frey and Speare's (1991) 'back to the future' scenario. This stresses the selective yet sustained character of future non-metropolitan growth, in which growth will be concentrated near the large metropolitan areas, in attractive, scenic environments and in tourist areas. There is clearly a strong inter-linkage here between employment-related migration, counterurbanisation and broader quality of life issues. Nonetheless, the sort of migration seen in eastern Australia poses environmental and general planning challenges (Caulfield 1993), which have to be met if the environment that many of the migrants are seeking is not to be devalued.

Source: Hugo 1986a, 1994; Murphy and Zehner 1988; figure reprinted from *Geoforum*, **24**, Sant and Simons, Counterurbanization and Coastal development in New South Wales, pp 291–306, © 1993 with kind permission of Elsevier Science Ltd, The Boulevard, Langford Lane, Kidlington, OX5 1GB, UK.

national migration from the developed world (Boyd 1989) and how such migration, especially for women, is usually a move to a less oppressive environment (Morokvasic 1984).

Finally, Charney (1993) noted how the state can play a key role in moulding location-specific migration through the localised activities of the public sector. For example, in the United States, welfare benefits vary between states, as do educational and other services. Recreational activities, too, can be promoted by the state sector in certain places, thereby creating the potential for location-specific migration. For example, the natural environment can be enhanced through anti-pollution and cleanliness drives and through the establishment of 'natural areas', while public provision of recreational facilities can be provided through organised activities and designations such as National Parks. This example demonstrates clearly the extent to which quality of life can be constructed 'on the ground' as well as more ideologically.

Conclusion

This chapter has illustrated the intersection of qual-

ity of life reasons for migration with those linked to employment, family and other concerns. Overall, therefore, Rudzitis's emphasis on the importance of quality of life factors is clear but must be placed firmly in the context of people's everyday lives. This comes across clearly when we contrast the relative importance of quality of life factors to different societies, or between different groups within individual societies. While jobs might be 'following' the migrants' residential quality of life choices in the north west United States (Rudzitis 1993), or in certain parts of the British countryside (Thrift 1987), this reflects the high levels of independence and economic and cultural power that the migrants possess. In many other places and for many other people, the potential of quality of life concerns to drive migration decision making is heavily constrained. Nevertheless, even many seemingly 'forced' migrations by the poorest of people can contain within them the potential to realise quality of life goals. Finally, emphasis on the links between quality of life and other factors when explaining migration emphasise how migration is a highly cultural action (Chapter 9). From this perspective, it bears repeating how quality of life is an extremely culturally constructed concept, which varies between and within different societies, across both time and space.

CHAPTER 7

Migration and social engineering

Introduction

This chapter considers the issue of state intervention in patterns of migration and mobility. Using examples, it outlines the reasons why some states elect to pursue interventionist strategies and it describes the impact of these policies upon people who are denied entry and those who are encouraged to leave. There is, of course, a thin line between vigorous encouragement on the one hand and compulsion on the other, and in many cases it can be difficult to distinguish between the two. While we acknowledge this difficulty, and the arbitrary nature of any division that might result, the current chapter focuses upon those examples of state intervention where the individual is still left with some realistic choice in whether to move or stay. It is left to the next chapter to consider cases where the element of choice is minimal or totally absent. We are therefore following Petersen's (1958) widely quoted typology of migration in which he distinguished between 'impelled' and 'forced' migrations, both of which result from state intervention.

Engineering migration and the place of human rights

Throughout history, states have sought to achieve particular objectives by manipulating all three population dynamics (fertility, mortality and migration). *Anti-natalist* and *pro-natalist* policies have been used to alter, selectively, the composition of a nation's population (for example, Berelson 1979; Helsinki Watch 1992) and, recently, *ethnic cleansing* has targeted minority ethnic or religious groups for localised extermination. Often such policies are implemented in conjunction with deliberate and direct intervention in migration, the focus of this chapter.

Dowty (1987) rehearsed the arguments used by the modern state to support intervention in the right to leave and to stay. He suggested that the modern state will control free movement, since such freedom would allow individuals to escape domestic law, has the potential to threaten public health, can undermine national security, may introduce 'contaminating influences', may lead to the abuse of unwitting citizens, can jeopardise national economic development and, ultimately, challenges the moulded homogeneity of the national character. Since few regimes welcome any of the above, a consensus has developed globally that it is not only in the best interest of states to manage immigration and emigration selectively, but it is actually imperative to do so. The protection and reproduction of national character has come to be seen as one of the central roles of the state.

Restrictions on freedom of movement not only have international legitimacy on their side, they also have practical advantages over other forms of population engineering. First, the legislation that controls migration can be changed very rapidly in times of emergency. The indecent speed with which the 1967 Labour government in Britain deprived East African Asians of the right to enter the United Kingdom is a case in point. Faced with the prospect of admitting 80,000 Asians who held full British citizenship and, therefore, rights of settlement at a time of public hostility towards further primary settlement of black people in Britain, the government introduced the 1968 Commonwealth Immigrants Act, by which only British citizens with patrial links to the United Kingdom were allowed entry. The effect was to leave white Kenyan settlers free to enter the United Kingdom, while denying this to Asians from Kenya. The change – including debate – took only six days. Second, there is a good deal of evidence to show that decisions about mobility are much more sensitive to changes in the external environment than are changes in fertility. During Castro's temporary relaxation of emigration restrictions in 1980 – the Mariel incident – over 125,000 people left Cuba for the United States in a six-month period, escaping on a flotilla of small boats that were either hired or provided by Cuban-Americans. Elsewhere, the ethnic violence that erupted in Rwanda in April 1994 prompted as many as 1 million people to cross the border into Zaire in only five days between 13 and 18 July. Similarly, the first opening of the Berlin Wall, on the weekend of 11 November 1989, produced the spectacle of what *The Times* estimated to be 'tens of thousands' of East Germans rushing to drive across the border in their Trabants with as many of their belongings as they could carry, often leaving behind vacant houses and even children in the rush to escape. By 17 November, some 9.5 million of East Germany's total population of 16.6 million had been granted an exit visa to visit the West.

Individual freedoms versus collective good

While the regulation of population movement is seen not only as a right by governments but also as a duty and an effective tool of social control, others see it as an infringement of a basic human right. Dowty (1987: 10) again summarised the arguments, noting the prevailing mind-set that has to be challenged by those advocating freedom of movement:

Modern ideologies ... have perfected the idea that the interests of the state ... should take priority over individual whim and caprice ... the burden is on individuals to demonstrate why they should be allowed to leave, rather than the state to show why they should not.

This view, he suggested, ignores five points. First, freedom to stay or go is a basic human right or, as he terms it, 'the forgotten right'. Second, the 'right of self-determination' is vital to the achievement of other basic human rights such as access to family life, marriage, an adequate standard of living, education, employment and freedom to practise a religion. Third:

the extent of a country's respect for personal self-determination says something very basic about how it is governed.

It is an index of a state's responsiveness to its citizens and, by extension, a measure of overall social health.

(*ibid*.: 226).

Countries that govern by consent must allow their citizens the right to leave, so that they can express their dissatisfaction with the performance of the government and with the 'contract' it is offering its citizens. Otherwise, consent is replaced by coercion. Fourth, there needs to be symmetry in intervention. Since the right to enter a new country is usually discretionary, there should be no right to expel people, thereby making them stateless. Fifth, freedom to leave a country also represents freedom to escape persecution.

Administrations have been slow to embrace the concept of *open borders*, and action groups have made little impact upon governmental obduracy. Dowty suggests a number of reasons for inaction and procrastination. Prime amongst these is that many activists believe that the prospects for change are negligible, since all states find the implications of open borders too disturbing. States fear that they will either lose population through emigration or gain unwanted population through immigration. In addition, those who are most affected by curbs on their freedom of movement are often those who are least able to articulate their complaints, either because of the regimes in which they live or their inability to access effective means of communication. For example, how many people in the West would have known about the 50,000 or so Nepali women trafficked across the Indian border to serve, against their wishes, in the brothels of Bombay, had it not been for the Human Rights Watch (1995) *Global Report on Women's Human Rights*?

Some commentators believe that there is a conflict between allowing freedom of movement and economic development, rooted in *mercantilist philosophy* which advocated the hoarding of population alongside the hoarding of precious metals. People were seen as not only producers of goods for export but also as consumers of domestic goods, which could themselves be manufactured more cheaply because of economies of scale. Countries should thus prevent their populations from emigrating. Modern perspectives are more sophisticated but still share some of these principles of mercantilism. Many developing countries argue that skilled personnel trained by the state and at the state's expense should not be allowed to leave and work elsewhere. Instead, they should repay the investment made in them by helping to develop the economy of their own country.

Lastly, progress towards open borders has been slow because, for those countries that attempt to control population movement, the issue is a highly sensitive one.

A logical fallacy behind all of these arguments is that they assume a huge unsatisfied demand for emigration and immigration. This assumption ignores the very powerful forces that tie people to their home regions: inertia, the desire to be amongst like-minded people, the pull of family and social networks, and the comfort of familiar surroundings and practices. All of these ties act to discourage potential migrants, despite the opportunities that migration might offer. As Dowty (*ibid.*: 223) puts it: 'For most people, deciding to leave requires a highly compelling set of circumstances, an unusually strong combination of push and pull.' Nevertheless, most governments retain a *laager* mentality as far as migration is concerned, regarding every immigrant or skilled emigrant as the thin end of a thick wedge (Robinson 1996a).

Engineering international migration

Emerging definitions of nation

Governments may choose to use their power of control to effect various forms of homogeneity or conformity, and they may engineer distinctive patterns of either internal or international migration, thereby regulating community membership. International migration is frequently easier to influence, since it is now regulated by national law and by internationally agreed conventions, definitions and arrangements. As shown above, legislation can be introduced by administrations relatively quickly and with little opportunity for effective dissent. Moreover, changes in legislation can have an immediate effect, with international migration finely adjusted to meet prescribed needs.

This was, of course, not always the case. In the era before Western nation-states, membership of a nation was derived from common descent and native culture, not from current place of abode. As Smith (1991: 12–13) puts it:

Genealogy and presumed descent ties, popular mobilization, vernacular languages, customs and traditions: these are the elements of an alternative, ethnic conception of the nation, one that mirrored the very different route

of 'nation-formation' travelled by many communities in Eastern Europe and Asia.

The significance of such a definition of nation is that it is exclusive not inclusive. Residence within such a nation does not entitle one to become a part of it; nor does emigrating to another location sever the primordial ties that make individuals members for life. Immigration and emigration thus represent no challenge to the nation's identity and, therefore, need not be regulated closely. Groups such as nomads and traders were left free to move around unhindered, administrative boundaries being largely irrelevant. In contrast, the Western or *civic model* of the nation has very different bases (Chapter 9). Smith (*ibid.*: 9) argues that

it is in the first place, a predominantly spatial or territorial conception. According to this view, nations must possess compact, well-defined territories. People and territory must ... belong to each other.

A fixed *homeland* is required, in which the historical national memory is located. Those who live within and are accepted as part of the nation receive certain benefits and are subject to certain obligations. They have exclusive rights to the land's resources and are afforded protection by the state. However, by becoming citizens, they are subject to the laws of the patria, they become part of a community of laws and institutions. The media and educational system ensure the transmission and reproduction of the common values and traditions that serve to bind the population together within their homeland. In sum, the Western civic model of the nation can be characterised as 'historic territory, legal–political community, legal–political equality of members, and common civic culture and ideology' (*ibid.*: 11).

The civic conception of nation has clear implications for migration. If membership of a nation is determined by place of abode, the benefits that derive from membership are allocated on the same criterion and individuals can be socialised into a new civic culture and ideology. Unregulated immigration poses a major threat to such a nation, since it implies a loss of control over who can and cannot become citizens and, therefore, enjoy the benefits of citizenship, and it also implies a continual challenge to the homogeneity of civic culture. Not surprisingly, then, the civic model also requires precise territorial borders, border controls and discriminating immigration criteria. In some cases, where migrant labour is essential for economic growth, but where nations are unwilling to offer newcomers membership of that nation, administrations fall back on a separate set of eligibility criteria, which replace residence with length of residence. Guest workers in Germany (Chapter 9) or the Gulf represent groups of people denied legal membership of a nation in this way (Castles, Booth and Wallace 1987).

Modern civic nations thus have little choice other than to become committed to selectivity in international migration. Such a policy serves to bolster national self-identity and homogeneity, through the admission of certain ethnic, racial and religious groups and the exclusion of others. *Ethnicity* is often the core and precursor of the modern nation, while differing from it in several important respects. Ethnic communities

need not be resident in 'their' territorial homeland. Their culture need not be public or common to all the members. They need not, and often do not, exhibit a common division of labour or economic unity. Nor need they have common legal codes with common rights and duties for all.

(Smith 1991: 40).

An ethnic community can thus be scattered across several territories, as is the case with the Indian (Table 7.1) or Jewish *diaspora* (Chapter 9), and this creates the possibility of flows of international migration between different satellites of an ethnic community. Furthermore, in the same way that ethnic identity is often nested within national identity, so

Table 7.1 The Indian diaspora: the total stock of people claiming Indian ethnicity by country of residence, 1987.

Country	Numbers
UK	1,260,000
Malaysia	1,170,000
Fiji	839,340
Mauritius	700,712
USA	500,000
Trinidad	430,000
Kuwait	355,947
South Africa	350,000
Burma	330,000
Guyana	300,350
Canada	228,500
Oman	190,000
Singapore	169,100
Surinam	140,000
Yemen	103,230
Netherlands	102,800
Global total	8,691,490

Note: only communities of over 100,000 included.
Source: Clarke, Peach and Vertovec 1990, reprinted by kind permission of Cambridge University Press.

race and religion often become components of ethnic identity. In attempting to control and homogenise national identity, states often wittingly or unwittingly become involved in engineering racial and religious groups, their spatial distributions, and their mobility.

International migration and ethnic engineering

Recent German history provides us with classic cases of international migration being manipulated to feed nationalism. The treatment of Jews during the Second World War, the resettlement of emigrant *Volksdeutsche* from 1940–44, and the repatriation of *Aussiedler* since 1945, each demonstrate how the German state has intervened in migration patterns in order to produce ethnic homogeneity, expand ethnic settlement back into historic homelands, or repatriate 'lost' Germans stranded by the political settlement at the end of the Second World War. All of these actions depend in part upon the concept of a special ethnic bond between *all* Germans (*Volkszugehorigkeit*), first given legal expression by the Nazis.

The new map of Europe after the war left many ethnic Germans outside the territory of either the Federal Republic (West) or the Democratic Republic (East). P. Jones (1990) estimates that, in 1939, over 8.6 million Germans were living outside the 1937 boundaries of Germany. The continuing 'repatriation' of these ethnic Germans to German soil since the end of the Second World War has created a migration of some 12 million people, with 8 million arriving in the years immediately following the war, and an average of 71,000 entering Germany each year between 1950 and 1993 (Jones 1996). Marshall (1992: 247) argues that the integration of these migrants and their subsequent contribution to the rebuilding of the German economy 'is considered one of the outstanding achievements of German post-war history'. The German government engineered the migration of these *Aussiedler* and its outcomes in three separate ways. First, under Article 116 of the 1949 Basic Law, all ethnic Germans were granted the right to enter and settle in Germany if they could prove that their forebears were German citizens, were of German ethnicity (usually through old census enumerator's returns) or had wished to become German at some point in the past (perhaps through being included on the Nazi *Volkslisten*). The presumption behind this gesture was that ethnic

---

Box 7.1 German attempts to engineer ethnicity through migration during the Second World War

The desire to retain ethnic purity and prevent racial intermixing was, of course, at the heart of why the Jews suffered so greatly during the Second World War. Managed emigration was a precursor to the Final Solution. Until 1941, the Nazi authorities encouraged and even orchestrated the emigration of Jews, first from Germany and later from occupied Austria and the Sudetenland. Individuals responded to harassment and increasing restrictions on their lives by seeking emigration opportunities. In 1939, 17,000 Jews fled to Shanghai, for example, where an *émigré* community was established with its own cultural infrastructure, largely funded by a Jewish voluntary body. Others sought illegal refuge in British Palestine or asylum in the United Kingdom. However, while individuals made their own legal or illegal arrangements a wider plan was taking shape involving the German authorities and other governments. Hitler suggested overseas 'reservations' for German and other Jews; in their haste to appease, other governments investigated the possibility of re-settling Jews in remote unsettled parts of the world. Suggestions included places such as Madagascar, Kenya, Northern Rhodesia (Zambia), the Dominican Republic and Cyprus. In reality, as Marrus (1985: 188) caustically comments, while 'government leaders ... warmed to the image of young Jews tilling the soil of sunny, distant lands, improving their own lives immeasurably in the process', three-quarters of all German Jews were aged over 40, most were from urban backgrounds and only 2 percent knew anything about farming. With the failure to find an international solution, Germany turned instead to its own internal solutions. While one official of the German Foreign Office pointed out that, in 1939, the 'ultimate aim of Germany's Jewish policy is the emigration of all Jews living on German territory', in the early years of the war the Nazis were content with the resettlement of Jews in prescribed areas where they could be held until overseas resettlement became possible after the war. After the capture of Poland, the south and east of that country was declared a governorship of the Reich and an experimental Jewish reservation was established there, near Lublin, to which 60,000 Jews were sent. The site was unprepared and unsuitable, and failed. In late 1941 the policy of internal resettlement was replaced by systematic extermination.

While plans existed to ensure the ethnic purity of the German peoples in their cultural heartland, other groups were simultaneously being encouraged to expand that heartland through large-scale and centrally organised migration and settlement. The German government created an organisation as early as 1938 to encourage the return of ethnic Germans, who since the Middle Ages had migrated east and south east in search of new opportunities. By the end of 1938, they were receiving 2500 returnees per month and in the following year this number rose to 4000. In late 1939 and early 1940, a series of bilateral agreements was forged with various countries, such as the Baltic nations, Romania and the Soviet Union, to allow the speedier return of the *Volksdeutsche*, who were then resettled in the north and west of newly conquered Poland, an area experiencing intense Germanicisation. Eventually, one and a quarter million ethnic Germans were channelled into this new territory and over half a million were resettled on farms vacated by 750,000 Poles forcibly expelled by the invading troops.

Source: Marrus 1985.

Germans living in the East faced persecution. Consequently, *Aussiedler* were granted rights of entry despite the continually repeated claim that Germany is 'not a country of immigration' (Chapter 9) and recently expressed views in Germany that 'the lifeboat is full' (*das Boot ist voll*). Thus, a third-generation ethnic German who has lived all her life in Russia, speaks Russian and has never visited Germany would be allowed entry, while German-born, German-educated and German-speaking children of Turkish guest workers have no right to citizenship or permanent residence.

The second way in which the German administration has intervened in the migration of *Aussiedler* is through inter-governmental agreements. From 1978 onwards, the Communist regime in Romania issued exit visas for about 12,000 ethnic Germans in exchange for payments by the German government of 8800 DM per head.

Third, in 1988, the government put in place a generously funded programme for the resettlement of the *Aussiedler*. This was in response to the sharp increase in the number of arrivals from 1987 onwards as the political reforms of *glasnost* took hold in Eastern Europe. Whereas, in 1986, nearly 43,000 *Aussiedler* arrived in Germany, the following year this rose to 79,000. In 1988, immigration from this source had climbed further to 203,000 and by 1990 the flow numbered nearly 400,000. The *Sonderprogramm Aussiedler* of August 1988 gave new arrivals complete equality with Germans, a right to German citizenship on arrival, and a free choice of where they wished to settle in Germany. Unlike asylum seekers, they are also granted complete freedom of movement within the country. Cheap loans were made available to assist with the search for accommodation (5000 DM per couple over ten years) and 300 DM 'integration money' was given for each family member to allow them to purchase furniture, clothing and household necessities. They were immediately entitled to all state welfare benefits and labour exchanges were told to find them work as quickly as possible. P. Jones (1990) notes that the total cost of this package of incentives in 1989 amounted to 1900 million DM, excluding the expanded housing programme begun in that year. Although the cost of the programme required it to be scaled down within twelve months and rejigged completely in 1991, it nevertheless represented a major investment by the German government and a considerable inducement to ethnic Germans to migrate to Germany. However, the programme has not been without its problems (Heller and Hofmann 1992). Many *Aussiedler* were unable to speak German and were therefore unable to work. In addition, the peak of immigration coincided with rising unemployment and an acute housing shortage. The problem became known as *Aussiedlerstau*, or the 'bunching' of ethnic Germans in the housing and labour markets. As a result, there have been calls for a shift in policy towards anchoring and assisting ethnic Germans within their eastern European adoptive homelands.

International migration and racial engineering

Governments also intervene to maintain perceived racial homogeneity. Such intervention often, but not always, takes the form of restrictions on immigration. The history of immigration control in the United Kingdom since 1945 demonstrates how and why a state might engineer international migration to fit a pre-existing racial template. It is also a story of hypocrisy. Hiro (1991: 200) notes how in public:

the leaders of both major parties regarded the Commonwealth as 'one of Britain's principal sources of diplomatic influence'. This source was impressive not only in its magnitude ... but also, more importantly, in its unique exemplary nature. As the head of the Commonwealth, routinely described as a 'multiracial society', Great Britain was seen as a first-rate moral leader, setting an example to the world at large.

Yet, as Britain was congratulating itself on transforming an empire into a club of nations, it also realised that it needed a public symbol both to bind the members of this club together and to demonstrate its moral leadership and responsibility for the group. This symbol was the 1948 British Nationality Act, which confirmed for most members of the Commonwealth their right to enter and live in Britain. By reiterating that black people were free to migrate to Britain and settle there, Britain was publicly making two statements: that the legacy of empire was capable of benefiting both the conquered and the conqueror; and that all citizens were equal regardless of their skin colour in the Commonwealth 'family'. However, both of these statements were made at a time when large-scale black migration to Britain was neither envisaged nor thought likely because of the high cost of long-distance sea travel.

The private behaviour of successive post-war governments in relation to non-white ethnic minority immigration demonstrates the hypocrisy of these public gestures. Immediately after the war, for example, Chinese pool seamen who had voluntarily crewed merchant navy vessels in order to allow British sailors to enlist in the Royal Navy were deported from the United Kingdom, even though some of them had married local women while living in ports such as Liverpool (Robinson 1992). The 1945–51 Labour government also began in private to look at ways of discouraging or stopping black immigration. Indeed, both this administration and the following Conservative administration used covert and illegal administrative measures to achieve this end, even though the 1950 cabinet committee established to look at the need for immigration control had reported that control was not necessary, saying that problems arising from unconstrained black immigration were too localised to merit national action. Even so, the tone of the covert debate about 'coloured immigration' was such that it was instrumental in associating immigration with racial and social problems in the areas of settlement. This association quickly became reduced to a shorthand of black immigration = problem. Such thinking both underlay and was used as a rationalisation for specific actions, such as Winston Churchill's attempt to have West Indians banned from serving in the Civil Service (Holmes 1991), and the general drift in government circles towards control in the mid- and late 1950s.

The association between black immigration and racial problems was 'confirmed' for many, and made a national public issue, by the 'race riots' in Notting Hill (London) and Nottingham in 1958. Although in the disturbances blacks had been victims of violence rather than its perpetrators, an opinion poll at the time showed 75 percent of those interviewed wanted immigration control (Hiro 1991). The national debate that followed the disturbances brought the true issues to the surface. There were few references to the brotherhood of the Commonwealth or the equality of its different racial constituents. Rather, when Cyril Osborne introduced a motion in the House of Commons in December 1958 proposing control, the debate that followed centred on racial purity. As Harris (1988: 53) noted:

when individuals like the Marquis of Salisbury spoke of maintaining the English way of life, they were not simply referring to economic or regional folk patterns, but explicitly to the preservation of 'the racial character of the English people'. We have developing here a process of subjectification grounded in a racialised construction of 'British' subject which excludes and includes people on the basis of 'race' skin colour.

The Times (03/09/58), too, voiced crude racist stereotypes when it wrote of the

three main causes of resentment against coloured inhabitants. ... They are alleged to do no work and to collect a rich sum from the Assistance Board. They are said to find housing when white residents cannot. And they are charged with all kinds of misbehaviour, especially sexual.

Thus, despite notions of moral leadership of a multi-racial Commonwealth, Britain introduced the 1962 Commonwealth Immigrants Act, which sought to control immigration. As William Deedes, a minister at the time, has subsequently admitted:

The Bill's real purpose was to restrict the influx of coloured immigrants. We were reluctant to say as much openly. So the restrictions were applied to coloured and white citizens in all Commonwealth countries – though everybody recognised that immigration from Canada, Australia and New Zealand formed no part of the problem.

(1968: 10).

The Act was operated in what appeared to be a non-discriminatory way. Those who had British passports issued by the British government, those born in the United Kingdom, and those named on the passports of adults in these two categories maintained the right of free entry to the United Kingdom. Others had to apply for an employment voucher, the issuing of which was closely regulated and inextricably linked to highly specific (and often professional) labour shortages. Decoding the official language, while

white Britons and white settlers (and their dependants) kept the right of free entry, black immigration henceforth came under tight political control, ostensibly driven by the needs of the labour market.

While some commentators have viewed the discrimination that underpinned the 1962 Act and all subsequent United Kingdom legislation on immigration as governments simply reacting to public opinion, and others have argued that it allowed capital to create a guest worker-type system, Solomos (1993) thinks otherwise. He claims that both political parties had realised, by the time of the Smethwick by-election in 1964 (when the Conservatives won against the odds using the slogan: 'if you want a nigger as a neighbour vote Labour'), the potential electoral benefit of being seen to be tough on immigration. As Richard Crossman (1975: 149) put it in *The Diaries of a Cabinet Minister*:

ever since the Smethwick election it has been quite clear that immigration can be the greatest political vote loser for the Labour Party if one seems to be permitting a flood of immigrants to come in and blight the central areas of our cities.

Such racially motivated immigration policies were not unique within the Commonwealth. From the Immigration Restriction Act of 1901 until the 1970s, successive Australian governments operated a White Australia policy. This sought to realise Australia's economic potential without sacrificing the racial purity of its population. Encouragement and exclusion were both used to achieve the desired end. The policy codified earlier actions taken by individual states against Chinese and black workers. The 1901 Act used coded language to exclude immigrants from countries other than the United Kingdom. Potential migrants were not excluded on racial or ethnic grounds *per se* but had their eligibility assessed by their ability to take down a 50-word dictation in a European language or, later, a 'prescribed language', usually English. Where immigration officials were particularly keen to exclude someone, Welsh would be used for this test. Conversely, both the individual Australian states and the federal government from 1920/1921 – when it took over responsibility for all immigration – did their best to encourage the immigration of white Britons and their descendants. Thus, in 1919, a settlement scheme was introduced for ex-servicemen from the United Kingdom and, in 1922, Australia collaborated with Britain in the Empire Settlement Act, designed to attract Britons to migrate to the

Dominions through the mechanism of subsidised transport. Over 282,000 Britons took a passage to Australia between 1922 and 1931, with over three-quarters receiving assistance under the Act.

While these twinned policies of exclusion and selective encouragement now seem both blatant and illiberal, Hawkins (1989) argued that they must be seen in the context of their times. She pointed to the fear of invasion by neighbouring Asian powers; an ignorance of other cultures, which made them seem threatening; a desire to secure for Britain a prior claim to Australia and its mineral wealth; a wish to build a new nation; and a belief that, by emulating Britain, Australia was emulating the pivotal economic and political power in the world at the time. Others have been less charitable. Collins (1991: 207), when discussing the continuation of the policy after the Second World War, commented that the architect of its re-affirmation, the Immigration Minister Arthur Calwell, 'remained a racist until his death'.

In their efforts to maintain racial purity, Britain and its Dominions did not only intervene in the international migration of adult workers. Between 1618 and 1967, United Kingdom-based charities and churches connived with the state to export 150,000 children from Britain to Canada, Australia, New Zealand and, to a lesser extent, South Africa, Southern Rhodesia (Zimbabwe) and the Caribbean (Bean and Melville 1990). While the children were described as destitute orphans, they were more frequently illegitimate, products of broken homes, or had been put into care temporarily by parents. Many parents were unaware that their children would be sent abroad and simply sought a better life for them. The Church of England, the Catholic Church, Barnardos and the British government abused their positions of trust and sent 4–14-year-olds abroad for reasons often unconnected with the children's welfare. Prior to the 1860s, children were exported because it was cheaper to pay their fare than keep them in the workhouse; while from 1870 to 1925, they were sent for emigration by the great philanthropists who, coincidentally, made five pounds per child profit by taking payments from orphanages and potential employers. In the post-war period, they were exported by the Catholic Church to sweep away the shame of illegitimacy and broken homes. Under all of these surface justifications was a common rationale: that it was Britain's task to populate its colonies with good British stock, partly to secure their strategic future and partly to create a viable workforce with which to exploit local resources. As a Catholic agency put it in 1938:

Those who are co-operating .. in this great and noble project of transplanting poor children who are without means, influence, and parentless in many cases, from congested and unpromising surroundings, to a land rich in natural, but undeveloped resources which are awaiting the correct type of people to render them productive, are doing much to strengthen and extend the Empire.

(quoted in Bean and Melville 1990: 6).

Child migrants offered the additional advantage that, being young, they could be moulded into good and obedient citizens. However, while Britain, Empire, employers, voluntary organisations and some children benefited from the schemes, most children did not. The majority suffered gross abuses of their human rights. In the case of those exported between 1945 and 1967, with the full cognisance and financial support of the government, these abuses took place when society had acquired a much more sophisticated knowledge of child welfare. At least one set of commentators (Bean and Melville 1990: 18) described the emigration of Catholic children to Christian Brothers' farm-schools in Australia during this period as 'one of the blackest spots in the whole history of child migration'. Thirteen-year-old boys were used to build dams, dormitories and chapels in the bush, with few tools, little thought for health and safety, and very little food. They were required to work for up to ten hours a day in all temperatures, while being ritually humiliated, sexually abused and beaten. Yet, it was not until January 1982 that regulations were introduced to prevent such gross abuse. The Secretary of State now has to give permission before a minor can be sent overseas.

International migration and religious engineering

States have also intervened in international migration for religious reasons. The partition of India in 1947 is an example from recent history (Ansari 1994: Kiernan 1995), but Israel provides one of the best contemporary examples, through its airlift of Falasha Jews in 1991 and its involvement in attracting and resettling Jews from the (former) Soviet Union. In the first case, Jews who shared neither ethnicity nor race with Israeli Jews were airlifted out of Ethiopia, then in the throes of a civil war. In the second case, Soviet Jews have been welcomed in Israel despite having no previous links with the country,

Box 7.2 The experience of twentieth-century Commonwealth child migrants

The following are some quotations from a series of conversations with adults who had been child migrants, selected so as to capture what it was like to be a child migrant, even in the latter two-thirds of the twentieth century:

We were put on a boat and shipped to Canada. The trip was appalling, horror laden, fraught with unconcern and disdain for those on board. We were all put in a large holding area in the ship and treated like cattle. The food was brought to the door and thrown to us. Whoever caught it, ate it.

There was this nun, she was a cruel woman. . . . I saw her beat 4 year old Cathy until she was black and blue, beaten to a pulp. . . . She enjoyed it.

I would get up at 4am and go to bed at 6 or 7 in the evening. You would work all day. . . . The farmer wouldn't feed me. I would steal stuff out of the barns, rhubarb and such.

We hated Australia when we first come out here. . . . I thought how am I ever going to see England again. . . . But when I got my first wages I went and bought a little miniature made in England. I never bought anything that was made in Australia, if I could help it. It took all my money. It was a tiny little English house all to remind me of England. I wanted to keep my identity with England. I felt if I had something English, there was still that hope.

I had a life of hell there. The man had a vicious temper and he used to beat me up if I forgot something or didn't do it right. . . . I was the hired man and a housemaid too. I worked my heart and soul out there for $3 per month. . . . It kept me with something to wear on my feet that's all, and people threw some of their old clothes at me.

Source: Bean and Melville 1990.

abling mechanisms were effectively latent due to the intransigence of the Soviet authorities and the willingness of the United States to take the few Soviet Jews who were allowed to leave.

For the United States, accepting what it labelled 'Jewish dissidents' was seen as a way of proving the moral superiority of the capitalist system. The number of Jews able to leave the Soviet Union fluctuated considerably in the 1970s and early 1980s, but in the late 1980s, as *glasnost* became a reality, numbers increased sharply. However, at the same time the dividend to be gained by the United States from accepting those fleeing 'communist oppression' fell and policy shifted, with Soviet Jews having to prove individual persecution as opposed to generalised group persecution. Israel and its envoys in the Soviet Union filled the void. From 1989 onwards, waves of Soviet Jews arrived in Israel. Their reasons for leaving were numerous and included the rising tide of ethnic nationalism in the former Soviet Union and the associated increase in the persecution of Jews (Cohen 1992). There was also a sense of ethnic affinity to Israel, fears about the future of the economy, concerns about environmental issues and their impact upon the quality of life, and a long-standing exclusion of Jews from positions of status and power (Birman 1979; Hiltermann 1991). Emigration, which had numbered 19,000 in 1988, increased dramatically (Table 7.2).

As US restrictions on entry came into force in 1989, an increasing proportion of emigrants headed for Israel. 24,700 arrived in 1989, and it was expected that over 120,000 would arrive in 1990 (Rowley 1990). The response of the state was immediate and

Table 7.2 Soviet Jewish emigration, 1968–92.

Year	Volume	Year	Volume
1968	229	1981	9,447
1969	2,979	1982	2,688
1970	1,027	1983	1,314
1971	13,022	1984	896
1972	31,681	1985	1,140
1973	34,733	1986	914
1974	20,628	1987	8,143
1975	13,221	1988	19,365
1976	14,261	1989	72,500
1977	16,736	1990	201,300
1978	28,865	1991	197,000
1979	51,333	1992	152,100
1980	21,471		

Source: Basok and Benifand 1995, reprinted by kind permission of Cambridge University Press.

being unable to speak the same language, and in some cases being enthusiastic advocates of Russian culture.

Israel long campaigned for the release of the 2.15 million Jews in the former Soviet Union, partly on religious grounds. In 1950, the Knesset (Israeli parliament) passed a law that gave Jews anywhere in the world the right to enter and settle in Israel. They were also entitled to citizenship immediately upon arrival. This Law of Return was based on the notion that members of the Jewish diaspora were simply returning to their spiritual home (Chapter 9). Two years later, the Knesset also formalised the status of the World Zionist Organisation, a body founded in 1897 to facilitate the migration and absorption of 'returnees'. However, prior to the late 1970s, these en-

> **Box 7.3 The evacuation of the Falasha Jews by Israel, 1991**
>
> The Falashas, or Ethiopian Jews, are a mountain tribe that first settled the Gondar region of north west Ethiopia in 300 BC. For most of their subsequent history the tribe led a very sheltered existence, eventually coming to believe that they were the only people in the world who practised the Jewish religion. While they celebrated the key Jewish festivals, practised circumcision and relied upon the *Talmud* (holy book), they were black and did not speak a Semitic language.
>
> The Falashas first entered public consciousness outside Israel when the Israeli government airlifted 12,000 out of Ethiopia in 1984, during Operation Moses. This followed the public declaration by the Chief Rabbi in 1973 that the Falashas – the 'lost tribe of Israel' – should be rescued from absorption and assimilation in Ethiopia and brought to Israel to be with fellow Jews.
>
> During 1991, the position for the remaining Falashas became precarious because of the civil war between the Mengistu regime and those seeking its overthrow. Foreseeing such an eventuality, the Israeli government had already made arrangements in 1989 to have the Falashas moved from the mountains to the capital, Addis Ababa, where they would be more accessible should a further airlift become necessary. Late in 1990, the Israeli government despatched a covert team to Addis Ababa to make preparations for the evacuation. Developments were hastened by the military successes of the rebels, and Israel began negotiations with President Mengistu about the airlift. Mengistu was keen to engage Israel in dialogue
>
> but he also regarded the Falashas as an important bargaining chip that would help him to retain American support. In the event, Mengistu gave way after a payment of £20 million was made by the Israelis to secure the Falashas' release. Between 26 and 28 May the Israelis put Operation Solomon into effect and 16,000 Falashas were airlifted to Jerusalem. The speed and secrecy of the evacuation was such that problems did arise. A contemporary press account describes how a family had unwittingly despatched their son to market to buy food just before the Israelis called to take them to the airport. Despite sending people to look for him, the family had to leave without their son, who was never evacuated from Ethiopia and presumably never learned what had become of his family.
>
> Once in Israel, the Falashas experienced considerable public support in the early period, with Israelis offering clothes, household items and jobs to the newcomers. However, press reports record that after the initial euphoria, the Falashas faced racism and discrimination, and also had difficulties because of their extended family structure, with Israeli housing being unable to accommodate the traditional household of over twenty members. An inability to speak Yiddish also made it difficult for Falashas to find work. The Falashas were also taken aback by the laxity of religious observance in modern Israel, since their form of the Jewish faith was very traditional.
>
> *Source*: *The Times* 27/05/1991, 28/05/1991, 29/05/1991.

on a scale commensurate with the size of the influx. In January 1990, the old absorption policy, reliant upon institutions, was phased out and replaced with 'direct absorption'. This involved giving a family of three, for example, a cheque for $2500 on arrival, plus $375 cash, six months' free health insurance, generous mortgage benefits, and free transportation from the airport to any destination within Israel. Once settled, a further payment of $8700 was made. The newly formed government body created to oversee absorption, known informally as the Aliyah Cabinet, also committed itself to a massive programme of housing construction, aiming to provide 45,000 new homes per annum between 1990 and 1993. Clearly, the Israeli state was willing to commit substantial amounts of its own (and American) money to attracting and settling Soviet Jews in order to strengthen the Jewishness of the Israeli state. Moreover, they did so despite both the economic

dislocation the programme created and the need to pressure other countries not to accept Soviet Jews. In March 1991, for example, the Israeli Foreign Minister asked the German government to stop issuing refugee status permits to Soviet Jews, and the Israeli ambassador to Austria requested posters be removed from transit centres informing Soviet Jews of resettlement opportunities in countries other than Israel.

International migration and economic engineering

For most countries, international migration policies are to a degree an adjunct of their labour policies. The direct link is often muted by additional migration components that arise from humanitarianism, foreign policy considerations or family reunion. However, in a number of cases the link between labour demand/supply and immigration/emigration is much more direct and economic imperatives are clearly driving state intervention in migration policy. States can either encourage the emigration or immigration of specialised labour, or they can seek to attract or export all types of workers. Equally, attitudes to illegal migration are also often shaped by labour needs.

Venezuelan migration policy in the 1970s provides an example of a country that sought, for economic reasons, to attract first any labour and later particular types of workers to fill labour market shortages. While Venezuela made use of legal migration, it chose not to move against illegal migration. The motor for such intervention was the dramatic rise in oil prices and revenue in 1972–74 and the greatly expanded development plan that was drawn up as a result. Even before this, private industry had complained of labour shortages and had lobbied the government on this point, backing up its arguments with research studies and making the issue one of national debate. The unions, too, had shifted from an anti-immigration to a pro-immigration position once they realised that national economic development was being stunted by the shortage of labour. In 1973, the government established an agency to recruit foreign labour (CORDIPLAN) and also signed bilateral agreements with the governments of Spain and Portugal, which had traditionally supplied it with immigrant labour. Agreements were signed with other Andean Pact countries to regularise the flow of illegals, who had added half a million workers to the population in the early 1970s alone (90 percent from Colombia). Those illegals who had already gained entry were offered an amnesty if they could prove they had a job. While there had been only 38,000 legalisations in the entire period 1960–72, the same number took place in 1976 alone, at a time when half the rural workforce were undocumented workers. Sassen Koob (1979) notes that, as a result of this raft of measures, the Venezuelan government hoped to add a net increment of 900,000–1,000,000 workers to the labour force over the period of the five-year plan.

Unconstrained immigration in the period between 1973 and 1976 did not however achieve all the objectives that the Venezuelan government had in mind. While the number of immigrants was certainly adequate, their distribution in the labour market was not optimal. The Latin American legal migrants often sought out professional work (many were refugees), the Mediterranean immigrants gravitated towards sales positions, and many of the former illegals remained working on the land. As a result, from 1976 onwards, the immigration policy was revised, making it more targeted (*ibid.*): the issuing of immigration permits was centralised rather than being delegated to overseas consular offices; the number of tourist visas was sharply reduced to prevent their misuse; work permits were issued to employers rather than potential employees; and employers were forced to consult with the government to ensure that no Venezuelan national was capable of filling the vacant post.

In Australia, incentives to immigrants were put in place less to fill specific and temporary labour shortages and more to increase the size of the total population beyond what was perceived to be the critical threshold of national economic viability. Australia's original search for immigrants goes back long before the Second World War, but many consider Arthur Calwell, Australia's first Minister for Immigration, to be its true architect. Calwell, appointed in 1945, was mindful of the forecasts of Australian economists who estimated that the country's population would peak in 1957 and then decline due to a falling birth rate. He knew that if such forecasts came true, dependency ratios would rise, as fewer and fewer young people entered the labour force, and the labour shortages that were already afflicting the coal, timber, steel and textile industries would become endemic. Calwell's response was to introduce a policy designed to increase Australia's population by 2 percent per annum, with 1 percent coming from immigration. This figure required net annual immigration of approximately 70,000 at a time when the Australian population numbered only seven million (the annual average for immigration for 1931–40 was just 3200). The policy of 1 percent growth through immigration was to remain in force until the 1970s.

Reflecting Calwell's cry of 'Populate or Perish', the Australian government waged an effective propaganda war to persuade the existing population that large-scale immigration was in everyone's best interests. For instance, it alluded to the possibility of invasion and noted the attacks that the Japanese had made on the north Australian coast during the Second World War. The policy was initially designed to attract Britons and other northern Europeans. To this end, subsidised transport was provided at £10 per ticket, with free transport for all British ex-servicemen and miners (Hawkins 1989). However, the inability of Australia to attract or even transport

Table 7.3 Net migration to Australia, 1921–75

Period	Average annual change
1921–1925	+36,654
1926–1930	+25,941
1931–1935	−192,177
1936–1940	+8,626
1941–1945	+1,562
1946–1950	+70,617
1951–1955	+82,765
1956–1960	+81,004
1961–1965	+79,978
1966–1970	+108,761
1971–1975	+68,901

Source: Hugo 1986.

70,000 Britons per year led Calwell to broaden the net in 1946/1947 to include people dislocated by the Second World War (displaced persons). Careful screening was introduced initially but pressure to fill the annual quota led to a retreat from selection, to the point where in the 1960s Turks were classified as honorary Europeans for immigration purposes. The success of Calwell's initiative was such that by the 1950s the net annual intake of migrants was 80,000; during the 1960s net migration rose to 109,000 per annum (Table 7.3).

The Australian population more than doubled between 1947 and 1982, from 7.5 million to 15.2 million, with post-war immigrants and their descendants representing 59 percent of the increase (Collins 1991). Between 1947 and 1985, five million people arrived in Australia as settlers (although more than one in five chose not to stay), with those from the United Kingdom and Ireland forming 39 percent of net migrants between 1947 and 1974 (Hugo 1986). The average migrant was five years younger than the average Australian, and post-war migrants took 61 percent of all newly created jobs over the period 1947–72, with immigrants forming 26 percent of the entire workforce in the 1970s (Collins 1991). Since then, however, there has been a realisation that rapid population growth through immigration does have consequences beyond the economy. For example, debate has developed in Australia about the impact of immigration upon the environment and the search for a sustainable population policy (*ibid.*). Blainey (1984) has been perhaps the most vociferous and controversial critic of the social/racial impacts of state-encouraged immigration. In his book *All for Australia*, he argued that the continued immigration of the 1980s was problematic

on two counts: first, it was taking place at a time of recession and could not therefore be justified on economic grounds; second, an increasing proportion of immigrants were Asians, when Australians had been led to believe that immigration was a means of maintaining an homogenous national identity. He noted the problems arising at a variety of scales from national to neighbourhood from taking immigrants who are culturally dissimilar and claimed that 'in many situations it might pay to bring in immigrants who are not necessarily the best in the world, but who adjust successfully to the existing society' (*ibid.*: 37). His views created a furore and what has since been termed 'The Great Immigration Debate' (Jupp 1995).

While all the above examples show how and why states might wish to encourage international immigration for economic reasons, other governments have sought to export labour. Bangladesh, for example, created a Bureau of Manpower, Emigration and Training (BMET) in the 1970s to service the growing international demand for cheap labour. BMET had responsibility for matching requests for labour with available workers and also took on the task of arranging transport, health checks and official permissions (Ali *et al.* 1981). An equivalent organisation in Pakistan, the Overseas Pakistani Foundation, even pro-actively sought to train people specifically for overseas appointments and at one point proclaimed a target of training 10,000 workers in 21 months (Shah 1983). Clearly, for states to become involved not only in the export of labour but also in the training of that labour, the returns must be considerable. Before the shift to South East Asian contract-tied labour in the early 1980s, the Gulf was a major destination for such migrants. The scale of South Asian labour migration to

Table 7.4 The scale of South Asian labour migration to the Persian Gulf, 1980.

Country	South Asian migrants	Percentage of non-national workforce
Saudi Arabia	65,500	8.4
Kuwait	84,700	22.4
United Arab Emirates	264,500	64.3
Oman	86,600	89.5
Bahrain	40,800	60.3
Qatar	38,000	47.3

Source: Robinson 1986b, reprinted by kind permission of Routledge.

Box 7.4 The advantages and disadvantages of further immigration to Australia

For

1 Growth *per se* is a desirable goal (the 'pursuit of numbers')
2 Expands internal market for Australian goods, especially housing – migrants have a greater propensity to consume than Australian-born.
3 Has multiplier effect on employment.
4 Expands skill base of labour force, relieves 'bottlenecks in labour market' (migrant labour force more mobile than Australian-born).
5 Expansion of refugee programme is a humanitarian act.
6 Migrants revive decaying parts of cities.
7 It is good foreign policy.
8 Ethnic/cultural heterogeneity and diversity enhance quality of life.
9 Has a 'younging' effect on age structure.
10 Enhances defence capability and international bargaining position.
11 Appeases ethnic lobbies.
12 Migrants facilitate structural changes.

Against

1 Puts extra strain on Australia's natural environment.
2 Exacerbates unemployment problems in an economy undergoing structural change and recession.
3 Prevents development of skill training within Australia.
4 Rapid population growth suppresses rather than promotes per capita economic growth.
5 Ethnic-based conflict could be encouraged by diversity.
6 Disproportionate costs are carried by poor and disadvantaged.
7 Concentration of migrants aggravates slum conditions, creates ghettos.
8 Pressure for expanded immigration comes from within immigration bureaucracy to ensure its own continued expansion.
9 Erodes Australia's export economy.
10 Causes increased competition, conflict and inequality within major urban areas.
11 Places strain on local government and welfare resources.
12 Migrants act against structural change by concentrating in declining industries.

Source: Hugo 1986.

---

the Gulf is clearly shown by Table 7.4, which demonstrates the size and proportionate importance of Asian labour by destination country.

While Table 7.4 is suggestive of the economic importance of immigrants for the destination economy, Robinson (1986b) has shown the return on migration for the source country. Here, the importance of seeing migration in a network context (Chapter 3) is clear. Robinson notes that, in 1982/1983, the Pakistani economy benefited from the inflow of £2.4 billion from expatriate remittances, representing 40 percent of its total foreign exchange earnings and 8 percent of its gross domestic product. In another study, Robinson (1986a) concentrated upon the value of remittances for the Indian economy, using Reserve Bank of India data to show that total personal remittances from Indians overseas in 1978–79 were of the order of £625 million.

The policies that governments introduce to combat illegal immigration, and the enthusiasm (or lack of it) with which these are enforced, can also have economic motives. Several authors have argued this

to be the case with illegal immigration across the Mexico–United States border in the 1970s (Bach 1978; Jenkins 1978; Portes 1978b). Adopting a structuralist perspective (Chapter 3), these authors believed that immigration was allowed to persist in the 1970s because of the needs of capitalism. Two possible explanations were proposed for the use of immigrants in agri-business: first, that there was a shortage of indigenous labour and, therefore, migrant workers were used as replacement labour; second, that immigrants were used because they can be 'super-exploited' in industries where the return on capital is so small as to be negated by the use of legal unionised labour.

More specifically, Portes (1978b) argued that agri-business in the south western states relied upon super-exploited labour, especially from Mexico. Welfare costs, world competition, labour organisations, the seasonality of labour demand, and very labour-intensive production methods all mean that the real rate of return on capital is slight unless extremely cheap labour is obtained and super-exploited. Portes also argued that migrant labour is not just super-exploitable because it is more docile or compliant but also because all migrants are vulnerable after crossing an international boundary. New to a country, they can be threatened with deportation or with having their work permits withdrawn, or they can be labelled as subversives or troublemakers. Furthermore, migrant labour offers benefits to the US economy overall, as well as to the individual employer:

immigrant workers have also regulated the level of class conflict directly by undercutting the collective actions launched by domestic worker organisations and indirectly by diverting the class hostilities rooted in the economy onto alien scapegoats.

(Jenkins 1978: 525).

Finally and more immediately, immigrant workers can be dispensed with when economic conditions deteriorate, avoiding the need to provide welfare or unemployment benefits. They are educated and pass through the non-productive phase of their early life in Mexico, and may well return there at the end of their working lives, so displacing any retirement and health care costs. Moreover, while this is true for *all* Mexican migrants to the United States, undocumented migrants offer even more benefits to the employer by virtue of their very illegality:

the act of surreptitiously crossing a political border places the worker in a still weaker position than legal immigrants, since it formally confronts him with the enforcement apparatus of the receiving state. ... This juridical weakness is 'appropriated' by employers in the form of a higher rate of return.

(Portes 1978b: 474).

Illegal immigrants are thus even more vulnerable than legal migrants, since they can be threatened with exposure or even sent home at will, and it is highly unlikely that they will seek to organise themselves publicly to campaign for better wages or conditions.

For these various reasons, agricultural employers prefer to use illegal Mexican immigrants than indigenous workers or legal migrants. Their problem, however, has been to maintain the permeability of the border, thereby ensuring a continuous supply of super-exploitable labour. Jenkins (1978) described how the political power of agri-business in the south west was such that local politicians both blocked measures which would have strengthened the border patrol and ensured that employers were exempted from legal sanctions for employing undocumented

workers (the Texas Proviso). Calavita (1995) recounted how the head of the Immigration and Naturalization Service in Arizona was regularly ordered to stop deporting illegal Mexican labour during the harvest period; Bach (1978) commented on the selective enforcement of immigration law; and Jenkins (1978) pointed out that, by targeting efforts on the border zone rather than the workplace, government was actually maintaining an illusion of threat, thereby heightening the sense of vulnerability of those who succeeded in getting into the United States and making them more open to super-exploitation.

International migration and political engineering

Selective intervention by governments to stimulate or discourage international migration can be undertaken to meet geo-political or strategic ends. Certain populations might be allowed entry or even encouraged to migrate as a way of confirming the unpopularity of the regime in the source country, or to strengthen political, military or trade alliances. Immigrants might be used as military personnel or as settlers to bolster occupation of a contested zone. Equally, other groups might be shunned, since to allow entry might involve either tacitly undermining a friendly regime or admitting people with views contrary to the prevailing norm.

An exemplar of such intervention in migration outcomes for geo-political reasons comes from the US government's contrasting attitude to migrants from Cuba and Haiti. Zucker and Zucker (1994: 1) summarise the situation as follows:

while many from the Caribbean migrate to the United States, the United States has singled out Cubans and Haitians for diametrically opposite treatment. Cubans who quit their island are assisted in coming to the United States, are called political refugees, and are given asylum, while Haitians who leave their island are labelled economic migrants, interdicted at sea, and returned to Haiti.

Explaining this apparent paradox is simple. Immigration decisions depend on the troika of economic costs/benefits, foreign policy considerations and the strength of the lobby groups supporting putative migrants. While there are undoubtedly differences in the lobby power and costs of resettlement for the two groups, the critical distinction for the Zuckers is foreign policy. American foreign policy has been dominated by an anti-communist stance since the Second World War and immigration policy has tended to follow suit. Those seeking to flee from a communist country are welcomed, since by leaving countries such as Cuba they condemn communism and validate the superiority of American democracy. In contrast, Haiti was until very recently a right-wing dictatorship with a tradition of undeviating support for the United States and condemnation of Cuba. Haiti even allowed its military bases to be used by US personnel during the Cuban crisis of the 1960s. As a result, when Haitian migrants started to flee the Duvalier regime from 1972, the US Immigration Service returned them. When numbers increased significantly in 1979, the authorities instituted a programme of detention and deportation without due process. Despite a court ruling which declared such a programme to be offensive, anti-constitutional and illegal, successive administrations continued to implement similar programmes until 1991. Indeed, in 1981, the US government held discussions about migration with the 'friendly' government of Haiti, during which it was agreed that the US Coast Guard would intercept all Haitians fleeing by sea and return those who had no credible claim to asylum. During the next decade, nearly 22,000 Haitians were interdicted at sea, but only 28 were allowed into the United States to pursue their application for asylum. In May 1992, the United States went even further when it announced the Kennebunkport Order, by which all Haitian migrants leaving by sea were to be interdicted and returned without any form of screening for suitability for asylum. All potential emigrants from Haiti had to apply through the US Embassy in Haiti, which was guarded by Haitian military and which found in favour of only 136 applicants in 1992. In contrast, between 1959 and 1994, nearly one million Cubans were resettled in the United States and most were given *proforma* asylum.

Politically motivated intervention is not just a recent phenomenon. Indeed, one of the most comprehensive attempts to influence migration flows was provided by inter-war Fascist Italy. While the state had previously encouraged emigration as a solution to unemployment, from 1927 onwards attitudes and policies changed, as Mussolini declared that 'to count for something in the world, Italy must have a population of not less than 60 millions when she arrives at the threshold of the second half of this century' (Ascension Day Speech, 26 May 1927). He drew unfavourable comparisons between Italy's population of 40 million and those of Germany and Britain (and the Empire). He saw a link between

population size and political, economic and moral influence, with emigration viewed as a loss of Italian subjects. Measures throughout 1927 and 1928 curtailed the emigration fund, abolished cheap rail fares to embarkation points, introduced a requirement for emigrants to have an official contract of employment from a foreign employer before departure, and made potential emigrants sign a declaration that they would not subsequently send back for their relatives to join them overseas. Punitive measures were also brought forward to discourage illegal emigration, including fines and periods of imprisonment for those caught, and increased powers for border guards to shoot escapees.

In parallel with these measures, renewed efforts were made to persuade overseas Italians to return, and to encourage others to maintain their Italian culture while abroad and their physical links with the homeland. The travel subsidy for returnees was increased, free transport was provided for pregnant women who wished to have their children born in Italy, and children were given free travel and holidays in Italy. Government money was used to establish Italian schools overseas, to subsidise Italian language newspapers and to found welfare organisations in the satellite communities. Glass (1967) charted the changing scale of Italian international migration but noted that it is almost impossible to separate the impact of government policy from other influences. Nevertheless, Table 7.5 documents the dramatic decline in net migration loss experienced by Italy between 1926 and 1936, while Fakiolas (1995) documented how the post-war period saw Italy become a country of emigration again.

Table 7.5 Population loss through net migration, Italy 1926–36.

Year	Net loss
1926	85,000
1927	78,000
1928	51,000
1929	34,000
1930	na
1931	58,000
1932	10,000
1933	17,000
1934	19,000
1935	18,000
1936	9,000

Source: Glass 1967.

International migration and social engineering

In some cases, selective migration can impact relatively quickly upon national characteristics, such as socio-economic characteristics. Nowhere is this more true than for small islands or recently inaugurated communities. The social experiment that produced the South Australian population is a good example of the latter (Hutchings and Bunker 1986), while the Isle of Man exemplifies the former. While the Isle of Man's history prior to the late 1950s is fascinating (Freke 1990; Kinvig 1975), its demography was essentially shaped by the island's peripherality and narrow economic base (Robinson 1990a) and

was traditionally characterised by emigration and population stagnation or decline. However, since the 1950s, the Manx Parliament (Tynwald) began to re-assert itself and chart an increasingly independent future for the island, often based upon policies with no counterpart in the United Kingdom (Quayle 1990).

Central to the island's future was the decision to engineer both an increase in the resident population and the attraction of wealthier migrants. Changes in legislation allowed the island to become a tax haven, attracting wealthy elderly immigrants and companies seeking relief from taxes. By the late 1960s, the success of the migration drive (Table 7.6) had been such that dis-benefits began to appear.

Box 7.5 Key interventions in population policy by the Isle of Man government, 1960–90

- The Isle of Man's first policy intervention, during the early 1960s, to encourage the immigration of wealthy immigrants was to reduce income tax, end surtax and abolish death duties. Collectively they formed the New Residents Policy. A low-tax regime has been maintained to the present day.

- A government office was created to stimulate immigration and advertise the advantages of life on the Isle of Man, to attract both inward investment and immigration. The unit has, for example, advertised the island's low crime rate, attended national and international exhibitions, produced a video about the quality of life on the Isle of Man, and organised briefings for potential migrants. The unit was particularly active during the period February 1985 to June 1987, during which a new New Residents Policy was pursued.

- A 1968 commission investigated population matters and recorded popular concerns about the threat to Manx culture posed by immigration, the impact of population growth upon the housing market, the changing nature of the labour market, and the financial repercussions of attracting elderly immigrants. However, it recommended further immigration and argued that the island's demographic problems stemmed from the emigration of the young not the immigration of the old. It did not demur from a view expressed elsewhere in the Manx government that the island was capable of accommodating a 50 percent increase in population.

- A 1973 report on immigration summarised the dichotomy of opinion that existed on the island towards the New Residents Policy. It also revealed first analysis of the 1971 census, which showed that the largest percentage of immigrants were

aged 20–24 years and that immigration had actually served to re-balance the island's age structure. The committee recommended no change in overall policy but suggested that immigrants should invest a lump sum in a government bond, the proceeds from which could be used to fund social housing for the indigenous population.

- A 1980 report on population growth and immigration raised two new issues: the notion of a two-tier society, and the impact of further population growth on quality of life (including environment). A committee had tested popular stereotypes of immigrants being elderly retirees against the findings of the 1976 census and discovered that the new residents actually included returning Manx men and women, that they were over-represented in the workforce, that they contributed a greater *per capita* share of the government tax take than did indigenous peoples, and that the island's average age had been lowered by the arrival of immigrants. The resulting report recommended further immigration up to a population target of 75,000 people.

- A 1982 report stressed that new residents had introduced skills, expertise and experience that were often lacking in the indigenous population, and that these qualities were now the driving force behind whole sectors of the economy. Local people had been offered broader economic opportunities through the arrival of immigrants.

- A 1985 report argued the case for a 75,000 population target.

- A 1987 policy document confirmed the wisdom of encouraging immigration but commented on the issue of regional equity and balance.

Source: Robinson 1990a.

Table 7.6 The Isle of Man's changing population since the introduction of policies to stimulate immigration, 1960–86.

Years	Population at start	Absolute change	Net migration	Net migration as percentage of change
1961–1966	47,166	+2146	+2815	131
1966–1971	49,312	+3916	+4677	119
1971–1976	53,228	+7268	+8346	115
1976–1981	60,496	+4183	+5606	134
1981–1986	64,679	−397	+1002	

Source: Robinson 1990a, reprinted by kind permission of Liverpool University Press.

There was a debate about the balance to be struck between increasing population size still further (to acquire economies of scale in providing infrastructure), quality of life and the maintenance of Manx culture. By 1976, Manx people noted that for the first time in recent history more than half of the island's population had been born elsewhere. Fears grew that the island was developing a two-tier society, which would disadvantage locally born people. Indicative of these strains has been the actions of FSFO (Financial Sector Fuck Off), which resorted to symbolic arson attacks on executive homes being built for 'incomers', the rebirth of Manx Gaelic as a living language and the decision of the Tynwald to support indigenous culture through the educational system and dual-language naming of streets. Clearly, manipulating international migration can have a dramatic impact upon small populations, with unintended and arguably deleterious impacts upon that society.

Engineering internal migration

Internal migration and ethnic/racial engineering

While state intervention in international migration is usually deliberate and is targeted at achieving a pre-determined and precise outcome, many interventions in internal migration are often indirect, unintended and vague. Nevertheless, national and local governments have intervened directly in patterns of ethnic or racial migration and distribution for four main reasons: to separate different racial groups in order to avoid racial conflict and disharmony; to maintain racially based power structures and the stereotypes that underpin them; to bring racial groups into closer proximity in order to increase mutual understanding and therefore racial harmony; and to defend or control valued resources by denying access to an out-group.

The former apartheid system in South Africa represents perhaps the most notorious example of many of these processes at work. While 'apartheid' simply derived from the Dutch word for 'separateness' or 'apartness', by the twentieth century it had acquired different connotations and more complex symbolic meanings. Christopher (1994: 1) described how apartheid came to mean 'a legally enforced policy to promote the political, social and cultural separation of racially defined communities for the exclusive benefit of one of those communities'. Much of this political, social and cultural separation was predicated on physical separation and racial zoning. When the National Party came to power in 1948 it either introduced new legislation to achieve these goals or strengthened existing legislation. The outcome was a comprehensive raft of laws that enforced separation at all levels, from national to household. Grand apartheid saw the engineering of separate tribal homelands and the exclusion of non-whites from the national political process; urban apartheid involved mandatory urban segregation of population and businesses; and petty apartheid created a parallel but unequal set of amenities for the different races and outlawed intimate social interaction between the races.

Grand and urban apartheid had the greatest impact upon internal migration, especially the 1950 Group Areas Act and the 1959 Promotion of Bantu Self-government Act. The latter sought to allow the white minority to retain their position of privilege at the pinnacle of the political and economic systems by fragmenting the black population into incipient nations, none of which had the economic or numerical power to prevail over the whites. A finely graded classification of all races, introduced in the 1950 Population Registration Act, became embedded into every aspect of official policy. The black population was split into ten separate 'national units' defined by language and, therefore, by supposedly pre-colonial tribal affiliation. These units were then allocated to their own territory (bantustan), which were carefully drawn both to exclude any mineral wealth and to form labour compounds for nearby white industry. Certain groups, such as the Zulu 'nation', were further fragmented by being allocated over 100 separate blocs of non-contiguous land. Within the bantustan, blacks were supposedly self-governing and had their own political systems with presidents, civil servants and state capitals, thereby obviating the need to give them representation in the parallel system of 'white' South Africa. The counterpoint to this contrived independence was that black national units should vacate the white urban 'homelands' and move to their own homelands. Relocation was engineered in a number of ways, but the net effect was that 1.7 million blacks migrated between 1960 and 1983. They faced varying conditions, with some moving to reception areas that had received little prior preparation, others moving to government-built resettlement villages, and others finding themselves in vast and under-provided resettlement camps. Such migration not only imposed psychic costs through the loss of homes and land and the break-up of communities, it also forced the migrants to bear economic costs. Many continued to work in the same place but were forced to commute much longer distances: the number of black commuters travelling into Pretoria by rail, for example, rose by 43 percent between 1969 and 1987.

The Group Areas Act sought to achieve similar separation of the races within cities, although such an objective had already been pursued in the colonial period (*ibid.*) and the inter-war period (Mabin 1992; Parnell 1988). The Land Tenure Advisory Board was given the job of partitioning South African cities so that zones were created for the exclusive occupancy of the different racial groups. Policies varied between cities. In some, internal zoning (Figure 7.1) and population exchange took place, while in others the entire city was declared white and other races were relocated to greenfield sites. Central business districts were almost always allocated to whites, who also enjoyed the better locations and sufficient land to live at low densities. The other races had to make do with peripheral lo-

cations, which were often too small for the population they were expected to accommodate. Implicit in this racial zoning was the fact that people would have to migrate to those neighbourhoods allocated to their race and leave those neighbourhoods that were presently mixed.

It is hard to calculate the number of people who were displaced by urban apartheid. Official estimates of 124,000 families displaced by the Group Areas Act are likely to be gross underestimates, since other legislation, such as the 1920 Housing Act and the 1934 Slums Act, was also used extensively to engineer segregation. Christopher (1994) suggests that three-quarters of a million blacks were relocated from inner urban areas alone, creating peri-urban townships such as Soweto and Crossroads (D. Smith 1982). The policy of urban apartheid was implemented throughout South Africa with such zeal that 'the administrators of apartheid planning ... had achieved virtually total segregation in residential patterns in most South African cities by the mid-1980s' (Christopher 1994: 132), with only 8.6 percent of the urban population living outside their designated areas by 1991.

Christopher (1987) uses the example of the city of Port Elizabeth to show the quantitative impact of population relocation upon one city and upon specific neighbourhoods. He describes how the black population of inner area Korsten was reduced from 17,200 in 1951 to only 2200 nine years later as the authorities cleared them to make way for 'coloureds'. In contrast, New Brighton, which had been designated a black area, saw its population of blacks rise from 35,000 to 97,000 over the same period. Christopher also shows the impact of changes such as these on the racial geography of South African cities (Fig 7.2) (page 172). Both Reintges (1992) and Western (1981) provide other examples.

While South Africa represents a well publicised and highly overt attempt to engineer racial settlement patterns in cities, it is not unique. Massey and Denton (1993), for example, make clear that American 'hypersegregation' is entirely intentional, Robinson (1997) has shown that the British government has manipulated the internal migration of refugees to produce desired geographical outcomes since 1973, and Flett, Henderson and Brown (1979) show that British local councils have also engineered desired patterns and levels of urban racial segregation.

An Ideal Apartheid City

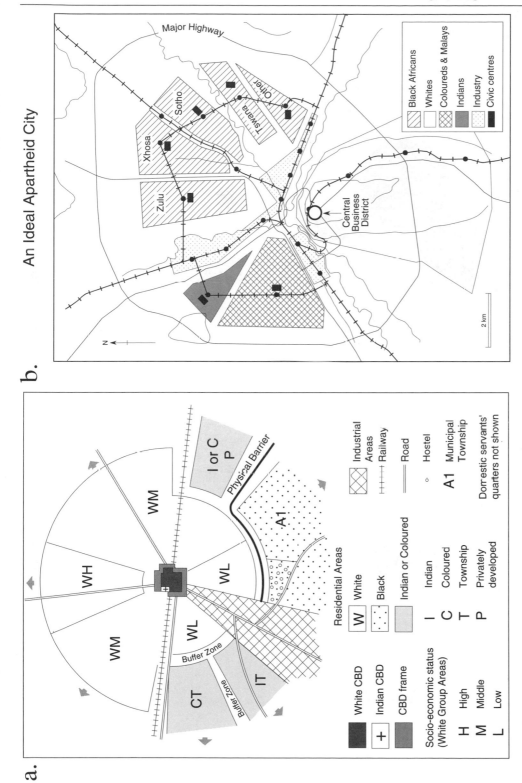

Figure 7.1 Models of the apartheid city (source: a. Davies 1981; b. Western 1981, reprinted from *Geojournal*, **Supplementary issue 2**, Davies. R, The spatial formation of the South African City. pp. 59–72, © 1981, by kind permission of Kluwer Academic Publishers.).

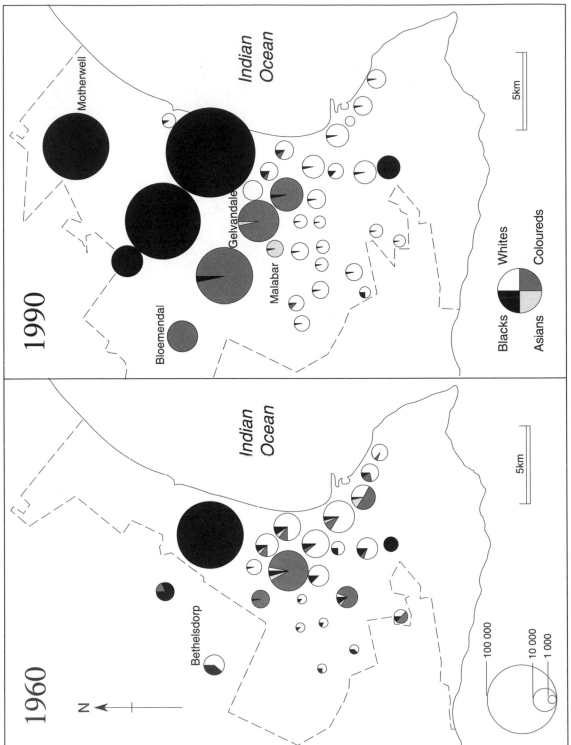

Figure 7.2 Distribution of racial groups in Port Elizabeth, South Africa, 1960 and 1990 (source: A. J. Christopher, Apartheid Planning in South Africa: the case of Port Elizabeth, *The Geographical Journal*, **153** (1987): 198. Reprinted by permission of the Royal Geographical Society).

Internal migration and economic engineering

Economically motivated internal population redistribution can be exemplified by the villagisation programme of President Mengistu of Ethiopia in the late 1980s and early 1990s. Described by a contemporary commentator as 'the largest government-organised movement of population in the world today' (Luling 1989: 34), its rationale was provided by Mengistu in a public speech in 1988:

I can create conditions conducive ... to controlling farmers. Then, having gained control over farmers, it is possible to decide and implement the level of productivity throughout the country, and to take disciplinary action against those failing to implement the decisions.

(quoted in Luling 1989: 34–5).

The programme sought to move the entire rural population of Ethiopia (38 million people) into planned centralised settlements. It was argued that this would improve access to agricultural extension services, thereby increasing productivity; encourage the rational use of land; conserve natural resources such as water; and give the rural population easier access to welfare and health services. Twelve million people had been moved by 1988.

Villagers were told to dismantle their (wooden) homes and transport them to new village sites laid out in a formal grid-iron pattern. Some of the new villages contained as few as 200 people, while the largest housed 2500. Each house was given an individual garden plot. Luling (1989) reports that little violence was used to get people to move and that few resisted the relocation, although Clay (1986) disagrees. There were a number of unintended economic consequences of the policy. Since most farmers continued to farm their old landholdings, the average distance walked each day before work could begin increased sharply to over 4 kilometres in the case of the Hararge administrative region. This had a negative effect on agricultural productivity, as did the need to drive animals from the village to the land each day. Furthermore, since the farmer was not always present to scare off predators, these took a greater share of output. Luling also records the social consequences of villagisation. Because farmers had to transport their own homes to the new villages, they tended to take the minimum of building materials with them and therefore erected less substantial houses. In addition, whereas traditional villages were constituted of kin groups, the new larger villages meant that neighbours were not necessarily kin and could not therefore be relied upon to help in times of need. Finally, farmers regretted having to leave their old hamlets and their rocks and trees, which had acquired sacred associations.

Indonesia's *transmigration* programme is another contemporary mass migration inspired by government wishes to stimulate economic development. Sixty five percent of Indonesia's 168 million people live on the island of Java at densities of up to 2000 persons per square kilometre, while the outer islands of Sumatra, Kalimantan and Sulawesi have vast tracts of land occupied at very low densities. The corrective to this imbalance has been described as 'the largest voluntary land settlement scheme in the world' (Leinbach 1989: 84). The main objectives of the programme are

the improvement of living standards and employment opportunities, stronger and more equitable regional development, achievement of a more balanced population distribution with the alleviation of population pressure in Inner Indonesia, and increased food and tree crop production. Mentioned also were the ability of the programme to foster nationalism and ethnic harmony.

(*ibid.*: 91).

Although policy has changed between the different Five-Year Plans that began in 1969, the net effect has been the same. The years 1950–87 saw 2.51 million people sponsored to relocate and over 6000 square kilometres of forest cleared to accommodate them. The World Bank had invested $439 million by 1989, the road network in the outer islands had been enhanced by a factor of nearly a half and airstrips had been built, as had numerous settlement clusters with their schools, health clinics, religious facilities and market places. The programme had created over half a million permanent jobs and as many temporary posts.

However, the relocation policy did not only involve sponsored transmigrants (*umum*), who were given extensive government support for up to five years, cleared land, uncleared land, housing and access to transport and social services. Three other groups were also involved, although numbers are less well recorded. First, local people living near the areas targeted for in-migration were also entitled to migrate and received some assistance to do so. There were also two groups of spontaneous transmigrants: the *swakarsa* were registered and, although they moved at their own expense, they also received land, access to services and credit (not direct subsidies); and unregistered spontaneous migrants were allowed to choose their resettlement destination and specific site but were granted no further assistance. Given that there is a four-year waiting list for official recognition as a transmigrant, many people choose to move as spontaneous migrants. Under the Third Five-Year Plan (*Repelita III*), spontaneous migrants formed 32 percent of all families moving, forming 60 percent of the total by 1987.

Although the programme has achieved some of its broad objectives, it has not been without problems or controversy. Leinbach (1989) noted the successes of redistribution, with Java experiencing a decreasing share of gross domestic product, agricultural production, and manufacturing activity. He also pointed to the failures. Some of these are environmental, such as the massive damage created by the clearance of tropical rain forest and the irreversible degradation of land after clearance (8.6 million hectares are now unusable). Other failures are ethical, such as the treatment of the indigenous population and their loss of land rights. For example, there is the link between migration to East Timor and the suppression of the independence movement there after the 1976 annexation by Indonesia. Still more failures are practical. Research has shown that many transmigrants end up on lower incomes than the average for the area from which they come. The government's choice of re-settlement model (for example, upland food crop, aquaculture and tree crops) has not always been suited to the natural environment and migrants have not been selected for their innovativeness or enterprise. Partly in response to these problems, the government is now putting much greater emphasis upon the long-term support of existing settlements rather than the creation of new ones.

Internal migration and political engineering

Administrations intervene widely in migration to support a particular ideology. Such intervention is characteristic of both ends of the political spectrum and, even within one country, successive religious/political ideologies can generate quite different migration regimes, as the example of China clearly demonstrates.

The traditional Confucian perception of internal migration in China is concisely summarised by Li and Li (1995: 149):

In traditional Chinese society migration from the ancestral birthplace was frowned upon. People were expected to stay

for life near their ancestors' graveyard. In addition, a high volume of interregional migration indicated social unrest. Chinese were indoctrinated with such concepts in their youth. For example, one of the most well known Chinese classics, *The Tale of Three Kingdoms*, repeated the famous passage, '*Liumin si-chuan, Tien-xia ta-rang*' which means that as migrants move in various directions, the country would be in turmoil.

As a result, what little is known about internal migration within pre-communist China suggests that permanent movement was unusual, with those who had undertaken such moves returning to their place of origin in later life. This followed a Confucian belief that 'falling leaves return to their roots' (*Ye lo hui gen*).

More recently, Maoist thinking stressed the need for an equal distribution of population and industry, a rapid reduction in urban population growth and, ultimately, the abolition of the distinction between urban and rural – a similar emphasis was given in Ceaucescu's Romania (Lambert 1989). Consequently, internal migration was tightly controlled, with migrants having to register their move in order to find employment and housing (Shen 1996). Few people succeeded in gaining permission and most population movement was inspired directly by the government for strategic or political purposes. For example, Han settlers were sent west from the seaboard to settle newly conquered Tibet and garrison its towns (Clarke 1994), and Li and Li (1995) describe how 'millions' of youths were temporarily sent out of towns in the 1960s to learn from peasants as part of the Cultural Revolution. In addition, during the Second Front period, industry and population was moved away from the eastern seaboard, where it was felt to be in danger from American seaborne invasion, and relocated in the west, where it could also help reduce regional disparities.

Following Mao's death in 1976 and his eventual replacement by Deng Xiaoping, market-led socialist ideology was introduced. Collective agriculture was replaced by household farms which have become increasingly market-oriented, industry was modernised and privatised, markets were slowly opened to foreign collaborators, cottage industry was established in many villages, and the policy towards regional balance and economic growth was radically overhauled. Lo (1989) noted how the Seventh Five-Year Plan marked a shift in thinking in which the eastern seaboard was to become the modernised economic powerhouse that would drive the Chinese economy. Special Economic Zones would become centres from which skills and innovations would diffuse. The growth of the largest cities was to be curtailed and medium-sized cities were to be developed, but the most active growth was to be directed at the small cities and towns, which would help eliminate the gradient that existed between urban and rural, and capture people seeking rural-to-urban migration before they moved to a major agglomeration. Population was to be encouraged to move permanently from large to small urban places and from small urban places to rural areas (Yang and Goldstein 1990). However, the major urban centres still required migrant labour to fuel their economic growth and transformation, so the restrictions on temporary migration were reduced. Relying on temporary migrants with no right of local residence has several advantages. For instance, the flow of such people can be regulated precisely in tune with urban and rural conditions, and

Within the political and economic system in China, these temporary residents will not burden the cities' educational facilities, employment and subsidized or rationed food supplies because they do not change their household registration and hence are not entitled to government assistance in these areas, as official residents are.

(*ibid*.: 531).

The consequence of relaxations on temporary migration have been profound. Approximately 15–20 percent of the population of most big cities are now temporary residents, with Beijing alone hosting over one million circulators (the 'floating population') in 1986 (*ibid*.).

Israel, too, has engaged in policies that encourage internal migration for political reasons and that are only made possible by a strong centralist state. In Israel's case, such intervention has been linked inextricably with international migration. We have already seen how Israel has courted overseas Jews. While one of the reasons for this has undoubtedly been to bring fellow Jews together in the same country, another has been the strategic imperative of defending the state of Israel, through land settlement. Cohen (1992) argues that, since the creation of Israel, the struggle for land has been at the core of the Israeli–Arab conflict, with Israel progressively annexing territory around it – the West Bank and Gaza were taken in 1967 and formally annexed in 1980 – in an effort to secure the future of a viable state, defensible from military attack. Newman (1985) even sees the policy of deliberate colonisation as a key factor determining the delimitation of the

Israeli state in 1948. The Alon Plan of 1967 sought to capitalise on these gains by shifting the demographic balance of the captured territories from being non-Jewish to Jewish. Central to this was the building and peopling of new Jewish frontier settlements and the progressive exclusion or relocation of the indigenous Palestinian populations (Figure 7.3) (overleaf).

Since 1948, the Absentee Property Regulations and Military Order Number 58 have been used to expel over two million Palestinians and expropriate their land, which was deemed to have been abandoned by its owners. As a result, by 1990, Palestinians were able to build on only 30 percent of the West Bank, and the percentage of Arabs of Palestinian birth or descent who lived in the West Bank had fallen from 34 percent prior to 1967 to 16 percent in 1992 (Abu-Lughod 1995). In their place, between 1948 and 1967, Israel attracted about one million Sephardic Jews from Morocco, Algeria, Tunisia and other Arab states (Cohen 1992). Since 1967, over 120,000 Israeli Jews have moved to 130 new settlements on the West Bank, thereby demographically strengthening a new eastern border for Israel (Kellerman 1993). Two sites have become especially significant for this demographic war. First, the Green Line – the 1948 international boundary – has been deliberately built over, with seven new cities planned to straddle the frontier. These will act as buffers between the Arab parts of new Israel and the Jewish parts of old Israel, and as bridges between old Israel and the settlements in occupied territory.

East Jerusalem has also become ideologically and demographically important. In 1990, the occupied sectors of east Jerusalem had a finely balanced demographic mix, with 120,000 Jewish inhabitants and 140,000 Palestinian inhabitants. Since that time Israel has done much to make Jerusalem its eternal capital by attracting Jewish migrants. Although in Jerusalem, as elsewhere, Israel has achieved many of its objectives by relocating the ultra-orthodox to new frontier settlements, existing citizens have been insufficient in number to maintain colonisation. The authorities have therefore turned increasingly to Soviet Jews as extra settlers. Indeed, in October 1991, the Mayor of Jerusalem declared:

the solution is to bring as many immigrants to the city as possible and make it an overwhelmingly Jewish city, so that they will get it out of their heads that Jerusalem will not be Israel's capital.

(quoted in Hiltermann 1991: 84).

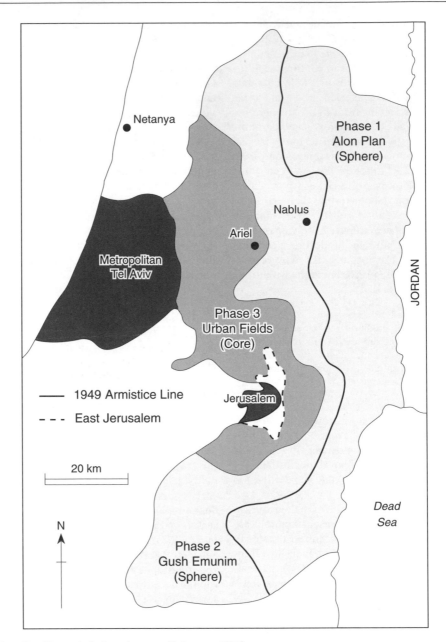

Figure 7.3 Israeli settlement strategy (source: Kellerman 1993).

To this end, Hiltermann (1991) describes how the city authorities established an Immigrant Absorption Project to absorb directly new arrivals in Israel, the Jerusalem Development Authority has broadened its remit to include absorption matters, and how the city approved six new Jewish neighbourhoods in east Jerusalem with 16,000 apartments to house 56,000 Soviet Jews. Soviet Jews have become critical to the expansion and affirmation of Israeli territorial hegemony. They have even been termed 'settlement fodder', with Cohen (1992) noting that the majority of the 15 percent increase in the number of Jews living in occupied territory in 1990 were Soviet Jews and Hiltermann (1991) recording

Box 7.6 The Westminster Council 'homes for votes' scandal

A recent example of an alleged attempt to engineer internal migration for political gain comes from the highly controversial case of the actions of certain members of an inner London borough council in Britain. The story is told here in a number of newspaper excerpts covering the subsequent investigation into the scandal:

The scandal has its origins in the 1986 local elections, when the Tory majority was cut to four. As the votes were being counted, Patricia Kirwan, the housing committee chairman, was confronted by Dame Shirley [Porter, the then Westminster City Council leader], and told: 'It's your policies which have nearly lost us this election'. Dame Shirley clearly thought that Mrs Kirwan's policies of encouraging developers to carry out part-private, part-council projects had increased the number of Labour-voting council tenants. She and her advisors devised the plan to reduce the number of Labour voters by selling council homes to potential Tory voters.

Council officials ... drew up plans to list 10,000 homes for sale, targeting properties in eight 'battle zone' wards, three held by Labour and the rest by the Conservatives. By the time the plan was presented to the housing committee in July 1987, it had been widened to cover much of the borough. Councillors agreed to designate 9,360 properties for sale and set a target of achieving 500 sales a year and to provide grants to existing council tenants to encourage them to move out.

Mr Magill [the Westminster district auditor] concluded 'My provisional view is that the council was engaged in gerrymandering.'

The Times 14/01/1994.

Westminster council Conservatives instructed their chief officers to 'be mean and nasty' towards the homeless to reduce the number of families they had a statutory duty to house.

A confidential Tory group strategy paper ... said action was needed to reduce the impact of the homeless families in the marginal target wards ...

Another group paper identified the factors 'contributing to the drop in natural support', including the provision of hostels for 'homeless/down-and-outs who are not our natural supporters'.

A paper by another councillor recorded the objective of the housing policy as ensuring that Westminster's residential mix enabled the Tories to keep control of the council and 'to retain the Conservative majority in the parliamentary seat of North Westminster'.

The Guardian 15/01/1994.

Efforts to build low-price flats for rent to poorer families in Soho, central London, have been frustrated at every turn by Westminster councillors and officials.

... internal files seized by John Magill ... suggest the council had effectively set up an exclusion zone barring housing associations from being able to build cheap homes in key wards in the centre of London.

The Guardian 20/06/1994.

The charge is that Westminster council neglected its statutory obligations to house the homeless in its zeal to see council houses sold to potential Tory voters. Mr Magill argues that by designating certain properties for sale, the Council kept the homeless out of them so they could remain available for any prospective buyer.

... but the whole matter requires more context than has generally been given. It is legitimate, arguably admirable, for any council to encourage home ownership in an inner-city area. There may have been a political benefit for Conservatives in increasing the number of homeowners but there was also a political benefit for the Labour leader of the London County Council, Herbert Morrison, in studding the city with council estates.

The Times 10/05/1996

---

the arrival of 5400 Soviet Jews in Jerusalem in the first six months of 1990 alone. However, despite all of these and other efforts, the 1990s have seen an ironic reversal of perceived causality. Some now claim that Israel is only expanding as a response to the needs of the unprecedented and uncontrollable flood of migration from the former Soviet Union to Israel.

Internal migration and social engineering

It is not difficult to find examples of where administrations intervene specifically in internal migration in order to bring about some desired social outcome. A fine example of this is provided by successive post-war British governments and their attempts to solve inner city problems through the decanting of population to new towns or expanded towns. Post-war policy was predicated upon a much older axiom that low-density living in semi-rural surroundings was socially, morally and physically preferable to residence in large cities (Chapter 6). Ebenezer Howard was by no means the first to express this view, but he was one of the first to suggest practical ways in which the urban populations might enjoy the benefits of rural living, and vice versa. His celebrated 'Three Magnets' diagram (Figure 7.4) (overleaf) demonstrated what he perceived to be the advantages of living in both urban and rural areas, and also suggested how these could be combined in a new rural–urban entity known as the garden city. Such cities would house populations of 32,000 at low densities on a site of approximately 2400 hectares, much of which was to be given over to green spaces, gardens, parks and productive agricultural enterprises (Howard 1946 [1898]).

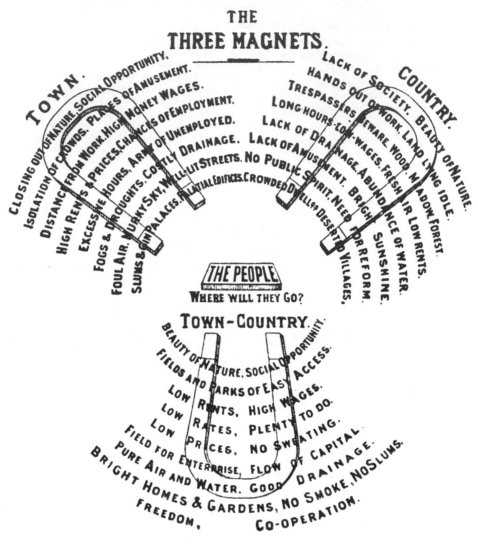

THE THREE MAGNETS

Figure 7.4 Ebenezer Howard's 'Three Magnets', illustrating the perceived benefits of rural, urban and garden city living (source: Howard 1946).

Garden cities were thought to solve many social problems:

the key to the problem how to restore the people to the land – that beautiful land of ours, with its canopy of sky, the air that blows upon it, the sun that warms it, the rain and dew that moisten it ... is the key to a portal through which, even when scarce ajar, will be seen to pour a flood of light on the problems of intemperance, of excessive toil, of restless anxiety, of grinding poverty.

(*ibid.*: 44).

Other commentators drew attention to different social problems that could be alleviated by mass voluntary migration to garden cities. For example, Mumford, writing at a time of concern over the falling birth rate in the UK, argued:

the task of our age is to work out an urban environment that will be just as favourable to fertility, just as encouraging to marriage and parenthood, as rural areas still are ... the sort of city he [Howard] projected was precisely the kind whose population will be biologically capable of reproducing itself and psychologically disposed to do so.

(1946: 38).

Even the chairman of the London County Council in 1891, who might have been thought to have been a champion of urbanism, described the capital as 'a tumour, an elephantiasis sucking into its gorged system half the life and the blood and the bone of the rural districts' (cited in Howard 1946: 42).

Urged on by such a climate of opinion, and in the spirit of post-war social experimentation, the 1945 Labour government accepted the need for decentralisation of population and industry from congested cities to new towns. The New Towns Act of 1946 paved the way for central government to achieve this. As Cherry (1988: 161) puts it:

This, after all, was the age of centralist planning; the state was the wise, beneficent steersman to a nobler future. New Towns would be civilised, attractive, agreeable places in which to live, with all the richness of community life which the new social order would bring.

Between 1945 and the early 1970s, 28 new towns were declared, many being clustered around London as recommended by Abercrombie's Greater London Plan of 1944, which sought to move almost one million Londoners to towns 30–80 kilometres from the capital (Schaffer 1972). By the close of the 1980s the British new towns housed 2.0 million people, perhaps 1.1 million of whom had moved from adjoining metropolises (Cherry 1988). They have been hailed as major successes. Ward (1994), for example, states that their residents have enjoyed better housing and employment opportunities in more convenient and conducive environments than would otherwise have been possible.

Whilst the new towns were exercises in grand social engineering 'well in tune with the psychological requirements for post-war reconstruction' (Cherry 1988: 161–2), they were also exercises in neighbourhood social engineering. Much was written about the need for new towns to have a 'social balance', by which was meant a mix of social classes, and this philosophy was even applied to the layout of the housing stock. As the chairman of Stevenage Development Corporation said in 1947:

we want to revive the social structure which existed in the old English village, where the rich lived next door to the not so rich and everyone knew everybody ... the man who wants a bigger house will be able to have it, but he will not be able to have it apart from the smaller houses.

(cited in Ward 1994: 105).

Judged against this criterion, the new towns have been less successful. Schaffer (1972), not surprisingly, notes that many of the in-migrants were young skilled workers – those most in demand in the local labour market – and that the elderly and less skilled were markedly under-represented amongst their populations. The same can also be said for ethnic minorities. Even within the new towns social segregation could not be prevented:

Most corporations started off with the intention of mixing the various types of houses and avoiding any suggestion of 'class segregation'. But it didn't work out. Harlow, for example, soon found out that 'middle-class' families ... like either to be somewhat isolated and to have big gardens or to have a large number of their neighbours drawn from similar income groups.

(Schaffer 1972: 184).

Thus, in this case, limits to the successful social engineering of migration were exposed.

Conclusion

This chapter has shown how and why states might choose to intervene in migration rather than in fertility or mortality when they seek to mould national characteristics. It has also provided a range of examples from around the world to demonstrate that such intervention is already commonplace under a variety of political systems in different geographical regions, as administrations the world over seek to homogenise the racial, ethnic or religious composition of their countries' populations or realise a particular national economic or social vision. It seems highly likely that as the power of the state and, increasingly, that of supra-national bodies strengthens, so the control of emigration, immigration and internal migration will be more comprehensively manipulated for ulterior motives. Nonetheless, such control is not always achieved, demonstrating clearly how migration remains an arena of social contestation and struggle.

CHAPTER 8

Forced migration

Introduction

The last chapter discussed international and internal migrations that were encouraged and facilitated by direct government intervention but where the individual was left with some choice about whether to acquiesce. While it is always difficult to decide whether discretion really exists in any particular case, the present chapter focuses upon examples where individuals have only a negligible choice over whether to move or stay, and where, if they exercise this choice, they may face life-threatening circumstances. We are therefore concerned with what Petersen (1958) defined as forced migrations.

Mention of forced migration conjures up images of political *refugees* trudging from one emergency feeding centre in Africa to another, or of *asylum seekers* fleeing from politically repressive regimes. However, while such individuals undoubtedly form a major proportion of the world's forced migrants, a central argument of this chapter is that we also need to consider other forms of forced migration that occur in the modern world and that are, too frequently, under-reported or ignored. Thus, while the United Nations High Commission for Refugees (UNHCR 1995) estimated that, at the beginning of 1995, there were 14.5 million legally defined refugees in the world, there was also another group of 5.4 million *internally displaced people* who did not qualify as refugees but had been forced to move by similar causal factors. Moreover, Cernea (1990) has calculated that a further 1.2 million to 2.1 million people are internally displaced worldwide *every year* by the construction of dams alone. Thus, while the suffering and trauma of refugees rightly attracts our attention, we must be careful not to ignore other less well publicised but equally important forced migrations.

Defining refugees

Why labels matter

One could argue that it does not matter whether someone forced from their home by persecution is labelled a refugee, a forced migrant, a displacee or an oustee. Rather, what is important is that coercion has taken place and individuals have had to uproot themselves against their wishes. Consequently, many people use the terms as if they were directly interchangeable. In reality, however, labels do matter. In particular, the label 'refugee' is highly potent, opening the way to a range of internationally guaranteed benefits but also carrying with it some negative connotations. As early as 1921, the possession of a League of Nations (the forerunner to the UN) passport – a Nansen Passport – meant that an individual was more likely to be recognised by national governments as being in need of refuge than someone without such a document. Subsequently, as the apparatus of refugee recognition and assistance became more formalised, so achieving the coveted status of *Convention refugee* (see below for explanation of this term) carried with it increasing benefits. At present, an individual who can persuade the authorities that his/her circumstances make them a Convention refugee is eligible for various entitlements. The UNHCR has a mandate that obliges it to offer refugees two things. First, in conjunction with other United Nations (UN) organs, it offers protection, although the concept of protection has changed dramatically over the last five years (UNHCR 1995). Second, they provide assistance, in the form of access to food, shelter and health services in the short term, and resettlement opportunities in the longer term. Host countries that are signatories to the

Convention also have obligations to refugees (Goodwin-Gill 1982). Most importantly, they should not return refugees against their wishes to the country from which they fled, a principle known as 'non-refoulement' (Goodwin-Gill 1986). National governments each have different arrangements for allowing Convention refugees access to their welfare services, but such refugees receive preferential treatment to those with other legal statuses (Black 1993).

While being awarded Convention status can thus change the life-chances of an individual dramatically, even making the difference between life and death, being labelled a refugee can also have deleterious side effects. Refugees are particularly vulnerable to the process of labelling, since forced migrations are often generated by crises that develop rapidly, usually in conditions of scarcity (for example, of shelter or food). In such circumstances,

those who define or assist refugees see their immediate task as providing mass emergency aid and refuge. This predisposes or even requires organisations to group people into manageable categories rather than treating them all as individuals. Groups are thus given a single *label* or identity ('refugee'), even though this might conceal a range of (contrasting) identities: political refugees versus environmental refugees, or the imposed label versus how refugees see themselves. Perhaps more importantly though, labels frequently become stereotypes, which carry with them assumed needs and prescriptions. Such stereotypes may well not come into being for malign reasons but simply as a coping strategy for staff or organisations under enormous pressure to achieve the impossible with limited resources. Even so, once such labels are in place, and especially under conditions of unequal power relations, stereotypes may become self-reinforcing, with refugees feeling that they have to conform to demonstrate their gratitude and merit further assistance.

Labels are not, however, just passive consequences of the actions of others but can create a powerful momentum for change. Zetter's (1991) study of the victims of the 1974 partition of Cyprus demonstrates some of the unintended but harmful consequences of labelling. To the outsider, the programme designed to resettle displaced Cypriots from the north of the island had been a resounding success, with 150,000 people rehoused in 40,000 new properties and an economy reshaped to provide them with employment. Closer scrutiny demonstrates four consequences of the dislocated being labelled refugees. First, the refugee housing provided by the state was on spatially segregated estates, which were highly visible and of a different type to those traditionally found in Cyprus. Residents consequently felt stigmatised, marginalised and dependent upon the state, which they felt controlled them. Second, refugees had to compete with each other for housing because of the allocation system imposed upon them. Consequently, a group of people who could have gained greatly from the solidarity of a shared misfortune were divided. Third, the very act of providing permanent housing for 'refugees' was perceived as weakening their continued demands for an international solution to the partition, which would allow them to return to their real homes and possessions in Turkish-controlled northern Cyprus. Fourth, the imposition of the label and identity 'refugee' undermined traditional cultural values and practices, such as the dowry system

---

Box 8.1 An example of the UNHCR's refugee assistance work

Below are selections from a brief account of an 'ordinary' day in the life of a refugee camp in Africa:

Refugees gather throughout the afternoon and into the night in a neutral strip of land between the Kenyan and Somali border posts near Liboi. By morning there are nearly 700 of them. A UNHCR team arrives at 8am to screen the new arrivals. They pick out those who are so sick or weak that they need immediate medical attention, issue ration tickets and try to prevent people who are not refugees ... from signing up for assistance.

By the end of the morning, 38 sick or badly malnourished people have been sent directly to the hospital in Liboi camp. ... Another 535 have been accepted for settlement at Ifo camp, one of the three sprawling settlements ... each of which shelters about 40,000 Somali refugees.

The journey ends at Ifo camp. After they get off the trucks, the children are taken aside by Médecins sans Frontières and given a cup of milk, a measles vaccine and a Vitamin A tablet. ... Meanwhile the heads of family line up to register and receive ration cards. More than 500 people are processed in little over an hour. The refugees then get back on the trucks and are taken to the distribution centre. ... They present their ration cards twice: first for non-food items including a tent or tarpaulin, blankets, a small stove and jerrycans for hauling water. Next they join the food line and receive flour, beans, oil, sugar, salt and a tin of fish.

By the time the last groups are taken to their allotted sites, dusk is falling.

Source: UNHCR 1993.

and the communal residence of extended house-holds.

Legal/operational definitions of refugees

While academics have argued at length over how to define a refugee, in practical terms only two definitions matter. Both are legal definitions and, of these, one is pre-eminent. This first and most enduring definition was drafted in 1951 by the United Nations and is contained in the Convention relating to the Status of Refugees. It refers to any person who

owing to a well-founded fear of being persecuted for reasons of race, religion, nationality or political opinion, is outside the country of his nationality and is unable or, owing to such fear or for reasons other than personal convenience, is unwilling to avail himself of the protection of that country.

(quoted in UNHCR 1993: 163).

There are several features of note about this definition. First, it was drafted very much with the problems of Europe in mind, since the Second World War had left that continent with 14 million refugees and 11 million displaced people (Robinson 1996a). Indeed, Convention status was restricted solely to those forced to move as a result of the war in Europe prior to the Bellagio Protocol of 1967, such was the international community's preoccupation with Europe. Second, a person must cross an international political boundary to become a refugee, thereby excluding those internally displaced by the same causes. This anomaly was deliberate, since the United Nations – like the League of Nations before it – could only operate through the consent of national governments and had to distance itself from any involvement in the internal affairs of member states. Third, the definition relates to individuals not groups. Each person seeking Convention status has to demonstrate that they personally are in fear of persecution. It thus denies Convention status to those who have been displaced by war or violence but who have not been singled out for individualised persecution (Ferris 1985). Fourth, the definition emphasises political rather than other forms of persecution (D'Souza and Crisp 1985). Fifth, while the definition guarantees the right of individuals to seek asylum, it does not guarantee that they will receive it (Gurtov 1991). Despite these weaknesses, as of 17 May 1995, 121 countries had become signatories to both the 1951 Convention and the 1967 Protocol.

In 1969, the Organisation of African Unity (OAU) proposed an alternative and broader definition of a refugee, based more upon the experiences of developing countries than that of war-ravaged Europe:

every person who, owing to external aggression, occupation, foreign domination, or events seriously disturbing public order in either part or the whole of his country of origin or nationality, is compelled to leave his place of habitual residence in order to seek refuge in another place outside his country of origin or nationality.

(quoted in UNHCR 1993: 165).

The definition thus has two parts. The first is closely modelled on that of the 1951 Convention, while the second considerably extends that definition. Thus, the OAU definition includes the victims of war, violence and civil disturbance and it introduces the notion that refugees should not be treated differently because of their colour, race, religion or political affiliation. It also moves away from requiring individual asylum seekers to perceive that they are the target of persecution to more objective criteria (such as unbearable and dangerous conditions), which can be experienced by entire groups (Nobel 1982). This definition is now used throughout much of Africa, and its more generous spirit accounts in part for the less exclusionary attitude towards asylum seekers in much of that continent until very recently.

Finally, despite the apparent clarity of these definitions, there is enormous temporal and spatial variation in how they are interpreted and implemented. One only needs to look at the secular trends in the outcome of refugee-determination procedures in one country to see refugee criteria shift under pressure from public opinion and domestic political ideologies. For example, Figure 8.1 shows how the proportion of asylum seekers in the UK granted Convention status has declined sharply, initially in favour of the lesser status of Exceptional Leave to Remain but now in favour of outright refusal.

The contemporary global refugee crisis

Numbers

The modern refugee crisis involves larger numbers of people than have hitherto been seen outside the confines of world wars, moving in more geographi-

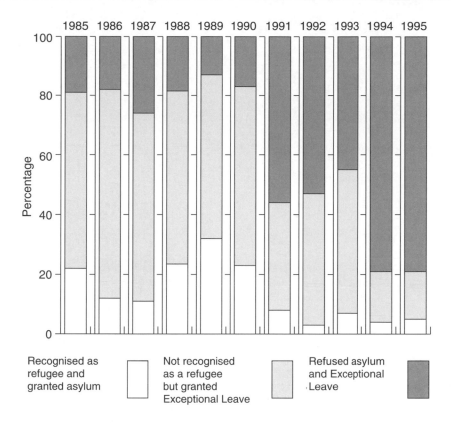

Figure 8.1 The proportion of people granted Convention status, Exceptional Leave to Remain and rejection in the United Kingdom, 1985 to 1995 (source: Home Office 1996).

cally and legally complex flows, with much greater re-sourcing implications over a longer period of time. Looking at the numbers first, it is difficult to quantify how many refugees there are in the world today. The organisations that collect and collate data do so in different ways and use different eligibility criteria, and they are often more concerned with dispensing aid than with undertaking censuses. Governments, which supply many of the basic figures, have a vested interest in either suppressing accurate data on departures – the number of people fleeing a country may be taken as a reflection of government failure – or exaggerating numbers of arrivals to garner extra international aid. Refugees themselves are highly mobile and are often suspicious of officials and officialdom. Despite these difficulties, the UNHCR suggested a figure of 14.5 million refugees in the world at the beginning of 1995 (UNHCR 1995), while the United States Committee for Refugees (1994) proposed a total of 15.5 million for 1 January 1994.

Figure 8.2 (overleaf) demonstrates how the world

has experienced a dramatic growth in the estimated number of refugees since the mid-1970s, with particularly pronounced increases between 1980–83 and 1988–91, albeit with a sustained reduction since the 1993 peak of 18.2 million. During 1991, when numbers were growing most quickly, the net number of refugees increased by 3300 people per day. The contemporary refugee crisis is thus different from most previous situations on numerical grounds alone.

Complexity

While numbers on their own make a compelling argument for the existence of a sea-change in the global refugee issue since the 1970s, other points reinforce this argument. Robinson (1993c) notes how changes in international communications have made it easier both for potential emigrants to acquire information about possible destinations and for those distant from refugee crises to learn about the horrors

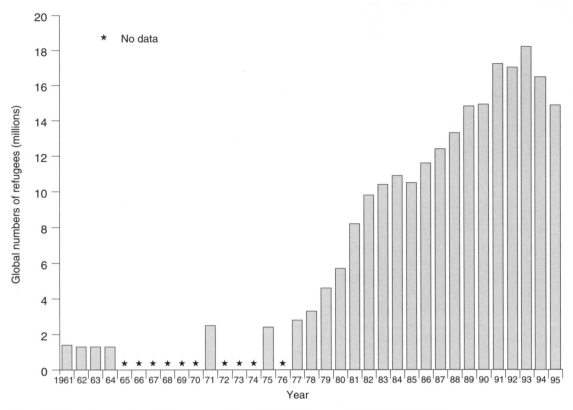

Figure 8.2 Changing number of refugees over time, 1961–95 (source: UNHCR 1993, 1995).

experienced by participants. Improvements in international mass transport have also made it much easier and cheaper for potential migrants to realise their ambitions. Scheinman (1983: 80) noted another change, namely how the 'granting/withholding of refugee status has become an instrument of the receiving state's diplomacy toward the sending state' rather than a simple matter of humanitarian considerations. Furthermore, as the number of independent states in the world has increased, more and more migrations that had hitherto been within the boundaries of one state have become international flows. The participants thus become eligible for refugee status but they are also put at the mercy of two, possibly rival, administrations. Marrus (1985) singled out another contextual factor that has profound implications for refugees, namely the development of the welfare state from the 1950s onwards. In earlier eras, refugee crises were short-lived, since refugees rarely survived more than one winter when separated from their traditional means of subsistence. These days, refugee concentrations can be sustained over the long term by state welfare provision and the intervention of the huge number of supra-national humanitarian bodies such as the UNHCR and UNICEF (United Nations International Children's Emergency Fund). Lastly, refugee migrations now take place within a world order increasingly dominated by xenophobic considerations and tainted by unemployment and sluggish economic growth, all of which change the climate of public opinion towards immigration and the admission of refugees (Hainsworth 1992b).

Consequences

How forced flight impacts upon participants, bystanders and specific localities will be considered later in the chapter. The contemporary refugee crisis has consequences broader and on a scale hitherto not seen. For example, Figure 8.3 looks at the geographical distribution of those refugees recognised by the UNHCR on 1 January 1995, while Figure 8.4 (page 186) shows their origins. The figures demonstrate how some regions and countries carry a disproportionate share of the costs of the refugee crisis.

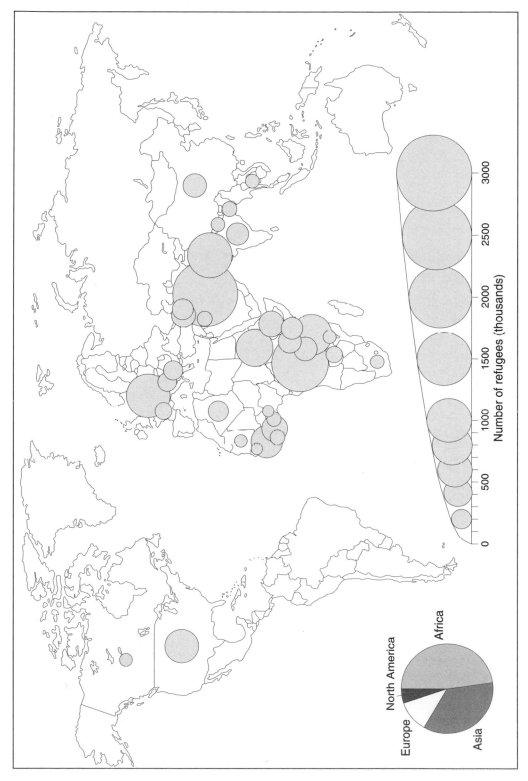

Figure 8.3 Location of the world's refugees, January 1995 (source: UNHCR 1995).

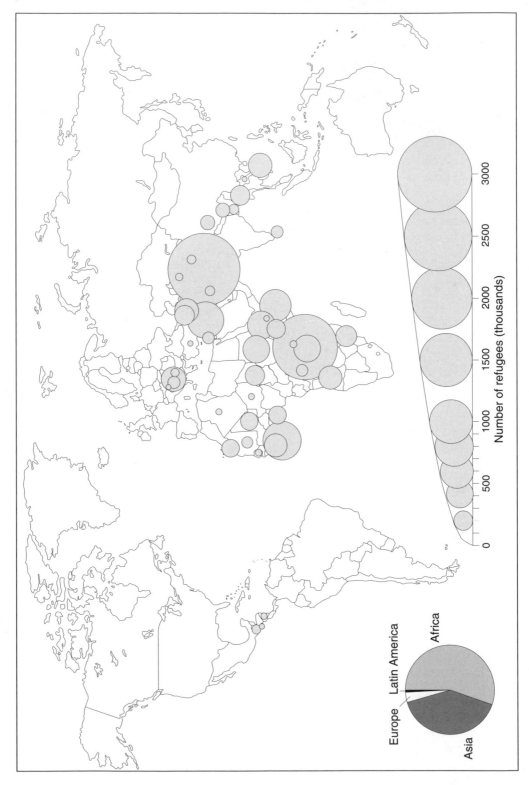

Figure 8.4 Origin of the world's refugees, January 1995 (source: UNHCR 1995).

Africa is the epicentre of the global refugee crisis, with 47 percent of the world's refugees. Asia is also important. Together, eight out of every ten refugees in the world today live in these continents. Differences are even starker at a national level. In January 1995, two countries alone (Afghanistan and Rwanda) accounted for 35 percent of all refugees, and the leading five nations contained nearly half of the total. Within Europe, Germany hosts over one million refugees, more than half of the continent's total. These major concentrations can be contrasted with Japan, which contains only 0.06 percent of the total.

These crude numbers need to be related to the size of the population and the affluence of the host country or region to give a more accurate picture of the impact of refugee populations. In 1995, recently arrived refugees formed more than 5 percent of the total population in four countries, and above 2 percent in a further ten (Table 8.1). For these countries, the contemporary refugee crisis is immediate and severe, requiring difficult decisions about the allocation of resources between different population groups and impacting upon overall levels of economic development and growth. As Table 8.1 also suggests, countries with the largest proportional number of refugees are often those least able to afford them.

Costs can also arise for other reasons. European governments, for example, spend a large amount of money on trying to decide who is a deserving refugee and who is a bogus applicant, and then denying access to the latter. Spain has spent 520 million pesetas on closed-circuit television (CCTV), barbed wire and monitoring equipment to strengthen the frontiers of its North African enclave, Melilla (Webber 1991). Indeed, it has been estimated that, within Western Europe, there are now 20,000 administrators and case-workers involved in refugee work, and that the cost to those countries of staff, legal aid, translators, reception centres, social aid and removal expenses had reached $7.5 billion in 1991.

National governments also bear the cost of the refugee crisis in other ways. The UNHCR spent $1.17 billion in 1994, the majority donated by individual countries and raised from their taxpayers. The geographical distribution of where UN funds come from is also highly localised. Figure 8.5 (overleaf) graphs the relative contribution of the top donors to UNHCR funds. Not only is the European Union (EU) the second largest contributor as a bloc, but ten of the fourteen individual countries mentioned are from Western Europe. When converted into donations per head of population, European countries fill the top eight places, with countries such as Norway and Sweden donating $9–$10 per head of population per year against the $1 per head of the leading non-European contributor, the United States.

Nation-states and the causes of refugee movements

Many authors provide checklists of the separate factors that they feel stimulate flight (Kliot 1987). Zolberg (1983), however, goes beyond this, providing perhaps the first attempt to develop an integrative explanation for refugee movements. He began by noting that much previous work in the social sciences dichotomised international mobility, with migration characterised as possessing a degree of regularity, since it was the aggregate product of individual responses to economic circumstances. It was thus amenable to theoretical analysis and could be 'explained'. In contrast, refugee flows had been stereotyped as 'singular', 'unruly' and 'unpredictable', since they arose from causes such as civil strife, changes of regime, war or government intervention. Each was seen as a unique event with unique causes. Zolberg (*ibid*.: 25) rejected this dichotomisation and suggested that

Table 8.1 Proportion of the population comprising refugees, by country, 1995.

Country	Percent of total population recently arrived refugees
Guinea	8.8
Armenia	8.1
Djibouti	6.0
Burundi	5.0
Liberia	4.5
Zaire	4.2
Croatia	4.1
Belize	3.9
Mauritania	3.8
Iran	3.5
Tanzania	3.2
Ivory Coast	2.7
Sudan	2.6
Guinea-Bissau	2.3
Kuwait	2.1

Source: 1993 UN Demographic Yearbook 1995; UNHCR 1995.

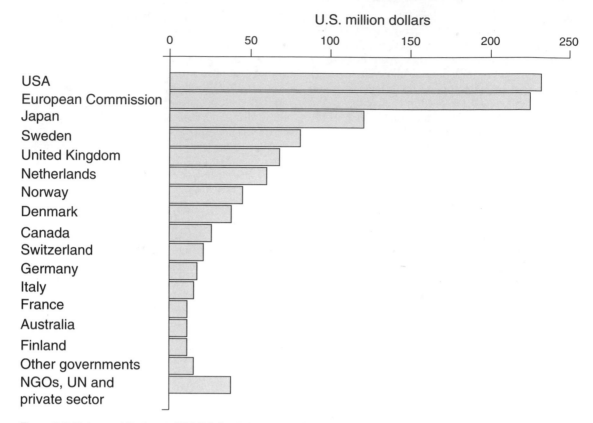

Figure 8.5 Major contributors to UNHCR funds by country (source: UNHCR 1995: 255).

it should be possible to view refugee flows in theoretical perspective, as particular instances of a general phenomenon that is as much a concomitant of world politics as ordinary migration is of world economics.

Zolberg argued that while migration had its roots in economic forces, so refugee movements had theirs in political forces, particularly in the emergence of nation-states from multi-ethnic empires (for example, the Austro-Hungarian and Russian empires). Central to the creation of nation-states was the purging of non-conforming groups (Chapter 7). In particular, two groups of misfits would be created by the dissolution of empire: *minorities*, or people of one identity finding themselves in a country with a different identity, thereby losing full rights and legal protection; and the *stateless*, or groups whose identity did not correspond to that of any established nation-state or minority, because of either history or deliberate legal exclusion. Both groups would eventually become refugees, with the stateless being expelled and minorities being persecuted until they chose to leave. Zolberg also noted how misfit groups

could be defined in a variety of ways, either by ethnicity, religion, or social status. For example, as Castile and Aragon sought to create the single unified nation-state of Spain, 150,000 Jews were expelled in the fifteenth century because they were unassimilated, 275,000 Moslems were transported to North Africa in 1609 because of their links with the Ottoman Empire, and Protestants were persecuted in Spanish-held Belgium between 1580 and the 1630s because of their refusal to follow a state-prescribed religion.

While Western Europe effectively resolved its problem of misfit minorities within two and a half centuries of the onset of state formation, with the dissolution of the colonial empires in the developing world during this century, the same processes have now been visited upon the former colonies. Here, the problems of misfit minorities and nation building have been exacerbated by the lack of resources at the command of new governments, which instead turn to authoritarian methods to achieve their objective of homogeneity. In some instances, certain forms of colonial experience helped to create the foundations for state building, with institutions being created which cross-cut society and therefore encouraged alliance building and negotiation after independence (Anthony 1991). Examples include Kenya, Tanzania and Zambia. In other cases, the colonial experience was very different. Three alternative models of colonial administration existed. Under 'radical separation', a bifurcated colonial administration distributed resources unequally between different ethnic groups, thereby creating divisions, enmities and different expectations that spilt over into the post-colonial period; for example, Sudan. Alternatively, an administration favoured one ethnic group during the colonial period, but gave power to another group on independence; for example, Burundi. Finally, 'paternalistic' colonial administrations deliberately suppressed indigenous political organisations, relied upon imposed Western European channels of administration, and discouraged independence. On decolonisation, the local population was divided and unprepared for power; for example, Zaire. In all three circumstances, internal war was highly likely to develop as a substitute for internal politics, with this factional violence propelling many people across international boundaries (*ibid*.). The impact of decolonisation and post-colonial wars on forced migration in southern Africa is shown in Table 8.2.

The nationalism that underpins many refugee movements can be expressed in different ways and will tend to produce different triggers to flight. The expulsion of the Ugandan Asians saw a newly emerging nation pursuing policies of racial persecution and exclusion to achieve the racial homogenisation of its population (Robinson 1995; Twaddle 1995). Asians in East Africa were a middle-man minority. They were traders and storekeepers who thrived under colonial rule, acting as intermediaries between local African producers and European exporters. In pre-war Uganda, for example, Indians controlled 90 percent of all trade (Delf 1963). After 1945, the Asian communities diversified and sought to gain occupational mobility through education. In Uganda, they were the first to establish schools and, in general, they achieved considerable success in entering the civil service and the professions. However, this economic success was gained at the expense of relations with local Africans, many of whom regarded Asians as uncommitted economic transients and exploiters, and unwilling to mix socially with Africans. This perception was not challenged by events after independence. Some Asians began to transfer their wealth out of Africa, others made plans to leave the continent, and many did not give up their British citizenship in favour of the citizenship of the newly independent states (Swinerton, Kuepper and Lackey 1975).

Table 8.2 Refugee movements created by decolonisation in southern Africa.

Country of origin	Duration	Peak years of asylum	Country	Numbers	Year of return
Liberation struggles					
Angola	1961–1974	1966–1974	Zaire	400,000	1974–1975
			Zambia	25,000	rarely
Namibia	1966–1989	1980s	Angola	74,000	1989–1990
Mozambique	1964–1974	1968–1974	Tanzania	100,000	1974–1975
			Malawi	25,000	1974–1975
			Zambia	15,000	1974–1975
Zimbabwe	1972–1979	1975–1979	Mozambique	150,000	1979–1980
		1974–1979	Zambia	43,000	1979–1980
Post-independence conflicts					
Angola	1975–1991		Zambia	103,000	
			Zaire	460,000	
Mozambique	1984–1992		Tanzania	100,000	
			Malawi	1,200,000	
			Zambia	30,000	
			Zimbabwe	200,000	
			Swaziland	75,000	
			South Africa	250,000	
Zimbabwe	1983–1988		Botswana	1,300	

Source: Wilson 1995, reprinted by kind permission of Cambridge University Press.

Consequently, once countries such as Kenya and Uganda had gained their independence it was not long before they intensified their Africanisation policies, with Asians being excluded from public employment, having limitations placed on their ability to own businesses, being restricted to trading in specified commodities, and being denied the opportunity to attend university. In Uganda, Africanisation was taken one step further by President Amin when, in August 1972, he ordered the expulsion of the Asian minority. Over the next three months, some 40,000 Indians fled Uganda, having been effectively stripped of their assets and suffering physical abuse as they left. Accounts of their flight (Marett 1989), resettlement (Bristow 1976; Robinson 1986a) and subsequent experiences (Robinson 1993d, 1996b, 1997; Mamdami 1993) detail their fate.

Elsewhere, the targets of policies designed to achieve homogenisation are political or ethnic rather than racial minorities. For example, nearly one million Vietnamese left that country between the end of the Vietnam War in 1975 and 1989, in two major waves. The first wave comprised those who feared the communist regime that had taken control of South Vietnam, either because they had associated with the previous capitalist administration or with the Americans during the war, or because they were intellectuals, business people or critics of communism. They were the victims of North Vietnam's determination to absorb the south and impose its own identity and ideology upon it. The second wave, numbering well over half a million, fled after 1978, as the government sought to integrate the south and the north of the country and homogenise the united population. The ethnic Chinese minority in former North Vietnam saw their liberties and rights gradually eroded, to the point where they could not meet in public or own a business. Deteriorating relations between Vietnam and China, culminating in the Third Indo-China War, did not assist their cause. Under growing pressure, a quarter of a million ethnic Chinese crossed the border into China and many others (the Boat People) sought to escape by sea to destinations such as Hong Kong, from whence most were resettled to Australia, Canada, France, Britain and the United States following the Geneva conference of 1979. The legacy of this internal persecution remains in the demography and social trajectories of the resettled groups in different countries (Robinson 1993b). Britain and Australia, for example, accepted Vietnamese refugees from the second wave, most of

whom had had little contact with the developed world, did not speak any English, and were either peasant farmers or fisherfolk. In contrast, the United States' involvement was much more weighted to the first wave, many of whom were educated, white collar workers, with prior contact with the developed capitalist world.

The Gulf War of 1990–91 demonstrated how, in its desire to strengthen itself internally and also gain territory and resources from others, a state might engage in war. The conflict was motivated by the desire of Iraq to regain what it felt was Iraqi sovereign territory and to appropriate natural resources to underwrite military expansion, making possible its desire to become a regional superpower. Iraq's incursion into Kuwait, and the military threat it posed to Saudi Arabia, eventually generated an outflow of four to five million refugees in several waves (Van Hear 1993). In the first wave, between August and December 1990, over a million foreign workers and professionals left Kuwait, fearing for their safety under Iraqi rule, while the Saudis harassed a further 850,000 Yemenis into leaving Saudi Arabia as a direct reprisal for Yemen's perceived support for Saddam Hussein (the Iraqi leader). In another wave (April–May 1991), Kurds began to flee from Iraq and persecution bordering on genocide. Two million tried to leave the country, with 500,000 entering Turkey and 1.3 million crossing into Iran. Many remained massed along the borders in makeshift camps, dying of exposure, malnutrition and disease, until the intervention of the United Nations' Operation Provide Comfort and Operation Haven. Finally, after the expulsion of Iraqi forces from Kuwait, Palestinians and others fled Kuwait following persecution by Kuwaiti militia and the reinstated government's decision to reduce its dependence on foreign labour by denying access to employment, health care and services. The nation-building activities of one country over a period of only ten months led to the flight of millions of people of more than fifteen different nationalities. For example, Van Hear (1995) has considered the impact of the conflict upon those Palestinians forced to return to Jordan.

Other examples of the way in which hegemonic territorial expansion can displace population groups across international boundaries are not difficult to find. For example, there was the flight of Afghans from the Russian invasion of 1979 and the civil war that followed (Findlay 1993c; Wood 1989). From an earlier era, Polish military personnel and government officials fled Poland after the German invasion of 1940 and went on to form the nucleus of the Polish community in Great Britain (Peach *et al.* 1988; Robinson 1997; Sword, Davies and Ciechanowski 1979).

Finally, regimes may chose to define national misfits in more strictly political terms. One-quarter of the population of El Salvador has been either displaced internally or forced to flee the country as a result of the long-running conflict between a right-wing government and a left-wing guerrilla movement (Stanley 1987). By 1981, 46 per thousand population were leaving the country, a rate eleven times higher than that six years previously (Jones 1989). Military sweeps by the government were designed to discourage peasants from supporting the left-wing guerrillas, while the latter used force to garner support, through, for example, forced recruitment (Stanley 1987). These activities led to the death of 1000 civilians in 1979, 8000 in 1980 and 14,000 in 1981. Rape, torture and crop destruction by the military were commonplace. Stanley sought to prove that there was a statistical relationship between the occurrence of such acts and the scale of flight from El Salvador to the United States. Multivariate linear regression (Chapter 2) was used to show that 72 percent of the variance in Salvadoran migration to the United States could be accounted for by political violence. While economic betterment might also have been a strong motivating factor for flight, political violence and economic disruption were often closely associated. Jones's (1989) subsequent analysis of similar data did not, however, produce supporting results.

The refugee experience

It is possible to generalise, to a certain extent, about the refugee experience, and the different temporal phases through which a refugee might pass (Hitch 1983; Stein 1981). Below we describe an eight-stage model, although not every refugee would be expected to experience all of these stages. Information has been attached to each stage about the mental health and outlook of refugees.

- *Perception that a threat exists.* This is often triggered by a specific event that forces people to realise the precariousness of their position. It might involve witnessing an event (the arrest of a relative), experiencing an event (having one's village attacked or business closed), or some official pronouncement (a government edict preventing a particular group meeting in public).
- *The decision to flee.* This is usually an especially difficult decision for refugees despite the threats facing them. Flight may be possible only for some of a group or family, so those contemplating it may have to weigh both the repercussions of their own flight on those left behind and the prospect of having their families permanently divided between countries or continents; they may even have to make the unenviable choice of who should be sent to safety and who should not. Besides breaking their place attachments, potential refugees will also almost certainly face considerable financial loss and great financial insecurity. There will be fear of what might happen if the decision to flee becomes public before the refugee has reached safety. Reflecting the complexity of the decision to flee, Kunz (1973) suggested that two groups will develop in many refugee flows. There will be *anticipatory refugees*, who are quickest to divine the threats to their way of life and make appropriate preparations at an early stage. They are able to liquidate assets, plan their means of flight and make arrangements in their country of destination. Such refugees are more likely to be educated and affluent and to travel as families. *Acute refugees*, on the other hand, are those either caught up by the speed of events or lacking the foresight to have predicted a future crisis. Their priority is to get to safety, which is accomplished without extensive planning or preparation, and may lead to the accidental break-up of family units. Even for acute refugees, though, flight might not be immediate. Because of delays, Kunz (1981: 140) introduced the concept of *vintages*, defined as:

> departure-and-transit cohorts uniting people with shared experiences before and during displacement, who because of their shared timing of departure often hold common views and attitudes.

- *A period of extreme danger leading up to and including flight.* As preparations are made for flight, so the risk of detection increases, as acts become more overt and the decision is shared with more people. Applications for visas might have to be made at foreign embassies, travel arrangements need to be made and resources gathered together for flight. Members of the family not going must be told and, as resources are expended and assets disposed of, it becomes harder to reconsider the decision.

- *Arrival at a safe destination*. This often induces a temporary feeling of euphoria at having escaped and survived, accompanied by feelings of great optimism about the future.

- *Reception camp life*. Many refugees begin their lives outside their own country in a reception camp created either by voluntary agencies or by the country of first asylum. Murphy (1955) argues that during this phase euphoria is replaced by profound depression, as refugees are confronted by a dual realisation. First, there is the enormity of their loss; of loved ones, their homeland, their previous identity and their former lives. Such loss might be sharpened by feelings of guilt at having left relatives behind or at having survived flight when others had not (Scudder and Colson 1982). Second, there is the devastation of prolonged camp life. This life and the dual realisation typically produce a sense of dependency and strong feelings of apathy arising from a perceived loss of control and uncertainty about the future. For others, however, a sense of perceived invulnerability (stemming from having survived persecution and flight) and displaced guilt produce aggressive instincts and behaviour, which are worked out on other camp inmates (Murphy 1955).

- *Onward migration to a third country*. Those who are fortunate to have been offered a resettlement place will leave the camp, embark upon a second migration and thence begin a new life in a third country. Stein (1981) suggests that this will be accompanied by heightened anxiety as the familiarity of the camp and its regime is replaced by the fear of what a new life in a new country might bring. Such feelings will coexist with feelings of relief at having escaped the unpleasanter sides of camp life (malnutrition, harassment, sexual harassment, disease and inactivity). In some cases anxiety will be heightened by the lack of information about a destination or even misinformation.

- *Initial resettlement*. This covers the period immediately after arrival and is characterised by disorientation and a recurrent sense of loss. Disorientation arises from trying to come to terms with a new way of life, new customs, a new language, new administrative procedures, and even family role reversals. For example, personal research revealed a Vietnamese family in Swansea in 1979 that had been given coal to heat their house but, since they had never seen coal before and did not know that it had to be ignited, simply put it in a fire grate, as they had been told, and waited for something to happen. The sense of loss apparent in the camp phase returns, but to this is added a sense of lost status in the community (surviving on welfare payments), lost culture, lost identity and even loss of respect from other family members (who might be quicker to learn new ways). Stein (1981) notes feelings of nostalgia, depression, anxiety, guilt, frustration and anger amongst newly settled refugees, many of whom feel isolated. Scudder and Colson (1982) suggest that the multidimensional nature of stress during this phase leads to an outlook of extreme conservatism, in which refugees cling on to as much that is familiar as possible and change only those things that are absolutely necessary. This might include 'cultural involution' and working in ethnic work gangs in particular industries. Such a phase often lasts two years or more.

- *Mid to late stages of resettlement*. Scudder and Colson (*ibid.*) suggest that the critical change occurring in this stage is a willingness to take risks, with Stein (1981) feeling that this period is marked by greater hope and a return to ambition and planning. There is a common drive to recover, to rebuild lives and to plan for a more certain future. This might involve the creation of a business or the acquisition of new occupational or educational skills, as was the case with the Ugandan Asians in Britain (Robinson 1986a, 1993d). Even though this phase is marked by greater optimism, control and some recovery in social and material status, the extent of de-statusing during flight should not be forgotten, nor should the adjustments that refugees have had to make to their life-aspirations. Yet, Finnan (1981: 308), in her study of Vietnamese in Santa Clara County, San Francisco, showed how refugees may react positively to de-statusing:

When refugees realise they cannot pursue their former occupations and must choose another, they adjust their self-images to accommodate occupations with less prestige than they formerly enjoyed.

Years after the original flight, the refugee experience still leaves an imprint. Refugees often have high expectations and feel that they deserve success after the traumas they have experienced (Stein 1981). When these expectations are not fulfilled, or individuals or organisations disappoint (for example,

---

Box 8.2 The life-course of the refugee flow of three Hmong from Laos

Phase 1: Perception of threat
Warfare provided the initial impetus for the refugee flow:

The day of the invasion, we heard dogs barking and knew something was wrong. The Laotians and Karen must have surrounded our village and waited until daybreak. ... Half of the village was still in bed. But after that first gunshot, everyone was running round like crazy. ... The soldiers burned all the houses, farms, and everything down, and killed all the pigs, horses and other animals. When we came back to the village we did not know how to live.

Phases 2 and 3: The decision to leave and the act of flight
The respondents lived homeless within the rain forest for seven years after their village was ransacked by Laotian troops, before eventually deciding to make the arduous journey to Thailand:

We made the decision to leave. ... We picked a Vang man to take charge and lead us to Thailand. ... Each carried whatever they needed. Besides rice, we carried knives, blankets, pots and a bag of wild potatoes. We only brought one set of clothing each, that which we wore. We did not have any money. ... We were always on the move. We walked quickly and kept as quiet as possible. On the trail we saw dead bodies, food and clothing left behind by those who went before us. We slept at two or three hour intervals and walked day and night. ... It took a month to reach Thailand, and we met with many Hmong on the way. I saw soldiers far away. They killed the group ahead of us, and I remember seeing the bodies as I passed by.

Phases 4 and 5: Life in a reception camp
Living in the camp proved to be extremely hard. One of the respondents, Kue, remarked that at first sight, the reception camp, Ban Vinai, 'looked so bad, I didn't know how we were going to survive there.' The respondents described the living quarters as both alien and dangerous. All complained about inadequate food, medical treatment and sanitation, and about periodically being robbed by the Thai:

In the camp, everyone ... did needlework all day every day. ... One person would get about 25 cents a week for the needlework. ... I felt like a slave. It was frustrating, depressing, sad and very maddening.

I didn't have any feelings about our misfortune. I thought I was going to die soon. I felt very much like a prisoner.

Phase 6: Onward migration to a third country
Although the respondents were told that Americans were cannibals, they began to believe that they had to leave the camp or die. They saw others leaving in buses, bound for the United States, so in desperation they decided to apply themselves. Eventually, the respondents left, with the trip being extremely arduous. The Hmong got little sleep and experienced much vomiting. They were not told how long they would travel and no-one provided them with food on the journey. On arrival at the airport in New York City, one of the refugees, Nib Yia, was very sick and the others thought she would die.

Phase 7: The early phases of resettlement
The respondents found the initial experience of life in the United States extremely alien:

It was so cold! All the trees outside looked dead so I went out to collect firewood. Then I realised that the wood was green ... I was afraid to touch the snow, afraid that something might happen to my hands.

When we first came, we were very sad because no one visited us. I knew everyone was busy, but I needed someone to talk to. I was so sad, so lonely.

Phase 8: Later stages of resettlement
In different ways, all three of the female respondents eventually regained control over their lives:

Xai actively distances herself from the past, and tries in her old age to live as much as possible in the present. She has developed a flexible optimism to deal with rapid change. 'I never dream about going back home. ... My real ties are here in the United States and this is where I want to remain. ... Living in Laos and Thailand were both equally hard times. But living in America now is like a breeze.

For her daughter Kue, religion has played an important role in self-empowerment. ... Her conversion ... is clearly more of a rejection of a traditional belief system ... than the embrace of a new one.

Nib Yia uses education as the means to gain some control. She avoids boys at all cost, and explains that she ... [is] working towards a skilled career.

Source: Monzel 1993

---

government agencies), feelings of bitterness can develop. Valeny (1996) describes how some of her Ugandan Asian respondents felt that the British establishment had somehow conspired against them and limited their business success. Nor can the experiences of life immediately before or during flight easily be eradicated; post-traumatic stress disorder can be experienced by many refugees, even years after flight:

Periods of relative equanimity in their lives are often shattered by a chance viewing of a television programme, reading an item in a newspaper, or realising that a particular day signals the anniversary of the event. Stimuli such as these trigger an episode of intense distress which may last for weeks or months, during which the person re-experiences his or her past in recurrent dreams and painful intrusive recollection, develops feelings of estrangement from others, loses interest in previously enjoyable activities, becomes distractible, distances him or herself from previously loved people and loses hope for the future.

(Beiser 1993: 216).

Several authors have demonstrated, through the use of art, how the vivid images of war, loss, and flight have been imprinted on the subconscious of children, frequently to return in the form of nightmares. Perhaps the best example of this genre is the study by UNICEF (1994), which contains striking and disturbing images drawn by children who have lived through the wars and their aftermath in former Yugoslavia.

The impact of refugees upon localities

Insufficient research has been undertaken to date on the impact of refugee flows on specific localities. Most research has tended to be refugee-centred and has focused upon the plight of refugees, their use of services, their access to housing and employment, and the government policies designed to assist them. Relatively little detailed work has considered the impact that concentrations of refugees have upon housing markets, labour markets, the built environment, or the attitudes and lives of their new neighbours. However, work in Australia has researched the scale and density of Vietnamese settlement in major cities and, by default, the process of residential transition experienced by other ethnic groups living in the same neighbourhoods (Burnley 1989; Coughlan 1989a; Hugo 1990a, b; Wilson 1990). More directly, Neuwirth and Clark (1981) have looked at how the attitudes of Canadians changed during the Vietnamese resettlement programme. Black (1993) provides a broad overview of the impact that refugees have had upon Western Europe, while Findlay (1994b) and Spencer (1994a) have listed the positive benefits generated by the arrival in the United Kingdom of refugees and immigrants. Nonetheless, such audits are necessarily too broad to be anything other than a first step.

In the developing world, the impacts of refugee concentrations are both more immediate and more profound because of the relative poverty of the indigenous populations and the general shortage of resources. Some detailed work has tried to identify and assess the impact of refugee settlement. Ghimire (1994) was concerned with the pressure that refugees put upon local woodland, which provides fuel for cooking and heat, a source of fodder for animals, a free building material and a commodity that can be traded and sold. The arrival of refugee populations was often associated with rapid deforestation, as wood was taken for specific purposes, clear-felled for sale, or stripped to produce extra grazing land. This had severe consequences for the refugees themselves and for local people: food supply was reduced, soil erosion was precipitated, water tables were altered, a valuable source of income was denied, and refugees had to forage firewood and building wood from further afield. Ethiopian refugees in Somalia travelled between 5 and 8 kilometres daily to collect wood, and Afghan refugees in Pakistan were spending up to three hours per day collecting fuel. Ghimire notes that disputes over the extraction of wood and the use of common pasture were the two issues that most frequently caused resentment and conflict between the local and refugee populations. The UNHCR (1995: 163) concurs with Ghimire's work, and notes that 'settlement sites are visibly surrounded by large areas of land which have been stripped of trees and vegetation'. It estimated that the one million Mozambican refugees in Malawi consumed 500,000–700,000 cubic metres of wood per annum for cooking and heating, and they drew attention to studies which showed that, after refugee deforestation around camps in Tanzania, soil structures broke down, allowing invasion by weeds and the eventual destruction of fertility. They also noted that refugees were responsible for water shortages, the pollution of existing sources of water, and the disruption of habitats, which previously provided locals with medicinal herbs and plants. Consequently, the UN for one (Black 1994a) now has its own environmental guidelines and an environmental database, and has committed itself to undertaking environmental impact assessments of all its projects. It is heavily involved in re-afforestation projects and is seeking to develop more efficient stoves for distribution in future refugee crises.

The UN has also argued that the negative impacts of refugee settlement in developing countries should not be exaggerated and that resettlement can benefit locals and refugees alike (UNHCR 1995). They describe the settlement of 20,000 exiled Mozambicans in the new township of Ukwimi in south east Zambia in the late 1980s. Prior to the arrival of refugees, the area was sparsely populated and experiencing economic decline but, as a result of their arrival, the town gained a major influx of development capital from overseas. Roads, schools, health centres, training workshops and an agricultural advice centre were provided and made available to the pre-existing population as well as to refugees. In addition, the Mozambicans managed to become self-sufficient in food production within two years of arrival, an example that was not lost on local people. As surplus production began to accrue, the refugees began to trade with locals and patronise the itinerant traders who had been attracted to the area by the expanding population. These traders introduced new goods and new agricultural inputs to the region. Eventually, when the Mozambicans were able to return home in December 1994, the entire settlement was handed over by the aid agencies to the Zambian government. In this case, local people benefited from the presence of refugees and the extra development resources that they were able to attract from overseas (Black 1994b). Wilson (1985) provides corroborating evidence of such synergies from another African case study, and Rogge (1986) notes how refugees have been responsible for bringing unused and under-used land into full production in Sudan, Tanzania and Botswana.

Policy towards refugees

Local integration

Within the refugee studies literature policies tend to be grouped into what are known as the *durable solutions*, whereby

the refugees ... become self-sufficient, enabling them to integrate and participate fully in the social and economic life of their new country, or their homeland if they repatriate.
(Stein 1983: 190).

Traditionally, the UNHCR and others have proposed three durable solutions, although recently a fourth has become apparent in the actions of particular countries and blocs.

Historically, *local integration* was the preferred policy. Refugees would flee their country of origin to a neighbouring country, where they would be resettled or would self-settle. The prime difficulty with such a policy has been alluded to above, namely that many of the countries that generate the largest number of refugees are found in the poorest parts of the world. As a result, neighbouring nations are unlikely to have the resources to welcome large numbers of refugees. Despite this, the spirit of the Organisation of African Unity definition of a refugee and the attitudes of many Africans towards refugees has meant that local integration has been widespread within that continent. Kliot (1987: 115), for example, observed that:

Africa has had to take responsibility for its own refugees, and this is what it has done. In contrast to Europe, where refugees are looked upon as aliens, the African refugees are seen as fellow men and the local population, most likely kinfolk of the refugees, is usually friendly and generous to the refugees.

After crossing an international border, a number of policy options can come into play: dispersal amongst indigenous peoples; semi-permanent residence in camps; and settlement in self-supporting agricultural communities (Rogge 1977). Kliot (1987) estimated that 60 percent of African refugees chose the first option but Hansen (1990) considered the relative merits of the first and third options. Studying the 140,000 Angolan refugees in the border zone of Zambia showed that

scheme-settled refugees were found to be materially better off in 1989 than self-settled refugees, but self-settled refugees were found to be more integrated and 'at home'. Overall many scheme-settled Angolans remain 'refugees' after more than twenty years in Zambia, whereas self-settled Angolans are no longer 'refugees' in their eyes or in the eyes of their local hosts.
(Hansen 1990: 150).

Kliot (1987) suggested that, where government resettlement is preferred over self-settlement, the prerequisites for success are the availability of suitable land, a permissive government attitude, the establishment of settlements away from border zones, and a degree of ethnic affinity between local and immigrant populations. This was exemplified by Gasarasi (1987), Kabera (1987) and Rogge (1987), who described the nature of government-controlled settlement in Tanzania, Uganda and Sudan, respectively.

Third country resettlement

During the late 1970s a different world view on the refugee crisis began to develop. Shaken by the Vietnam War and its aftermath, shocked by government deceit, and realising both the possibilities of people power and the need to think in global rather than local terms, the public showed itself willing to become directly involved in a resolution of the refugee crisis. The preparedness of developed world populations to participate in government and private refugee sponsorship schemes (Indra 1993; Lanphier 1993) came at a fortuitous time. The flood of refugees from the Indo-Chinese wars was threatening to overburden the many neighbouring South East Asian countries that had agreed to allow refugees temporary first country asylum, but only as a prelude to longer-term resettlement elsewhere. The outcome was the growth of *third country resettlement*, in which refugees were transported to countries such as the United States, China, France, Canada and, to a much lesser extent, the United Kingdom for permanent resettlement. The years 1979–82 marked the peak for this policy (Stein 1983), with over 325,000 refugees from the developing world being resettled in the developed world in 1980 alone (Table 8.3).

For Kunz (1981), the critical issues for successful third country resettlement are whether a receiving nation is a traditional country of immigration, whether there is any cultural compatibility between the receiving population and the incoming refugee group (especially in language, values, traditions, religion and politics), what the attitudes of the host population are towards different types of immigrant, and whether the receiving society is seeking to achieve assimilationist or pluralist goals. Additionally, Neuwirth (1988) suggested that the occupational and social skills possessed by refugees are a major determinant of their successful resettlement. She also commented on the 'calculated kindness' of some recipient countries, which weighted their refugee selection criteria towards those who would

Table 8.3 The distribution of Indochinese refugees worldwide, by country, 1976–82.

Country	Number resettled
United States	486,778
China	262,853
France	86,640
Canada	85,139
Australia	70,735
West Germany	21,256
United Kingdom	16,036
Hong Kong	9,598
Switzerland	7,746
Netherlands	5,240

Source: Rogge 1985.

be most valuable in the labour market rather than towards those in greatest need. The pre-existence of a concentrated community of fellow nationals can also be highly beneficial to newly arrived refugees (Rogg 1971), while the degree to which the receiving government provides long-term support, well beyond the reception phase, is also vital.

The relative novelty of third country resettlement has seen the process attract a good deal of academic interest. Two main research thrusts, each with a considerable literature, have developed. The first has charted the progress of resettled refugee groups, especially during their early years in a new society, and the second has reviewed government policies and programmes put in place to assist refugees. Looking at the progress of resettled refugees, Neuwirth (1993) considered the fate of South East Asian refugees in Canada and their marginalisation within the labour market. Coughlan (1989b) did the same for Indo-Chinese in Australia and Robinson (1989; Robinson and Hale 1989) charted the secondary migration of Vietnamese households within Britain after government policy scattered them in dispersed clusters of between four and ten families (Figure 8.6 shows one component of this redistribution). Desbarats (1985), undertaking work in the United States, showed a similar pattern. From the policy perspective, Hammar (1993) reviewed the Swedish government's attempts at dispersing refugees, Robinson (1993e) compared the policies of Britain and Canada towards their respective quota refugees, Indra (1988) and Neuwirth (1984) reviewed the success of Canada's private sponsorship schemes, Lanphier (1993) assessed the effectiveness of Canadian public hosting policies, and Robinson (1997) reviewed critically the historical development of resettlement policies in the United Kingdom.

However, although third country resettlement in the developed world was a preferred policy option for a short period of time, it had fallen from favour by the mid-1980s. There were growing fears within the developed world that sovereign states were losing control of immigration (Scheinman 1983), with fears about illegal immigration making governments feel 'overwhelmed' (Stein 1983). A desperate response ensued, with negative consequences for refugees:

the protectionism to which our frustration over illegals has given rise risks catching refugees in the backwash. It is almost as if we believe we can make up in control over refugees what we fail to achieve with illegal entrants.

(Scheinman 1983: 81).

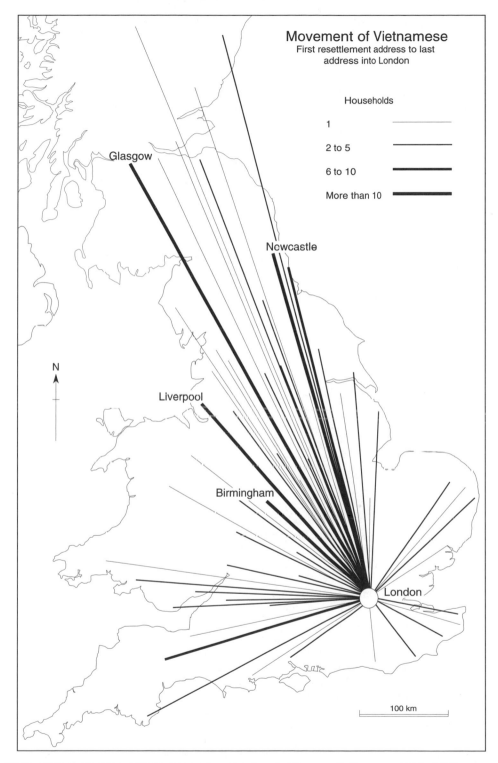

Figure 8.6 Secondary migration of Vietnamese refugees to London (source: Robinson and Hale 1989).

Moreover, the public began to question whether those who had been resettled were 'genuine' refugees or economic migrants. Consequently, many countries became far more introspective, a trend exacerbated by the global recession and competition for jobs. Critics pointed to the difficulties of integrating refugees from developing countries and the problems faced by refugees transplanted in alien cultures. *Compassion fatigue* set in, and the era of mass quota resettlement came to an end. However, despite the undoubted problems with third country resettlement, between 1975 and 1986 two million people were resettled in developed countries at a time when many regions of the world were in turmoil, with 10 percent of all post-1945 refugees experiencing such resettlement. Nor has the policy been completely abandoned. The UNHCR still argues that it has validity:

as an exceptional measure to be pursued ... for compelling humanitarian reasons or where the alternatives of voluntary repatriation and local integration do not exist.

(quoted in Stein 1983: 191).

Thus, Scheinman (1983) proposed that the objectives of third country resettlement should be adjusted so that policy sought to match fleeing refugee populations with countries that had pressing labour needs.

Repatriation

As developed countries closed their doors on refugees from the world's poorer countries (less than 0.5 percent of refugees now receive permanent resettlement in a third country), policy responded by advocating *voluntary repatriation* as the preferred option. It was argued that refugees would recover from their ordeal most rapidly and effectively if they were reintroduced to their region of origin, with its known language, way of life and means of subsistence. Attachment to place and people provides an important pull factor, as do a desire to regain citizenship and citizenship rights and a wish to reclaim property or land. Push factors encouraging repatriation include a worsening situation in the country of temporary asylum. For example, in 1989, 80,000 Ugandan refugees in Sudan returned after they had been attacked by rebel Sudanese troops. Two forms of repatriation exist; *organised repatriation* and *spontaneous repatriation*. In the former, governments and supra-national bodies negotiate a return to a precise plan, with responsibilities established

clearly in advance. Such returns might be voluntary or *involuntary*, depending upon the degree of coercion required to get refugees to participate. In spontaneous repatriations, refugees make their own decisions about when and how to return, and repatriation is often undertaken without external financial or material assistance. Both require an end to the conflict or persecution that caused initial flight and a belief that improved conditions can be sustained. The UNHCR (1993) noted, however, that such ideal preconditions are rarely fully met, with refugees returning while conflicts still rage and experiencing continuing violence after their return. Nevertheless, because of an easing of political tensions and the resolution of some previously intractable local wars, repatriation has become much more common in the 1990s, with spontaneous repatriation accounting for 70 percent of all return in 1992. In that year, over 2.4 million refugees were repatriated, an unprecedented flow of 6,600 per day.

Simon and Preston (1993) describe the organised repatriation of 42,000 Namibians from Angola and Zambia in 1989. The UNHCR took responsibility for registering those who wished to return, organising their travel arrangements and shipping their belongings. They were then handed over to an autonomous but UNHCR-funded body – the Repatriation, Resettlement and Reconstruction Committee – who handled the reception and dispersal phases within Namibia. The UNHCR spent $55 million on this one programme alone over two years. On the other hand, Findlay (1993) and Wood (1989), speculated about a spontaneous return of Afghans in Pakistan, an event that subsequently took place on a large scale after the fall of the Najibullah regime in 1992. The UNHCR estimated that 2.83 million Afghans self-repatriated between 1990 and 1995, with only limited assistance from that organisation. Returnees from Pakistan were given $130 and 300 kilograms of wheat when they handed in their ration cards.

Although repatriation has not been without its successes, the policy is fraught with difficulty. Refugees are often returning to a war zone, where they may face the possibility of attack or disablement from land mines. Infrastructure has often been devastated, land will have fallen into disuse and may have been usurped by others who did not flee, and social and kin networks may have unravelled. It has been estimated that 15,000 of Afghanistan's 22,000 villages were partially destroyed by armed conflict and the country's productive capacity had effectively been annihilated. Furthermore, agricultural production had fallen by 47 percent between 1978 and 1987 (Findlay 1993), irrigation systems had been left unmaintained, and over half of the farms in the eastern border region had been abandoned (Wood 1989). Perhaps as many as ten million land mines had been sown indiscriminately, and two million Afghans had been disabled during the war. Indeed, it was estimated that it would take six years to clear mines in priority areas alone, such as alongside main roads. Nor is collateral war damage the only problem associated with return; as the well-publicised return of Vietnamese from Hong Kong illustrated, not all organised returnees are repatriating of their own free will.

Containment and exclusion

While voluntary and involuntary repatriation have continued apace, so developed world governments have formulated and implemented a fourth strategy for coping with the refugee crisis. With electorates tiring of mass immigration and immigrants – witness the racist attacks on refugee hostels in former East Germany – and refugee crises erupting on their doorsteps (former Yugoslavia and Haiti) and spontaneous refugees now more able to travel and present their claims for asylum on the soil of their destination countries, so the developed world has moved to both *exclude* asylum seekers and *contain* potential refugees in their place of origin. Western European governments have taken various measures to deny asylum seekers access (Robinson 1996a). These range from fining airlines that carry them to erecting physical barriers, insisting on visas for travel from an increasing number of developing countries, legislating on the location where asylum claims must be made, deporting illegals or overstayers, and equipping neighbouring countries with technology to prevent illegal entry. The tendency to containment, or 'prevention' as it is euphemistically described by its supporters, has been advocated both by European governments and by the United States. The United States, for example, interdicted Haitian refugees at sea and returned them against their wishes. More subtly, Britain, France and the United States used UN Resolution 688 to create 'safe havens' for Kurds in northern Iraq after the Gulf War and established Operation Provide Comfort to supply and feed this population *in situ*. Both initiatives were designed to forestall the efforts of Kurds

to flee persecution and cross an international border into Turkey, a NATO ally, which did not wish to see its own Kurdish 'problem' intensified by refugees. In Rwanda (Rutinwa 1996) and Bosnia, too, 'safe' areas were created. In the latter theatre, as civilians in 'safe' areas were mortared and sniped at, the concept was further degraded by the admission that international peacekeepers could only secure safe corridors for the transmission of emergency aid. US President Clinton later admitted the failure even of safe corridors when he resorted to airdrops of supplies. A low point of current efforts to assist those in fear of violence occurred when the UN Protection Force in Sarajevo forcibly prevented people escaping to safety in exchange for a Serb assurance that aid would be allowed through.

Refugee policy has come full circle, with the *exilic bias*, which argued that those in fear of violence should be encouraged to flee their country of persecution and seek out residence in another, being replaced by a presumption of immobility and a *'Hollywood' bias*, in which refugees are now told to remain where they are and wait for the United States/UN 'cavalry'. Not all commentators have accepted such changes as *fait accompli*, however, with Gurtov (1991) proposing a new approach to the refugee crisis based upon global humanist principles.

Not all forced migrants are refugees

Other types of forced migration

So far, this chapter has discussed those who are seeking and are likely to be granted refugee status as defined by the 1951 UN Convention. However, while refugees are both unusually at risk and also prominent in the public eye, they are not the only migrants being forced to flee. Refugee studies is a relatively youthful field of enquiry but one which has grown rapidly since its inception (Robinson 1990b). In line with the growing volume of work has come greater definitional and terminological precision. While early works had perforce to use the UN definition of refugees, as people who were being individually persecuted and forced to flee across an international boundary, as early as the 1970s workers noted that such a definition was too restrictive. Subsequently, several typologies of forced migrants have been developed, each emphasising or introducing new causal categories which have spawned new types of forced migrant.

Ferris (1985) drew attention to *quasi-refugees*, who were suffering some degree of pressure to leave but who were also attracted to becoming migrants by the aid and assistance available in a new location. Following Scudder and Colson (1982), Cernea (1990) focused upon *development-induced displacees* or *oustees*, namely those forced to leave their homes by development projects, such as the construction of dams, but who do not leave their country of origin (see below). He subdivided oustees according to whether they lose their homes, land or both as a result of the forced move. Gurtov (1991) observed that many analysts mistakenly excluded all *internal refugees* from their work, not just those affected by development projects; internal refugees being people driven to flee by the same causes as refugees but unable or unwilling to cross an international border. Elsewhere, Oliver-Smith and Hansen (1982) distinguished between *allocatees*, *slaves* and refugees. They emphasised the power of the state to influence the two key decisions involved in any migration, namely the decision to leave an old place and the decision to relocate in a new place. Where the state determined both decisions, slaves would be produced; where the state influenced the decision to leave but determined the destination, allocatees would be produced; and where the state determined departure but only influenced destination, refugees would result. They also saw the same element of powerlessness in forced migrations caused by natural disasters. Scudder and Colson (1982) highlighted the same point by contrasting refugees fleeing life-threatening human action with those fleeing life-threatening natural disasters. The latter have more recently been considered by Wood (1995), who further distinguished *ecomigrant* and *environmental refugee*, arguing that the latter is a sub-set of the former. Ecomigrants are forced to move by rapid environmental change, such as an industrial accident in an urban area or the depletion of a natural resource that was the *raison d'être* for the settlement. Environmental refugees, on the other hand, are forced to flee for their lives from a cataclysmic natural disaster, such as a drought or an earthquake.

While authors such as those mentioned above have alerted us to the existence of different motivating factors behind forced migrations and have therefore proposed contrasting ways of differentiating those who have little choice to migrate, two have produced integrative typologies worthy of repetition. First, Richmond (1988) focused upon international migrants and argued that it has become increasingly difficult in the contemporary world to distinguish between migrations supposedly motivated by economic reasons and those supposedly driven by socio-political factors. In most cases, the motivation behind movement is a combination of the two. He therefore suggested that a continuum existed between the two polar extremes, with specific migrations located at any point along this continuum depending upon the balance of causes. In parallel, he also noted that it is impossible to dichotomise migrations into voluntary and involuntary. Again, a continuum exists with an infinite number of points upon it. The combination of these two axes produced a diagram similar to that shown in Figure 8.7. Any point within the diagram denotes a unique form of migration, with a unique mix of economic and socio-political motivations and a unique degree of autonomy in decision making. We are thus able to identify an infinite number of types of migration, in contrast to conventional typologies, such as that of Petersen (1958). Where a particular point on the diagram denotes a commonly occurring type of migration, we are able to allocate that point a descriptive label (for example, draft dodgers).

However, Richmond's continuum approach is not without its weaknesses. Unless the diagram were redrawn in three (or more!) dimensions, it is impossible to have more than two active continua, which assumes that the degree of voluntariness and the economic and socio-political dimension are the two key factors in defining migration types. Where, for example, would one locate ecomigrants or environmental refugees? Given the growing proportion which these groups form of all forced migrants today, and the hypothesised changes in migration occasioned by the impact of global warming, their omission is serious. We are also unable to distinguish between those forced migrants who can cross an international border and those who cannot. Furthermore, how does one capture the significant distinction that Cernea (1990) makes between those oustees who are fully compensated and those who remain uncompensated? Finally, the introduction to this chapter noted the power of labels and, in particular, the power of the label 'refugee'. This is not captured in the diagram at all.

Rogge's (1978) alternative (complementary?) typology related specifically to refugees in Africa. Several of the underlying principles of his schema are similar to those later adopted by Richmond.

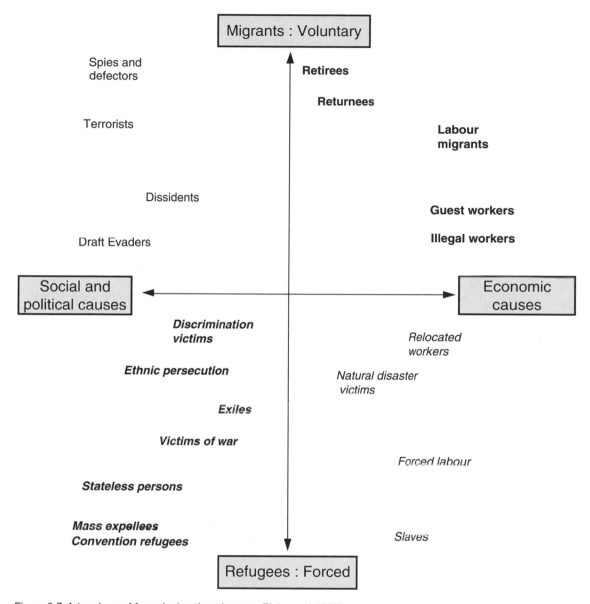

Figure 8.7 A typology of forced migrations (source: Richmond 1988).

First, varying degrees of coercion are involved in forced migrations and these need to be treated separately. For example, Tutsis fleeing death and violent persecution in Rwanda form a different sub-type to South Vietnamese fleeing because they feared their privileges would be undermined after the departure of the American-supported capitalist regime. Second, Rogge noted the point, already made above, that crossing an international border is an important discriminating factor between types of forced migration. Third, the attitude of the receiving country is also important. If that country fails to recognise (label) a group as refugees then it is unlikely that the international community will do so. Rogge cited the contemporary case of Senegal and Ivory Coast, both of which chose not to recognise Guinean forced migrants as refugees so as not to upset the Guinean government. Lastly, forced migrations can occur for a variety of reasons, some of which are political but many of which are not.

Box 8.3 Rogge's typology of African refugees

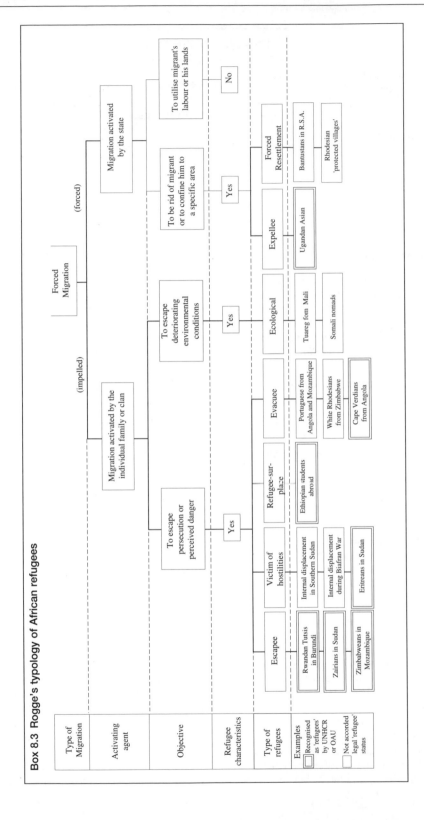

Box 8.3 (cont.)

The main types of forced migrant identified with the above diagram were:

- Escapees: anticipatory; politically motivated; invariably given international recognition; a common group;
- Victims of hostilities: people displaced directly by warfare or violence; escaping a war zone; often do not cross an international boundary;
- Refugees-sur-place: occur less commonly; international migrants; migrate for non-political reasons but find that, while abroad, political circumstances change at home, preventing their return;
- Evacuees: usually international migrants; rarely recognised as refugees; often acute refugees; often educated and of a higher social status; often a colonial minority; refugee-like (they have to leave in a hurry often without material belongings);
- Ecological refugees: not motivated by political factors, therefore not recognised as refugees by international bodies; refugee-like; may be political overtones/causes to the migration;
- Expellees: given no choice; expelled by government; may be prevented from taking belongings; often recognised by international agencies as refugees; often expatriates;
- Forced resettlers: very refugee-like; denied recognition by host government.

Source: Rogge 1978.

Development-induced migration

The study of those forced to migrate by development projects has a tradition stretching back into the 1960s (Oliver-Smith and Hansen 1982) but one which has taken academic research in a divergent direction to that of refugee studies. Cernea (1990: 321) was thus able to write:

a literature on refugees co-exists side by side with the literature on development-caused involuntary resettlement. The two literatures do not usually speak to one another. With some exceptions, most of the writings on refugees ... omit altogether oustee groups from the typology of displaced populations. And in turn the anthropological literature on oustees bypasses comparative analysis with refugees. As a result, the chance for more in-depth treatment by examining commonalities and differences is being missed.

Consequently, Cernea argued that the study of the

two different forms of forced migrants should be recombined as similarities between the two groups warrant this. They both lose their land, homes, jobs, assets and possibly their ability to feed themselves. In both cases, production systems are dismantled, social networks are often partially unravelled and relocation occurs in unknown areas, perhaps against the opposition of local people.

The current scale of the problem of development-induced displacement can be illustrated using the example of dam construction (*ibid.*; Weist 1995). As Table 8.4 indicates, 3.2 million people are currently, or shortly will be, dislocated by *major* dam constructions, with a further seven million entering the cycle of displacement each year as a result of other types of development project. Those affected are often the very poorest and least powerful in society: illiterates, people with low incomes, and those with an utter dependence upon the land (Scudder and Colson 1982). Cernea (1995) sketched a 'resettler's income curve', which illustrates the potential for downward mobility but which also shows how well-conceived and well-timed policy initiatives can forestall this. He stressed the importance of planning in advance. For example, at China's Shuikou hydro-electric power project, orange orchards were planted several years

Table 8.4 Number of people affected by major dam projects.

Dam	Country	Number of people
Already built		
Tarbela	Pakistan	86,000
Srisailam	India	100,000
Akosombo	Ghana	84,000
Kossou	Ivory Coast	85,000
Kainja	Nigeria	50,000
High Aswan	Egypt	100,000
Nangbeto	Togo/Benin	12,000
Saguling	Indonesia	55,000
Danjiangkou	China	383,000
Sobradinho	Brazil	60,000
Mangla	Pakistan	90,000
Under construction		
Shuikou	China	68,000
Narmada Sardar Sarovar	India	70,000
Almatti	India	160,000
Narayanpur	India	80,000
Itaparica	Brazil	45,000
Yacyreta	Argentina/Paraguay	45,000
Kayraktepe	Turkey	20,000

Source: Cernea 1990.

before commencement so that those dislocated would be able to harvest fruit from their new land in their first season. Moreover, Cernea argued that displacees need to be compensated at replacement value not market value, with this cost being internalised within project costs rather than being left to others to provide. He also described how mass displacement is not inevitable in major projects such as dams, the volume of displacement being heavily affected by design parameters. For example, reducing the height of the Saguling Dam in Indonesia by only 5 metres (from 650 to 645 metres) decreased the number of people who were forced to leave their homes by 45 percent. Scudder (1981) took such proposals a stage further, with a four-stage model of successful population displacement, while Oliver-Smith (1991) demonstrated the consequences of failed planning through his discussion of settler resistance and political mobilisation. Oliver-Smith exemplified this resistance with a number of case studies. One showed how the construction of the Kaptai Dam in Bangladesh ignited a resurgence in ethnic solidarity amongst local tribal peoples, who took to armed insurgency, and ultimately claimed cultural autonomy and indirect rule.

Industrial disasters

One of the best-documented industrial accidents that prompted mass forced relocation is the Three Mile Island incident on 28 March 1979. A nuclear power plant south east of Harrisburg, Pennsylvania, in the United States suffered a series of minor accidents and failures, which combined to produce a disaster in which thousands of gallons of radioactive water were spilled within the plant and radioactive gas escaped from it. Contradictory information was released, with different levels of government offering different advice to those who lived nearby, and it was not until two days after the accident that a public announcement confirmed that radioactive material had actually escaped. State authorities had considered evacuating a 8–16 kilometre zone around the plant but, in the event, only 2500 pregnant women and children near the plant were recommended to leave the area. Nevertheless, 144,000 people who lived within 24 kilometres of the power station spontaneously evacuated (39 percent of the potential population) and two-thirds of those who did not evacuate seriously considered this option (Zelinsky and Kosinski 1991). Evacuation had a degree of selectivity to it, with women, children, famil-

ies with small children, families of above-average income, and families containing a pregnant woman being most likely to evacuate. The same groups were also more likely to have remained outside the area for a longer period of time. Most evacuees left as a family. The most common duration for evacuation was five days, but 52 percent of people remained away for six or more days, with twenty-one states receiving evacuees, who had travelled a median distance of 135–160 kilometres. A clear distance-decay effect existed in the decision to evacuate, although Zeigler, Brunn and Johnson (1981) found a sharp discontinuity approximately 20 kilometres from the plant, modified according to prevailing wind direction. Many analysts, though, were surprised at the size of the area from which people evacuated. Zeigler, Brunn and Johnson consequently coined the phrase *evacuation shadow* for that area from which evacuation occurred but which might not have been expected. Clearly, people's fears about the consequences of nuclear accidents exceeded those about other types of accident, a finding confirmed by a later hypothetical study (Zeigler and Johnson 1984).

Subsequent work on evacuation from other industrial accidents (such as Bhopal in India) is extensive (Zelinsky and Kosinski 1991). Central to all of these studies is the importance of information in decision making and action taking. In the Three Mile Island incident, the lack of accurate and timely information on the threat was one of the reasons why even more people did not evacuate; while in Bhopal, the misguided recommendation for people to return to their homes near the plant shortly after the accident probably cost many lives. Khoser (1993) has looked in more detail at the role of information in the decision of refugees to return to their region of origin. Clearly, there are valuable parallels to be made between these different forms of forced migration in this context.

State slavery

The best recent example of mass state slavery is the policies pursued in the Soviet Union in the middle years of this century. Tolstoy (1981: 254) describes the mentality that underpinned them:

Stalin and his advisers entertained deep fears ... that Soviet Russia was a house of cards, held together only by the bonds of the NKVD, which would require only one determined push to collapse as suddenly as had Imperial Russia in 1917.

Stalin's fear of his own subjects prompted him to use the NKVD (secret police) to spy upon, denounce and punish 'subversives'. While others might simply have persecuted those deemed to be opposed to the state, Stalin also extracted forced labour from them. Since they were prisoners of the state, they could be directed to locations at which free labour would never countenance working and could be used for tasks beyond those that 'normal' workers would do. By 1941, it has been estimated that slave labour was responsible for fully 25 percent of the output of the Soviet economy, with perhaps 18–25 million slaves being employed at any one time. Conditions were such that mortality rates of 50 percent per year were not unknown in the early part of the war, rising to 60–70 percent by its end. Tolstoy described how one detachment of 7000 slave labourers, working in the Kolyma goldfields, had completed their construction task but were too weak to be marched to the next site. Rather than transport them, the NKVD simply herded them under a cliff, which was then dynamited above their heads. There were no survivors. Because of such losses, there were always 1.5 million slaves in transit at any point in time, being shipped thousands of kilometres to labour camps. Many died *en route*; 1650 froze to death in one train alone in 1941 and on another the mortality rate was 92 percent.

Such atrocities were not only visited upon proven 'dissidents' or Soviet citizens. When the Soviets took the Baltic states in October 1939, all 'anti-Soviet elements', including children, were deported to labour camps or *gulags*; 34,250 vanished from Latvia alone, representing 2 percent of the pre-war population. When the Baltics were formally annexed in August 1940, the NKVD drew up a list of those to be deported, which ran to 29 categories. It included 'mystics', freemasons, those previously employed by the state and anyone of aristocratic descent. Their seizure followed the usual pattern, with dawn raids on houses, followed by the splitting of husbands and wives, and their separate transportation to rail heads from whence they were despatched to different labour camps. Twelve million people died in the *gulags*.

Environmental disasters

Although there is now a huge literature on environmental disasters, many of which have prompted forced migration (see, for example, the journal *Disasters*), one example will suffice to demonstrate

---

Box 8.4 The treatment of political dissidents in the Soviet Union under Stalin

Some appreciation of the harsh treatment meted out to political dissidents in the Soviet Union under Stalin's rule comes from the following quotes:

Treatment of Soviet slaves was rendered more than habitually cruel because they were not regarded as merely a supply of cheap workers, but also as enemies of the nation.

(p. 11).

In the goldfields ... a man, loading a barrow, prodded by the shouts of a foreman or a guard, unexpectedly would sink to the ground, blood would gush from his mouth – and everything was over. From the chimneys of the Lubianka daily arose smoke from the corpse incinerator. ... Until 1948 every corpse was brought to the camp guardroom for registration. A sentry thrust his bayonet through the silent heart, to ensure that no one living left the camp, and the naked bodies were piled on to oxcarts for transportation to mass graves.

(p. 15).

When the death rate in the camp was so high that there was no room for the corpses in the shed ... they had to be stacked outside against the wall in roughly-covered piles, from which a leg or a ghastly, grinning face would stick out.

(p. 15).

It was not only the victims themselves who endured suffering impossible to describe effectively, but all those who were left behind. An Estonian woman described how she 'walked the streets next day, and met everywhere people with tear-stained faces. Fathers, mothers, sons, daughters, friends, acquaintances – everybody had someone to mourn after'.

(p. 218).

For a whole week (in June 1941) the Soviet transport system from the western frontier regions to the interior was crowded with slow-moving railway convoys shifting slave shipments east, through Leningrad, Minsk, Moscow, Kiev. ...

(p. 219).

Source: Tolstoy 1981.

---

how such movements might differ from those of legal refugees. The evacuation of the West Indian island of Basse-Terre, which represents half of Guadeloupe, occurred in 1976 in response to a feared volcanic eruption from La Soufrière mountain. The mountain had begun to emit cloud and ash early in July, and went on to experience 347 seismic shocks on 12 August alone. Initial evacuation was spontaneous, with 20,000–25,000 people leaving on 8 July when the local prefect declared a 'pre-alert'.

However, within two days, the same official had recommended that the population return home, which
most did, although renewed heightened activity
prompted an invitation for non-essential personnel
to leave again on 13 August. By 15 August, a total
evacuation order had been made and 18,500 people
left for the neighbouring island of Grande-Terre on
that day. Others subsequently followed, bringing the
total in the government's care to 33,000, although
an additional number moved in with relatives or
rented hotel rooms. In stark contrast to the previous
discussion of the *gulags*, Zelinsky and Kosinski
(1991: 73) described the Guadeloupe government's
arrangements for accommodating evacuees in positive terms, and concluded that

the Soufrière episode of 1976 did demonstrate that with
careful planning, preparation, and handling of information,
and the allocation of adequate resources, it is possible to
remove large numbers of persons from disaster-prone
areas quickly and safely and then restore the pre-
emergency pattern of existence reasonably well even in a
Third World community.

Conclusion

A consistent theme of this chapter has been the issue
of labelling and the importance of assigned labels. It
has been argued that forced migrants are given a
variety of labels, either by organisations, by governments, or by academics. Which label a group re-
ceives may well subsequently determine its real
world and academic fate. Those accorded the prized
label of 'Convention refugee' can expect the protection and assistance of governments and the attention

of a growing body of academic researchers and
media reporters. Those assigned other labels can expect much less. It is no coincidence that, when
writing this chapter, the type of forced migration
that proved most difficult to exemplify was state
slavery. Perhaps this was because so few participants
survive institutions like the *gulags* to tell their
stories, or perhaps there are insurmountable difficulties for researchers wishing to study such forced migrations. Even so, it is noteworthy that some of those
who have received the worst treatment have been
given the least attention. What is also clear from the
account above is that labels have led to a fragmentation of academic effort into the study of forced migration. While Cernea (1990) is right to point to the
dichotomy that exists between those who study
'refugees' and those who study development-
induced oustees, the problem is much wider and
more severe. Convention refugees are studied by
those in refugee studies, development-induced oustees by those in anthropology, ecological migrants
by those in the field of disaster studies, ethnic and
racial 'misfits' by sociologists, and state slaves by
social historians, while geographers tend to range
selectively over the spatial dimension of all of these.
Potentially, such a variety of disciplinary perspectives undoubtedly has much to offer, but equally it
means that research effort is dissipated or dupli-
cated, with important parallels, contrasts and policy
lessons missed. The challenges that will most tax
both concerned academics and governments in the
late 1990s are questions of labelling, namely how to
break down academic boundaries and ensure that
the whole is greater than the sum of the parts, and
how to wrestle with the moral and practical issue of
distinguishing between economic migrants and 'genuine' refugees.

CHAPTER 9

Migration and culture: some illustrations

Introduction

The previous chapters in this book have demonstrated some of the many different perspectives that can be brought to bear in the study of migration. We have seen migration described as a response to economic factors, as a quality of life decision, or as a desperate attempt to avoid persecution. However, although the rationale for migration may come from these varied directions, the specific act of migration itself is always part of an individual's life-course experience. This is the emphasis suggested by the biographical approach, introduced in Chapter 3, which argues that rather than see migration as a discrete event, carefully calculated at a specific time and in a specific place, the action must be seen as being embedded firmly within an individual's overall daily existence. Migration events relate to an individual's whole life – both past experiences and projected future expectations – and tend to have a wide variety of causes – some highly prominent and others more hidden but still essential to understanding the precise form that migration takes. This sense of embeddedness makes migration a very cultural event: migration is both a reflection of culture and a constitutive element of culture.

Before we can discuss the extent to which migration is a cultural event, however, we must be clear on what is meant by 'culture'. This is by no means straightforward. In recent years, academic understanding of the term has shifted from an emphasis upon highly skilled and distinctive human activities to a concentration on overall ways in which groups of people live their everyday lives – their actions, their motivations, their feelings. To study culture is to deal with people's 'maps of meaning' (Jackson 1989) or 'structures of feeling' (Williams 1973).

Cultures belong to both individuals and groups, existing through the shared and negotiated practices of everyday life. Cultures are both deeply felt and taken for granted; they are dynamic; and there are a wide variety of them, even within one small country. Finally, migration is intimately a part of many cultures, being a key practice in people's lives.

Reflecting the biographical emphasis on migration as a highly contextualised event, this chapter illustrates the extent to which migration is both infused with cultural values and infuses individuals with such values. It also discusses some of the cultural impacts caused by migration flows on the attitudes of those living at the migrants' destinations and expresses some of the political ramifications of migration. Therefore, this chapter re-integrates some of the migration experiences described in Chapters 4 to 8, both with the broader lives of those involved and with other activities and experiences with which people are involved. In particular, it does this by showing how the selectivity of migration by cause and type of person generates distinct *cultures of migration* and by demonstrating how such cultures feed back into reproducing these migration experiences over time and across space. The concept of cultures of migration expresses how both the association between migration and key events and experiences in people's lives and the selectivity of any specific migration process are reflected in the relative position that migration holds within specific societal groups. Moreover, the feedback mechanism between culture and migration serves to embed and make relatively permanent these distinct migration experiences and expressions, allowing us to map and model patterns and processes. In other words, cultures of migration serve to make selective migration experiences appear both fixed – they are *reified* – and expected – they are *normalised*. Finally, for parsimony, attention concentrates on groups of migrants rather than on individual participants, but we must recognise considerable variation within the groups discussed.

Only relatively recently has the distinct cultural dimension of migration been acknowledged explicitly in the academic literature (Chapter 3), with Fielding observing considerable unidimensionality within migration research:

There is something strange about the way in which we study migration. We know, often from personal experience, but also from family talk, that moving from one place to another is nearly always a *major event*. It is one of those events around which an individual's biography is built. The feelings associated with migration are usually complicated, the decision to migrate is typically difficult to make, and the outcome usually involves mixed emotions. An anticipatory excitement about life in the new place often coexists with anxieties about the move; pleasure at leaving the old place is often disturbed by the feeling that one has almost betrayed those remaining behind. Migration tends to expose one's loyalties and reveals one's values and attachments (often previously hidden). It is a statement of an individual's world-view, and is, therefore, an extremely *cultural* event.

(Fielding 1992a: 201).

The cultural dimension of migration has been appreciated best in the anthropological literature (Taylor and Bell 1994), notably in the strong community focus of the work, most of which has been undertaken in the developing world. In contrast, work in the developed world, more often undertaken by geographers, sociologists and economists, has rather overlooked culture. These two traditions have been examined and illustrated by Taylor and Bell in the case of Australia. For indigenous Australians, migration is typified by circulatory movements, reflecting a marginal attachment to waged labour, a difficulty in obtaining services and a nomadic tradition. In contrast, studies of the non-indigenous population displayed a narrower emphasis on migration's linkage to changing labour-market considerations. While the cultural aspect of Australian migration has been explored in the former case, it has been neglected when the non-indigenous population has been studied.

Bottomley (1992), again from an Australian angle, stressed the cultural experience of migration from the point of view of the migrant. She noted the experience of cultural change and discontinuity that can accompany migration:

By this very movement, migration challenges the idea of a distinct way of life. . . . Migration implies a radical change in objective circumstances. Migrants move into different political and economic systems where they must come to terms with already existing schemes of understanding and of power relations. Neither a subjective nor an objective account of this encounter is adequate by itself.

(*ibid.*: 3, 39).

Bottomley went on to draw attention to the writer John Berger's account of migrant workers in a book produced with the photographer Jean Mohr (Berger and Mohr 1975). She praised the way in which Berger, in part out of the empathy he felt as a marginalised English academic in self-imposed exile in mainland Europe, interweaved personal migratory experiences with various statistical and historical background information on the stimuli prompting emigration, such as poverty. Common roots of exploitation were seen by Berger in both south European migrant labour and the former slave trade in the Caribbean. Thus, there was much value in adopting such a culturally sensitive perspective.

The remainder of this chapter explores the idea of cultures of migration from five perspectives. The first four of these examine groups of people for whom the experience of migration is integral to their cultural identity and everyday experience: migration as spiralism, migration as escape and resistance, migration as nomadism, and migration as national diaspora. The final perspective takes a broader view through reference to the interactions between cultures of migration and the population at the migrants' destination, by exploring how migration can be experienced in terms of conflict. From all of these perspectives, we have concentrated on a number of relatively distinctive case studies, which we feel express some of the cultural aspects of migration in either very striking or especially interesting ways. It is largely left to the reader to translate these experiences to their own circumstances and to the more mundane occurrences of migration within everyday life in the contemporary world.

Cultures of migration: some illustrations

Migration as spiralism

Much has been said in this book, notably in Chapter 4, about the association between migration and employment factors. Thus, it is unsurprising that this association can generate feelings that link migration with economic success or betterment. Of course, we

Box 9.1 Researchable problems in the study of migration and culture

Ways of seeing places
How culture affects migration through the way in which places are seen and understood.

- The relationship between the strength of place identity and migration. A culture with a strong sense of place is likely to suppress migration to dissimilar places.
- The relationship between tradition of, or need for, co-location and migration. Cultures that tend to cluster spatially are likely to generate very specific migration patterns.
- The relationship between the balance of dominant and subordinate cultures and migration. Societal norms suggest net migration towards places where the dominant culture is strongest.
- The relationship between a mix of cultural identities and migration. Cultural conflict is likely to promote net out-migration.
- The relationship between dominant political values and migration. Individualist political values are likely to promote migration more than collectivist values.
- The relationship between conformity with the dominant culture and migration. Where conformity is high we might expect less specific migration flows than those to a place where conformity is lower.

Ways of seeing migration
How migration is experienced.

Group A: 'Stairway to Heaven' – the excitement and challenge of migration.

- Freedom: from boredom, the familiar, restrictions, social norms.
- A new beginning: 'wiping the slate clean'.
- Joining in: social life and activities.
- Opting out: getting out of the 'rat race', leaving stress behind.
- Going places: taking a step up in the world.

Group B: 'Crippled Inside' – the rootlessness and sadness of migration.

- Rupture: emotional break-up.
- Loss of contentment: unfamiliarity, nostalgia.
- Facing the inevitable: the only way to go.
- Failure: giving up, running away.

Ways of seeing migrants
How migrants are seen and understood within changing cultures.

- The relationship between mobility histories and local cultures. Does high mobility produce place-less cultures?
- The relationship between migrants' cultural characteristics and the class cultural nature of places. Do areas with high in-migration become more middle class?
- The relationship between migration and community development. Is community development impeded by high rates of population turnover?
- The relationship between migration and political geography. Are net migration flows altering countries' political geographies?
- The relationship between the cultural content of migration streams and cultural conflict. Does in-migration by people from locally subordinate cultures prompt anti-in-migrant hostility?
- The relationship between migration and the status of places. Are places experiencing high net in-migration held in higher esteem than those where net out-migration is occurring?
- The relationship between migration and cultural diversity. Are places that attract large numbers of migrants more culturally diverse?

Source: Fielding 1992a.

---

can be highly critical of this association as, for example, many married women move as tied migrants, actually harming their own career for the benefit of that of their spouse (Halfacree 1995a), but in dominant accounts, as given by politicians and other influential figures, migration is seen to be a 'good thing' economically, for both individuals and the nation. This linking of migration with economic betterment in a positive light is reflected in, and also can be qualified by, what can be described as cultures of *spiralism* (Watson 1964). Many professional and managerial jobs require the employee to migrate frequently, especially in the early years of his or her career, with each migration either directly or indirectly leading to economic betterment and potential promotion within the workplace. This spiralism, also represented in ideas such as Whyte's (1957) notion of a 'transient organisation man' sees geographical mobility overlapping with (upward) social mobility (Box 4.4).

Within developed nations, a key group that has been grappling with the culture of spiralism has been the *service class*, the professional and managerial fraction of the middle class, typically associated with high levels of academic and other qualifications. Savage and his colleagues have studied this lifestyle in the south east of England (Savage *et al.* 1992), developing insights from earlier studies (Johnson, Salt and Wood 1974). Migration is seen as central to the formation and reproduction of a *distinct* service class, since in many other respects the service class experience is very diverse. Service class members show a qualitatively different relationship to place than members of the working class, with their frequent migration leading to an 'expectation that the relationship of [a service class] individual with ... place or region of residence is a contingent one' (Savage *et al.* 1992: 33). For a successful service class career, a 'responsible' individual must be able to deal with 'non-locally-based information, codes, rules, and systems of thought and action' (Fielding 1992c: 15). This information source provides the individual with a level of *cultural capital*, which confers upon him or her a degree of social confidence. Such an accumulation strategy contrasts markedly with that of the *petit bourgeois* small shopkeeper or local self-employed artisan, for whom the cultivation and maintenance of local contacts can be critical to business success.

The frequent migration of service class persons leads to a de-localisation of their biographies, as the association between an individual's lifeworld (Buttimer 1976) and a strong local sense of place is unable to develop. This usually begins with the person moving away from home for higher education, although it may be an earlier feature of everyday experience if the individual comes from a mobile service class household. Leaving higher education, the newly qualified person then migrates to his or her first job and, if they are successful, their spiralism begins. The resultant non-local experience of the spiralist service class is reflected in their consumption practices (Fielding 1992c; Savage *et al.* 1992). For example, they show a positive association with such badges of experience as unusual foreign foods and clothes, second homes and exotic leisure travel.

While spiralism may be beneficial for an individual's career, much attention has been paid to the negative side of this experience. Taking account of the gendered character of this migration is critical (Halfacree 1995a), especially when linked to suburban living. Thus, although Pahl and Pahl's (1971) study of (male) managers stressed the importance of migration for occupational development, especially in the early years of a career, the Pahls also drew attention to the consequences of this migration for the managers' wives. For these women, issues of social life, education and other services were of at least equal importance to economic criteria in evaluating migration. However, such non-economic factors could be compromised severely with a move. Moreover, since wives are often well aware of the potential detriment to their own career of following a migratory lifestyle beneficial to their husbands, we can appreciate how the cultural position of migration is very different for males and females.

More generally, adopting a migratory lifestyle can prove stressful for both partners. Pahl and Pahl drew attention to the geographically dispersed friendship networks of managerial households and the very real efforts that often have to be made to create new friends after migration. Again, this issue is likely to be particularly significant for the wives, since they were largely the isolated suburban housewives, often looking after small children and unable to engage in paid work themselves (Oakley 1974). Elsewhere, Sennett and Cobb (1973) noted how a feeling that upward social and geographical mobility is somehow 'required' can cause stress, anxiety and identity problems.

The image of the service class as being extremely migratory has recently been challenged by changing modes of economic organisation. It has been argued that the spiralist model of career advancement, typified by the internal labour market (Chapter 4), is in decline (Savage 1988), especially in the south east of England (Savage, Dickens and Fielding 1988). Service class occupations have increasingly been located in smaller firms outside the large multilocation companies. In addition, Thrift (1987) has noted how service class members do not just follow job availability but also attract employment to the areas in which they live. All this would suggest a waning of the migratory lifestyle and culture of the service class. However, this point needs qualifying. First, the tendency for the service class to be more immobile seems much less clear-cut outside the south east. Second, the concept of the south east as an *escalator region* sustains the association between the service class and geographical mobility.

The escalator region hypothesis, with the south east region acting as a 'machine for upward social mobility' (Savage *et al.* 1992: 182), was developed through analysis of longitudinal study data (Chapter 2) from the 1971 and 1981 censuses (Fielding 1989, 1992c). Analysis has shown that the south east acts as a magnet for young, highly qualified individuals at the start of their career. This attraction can largely

Box 9.2 Spiralism and the affluent home-owner

A study of the housing histories of affluent home-owners located (when interviewed) in the city of Bristol in south west England reveals something of the complexity of the process of spiralism when played out within the context of everyday life. Although even the migrations for these largely upper middle class households were still largely constrained by employment factors – predominantly in the interest of the male's employment – the detailed processes of moving house are shown to incorporate a wide range of interests and priorities. Movement along 'an exclusive line with first class compartments only' also involves 'issues of culture and class and specific preoccupations, aspirations and constraints' (Forrest and Murie 1987: 358). Moves are seen to be strongly historically specific in their precise delineation, typically incorporating concerns with minimising family disruption, obtaining good schools for children, maintaining the investment value of property and general quality of life factors. This complexity is illustrated below, with respect to just one of the moves in the housing history of a senior executive of a multinational company. This narrative also demonstrates the importance of the range of subsidies and benefits that have the effect of cushioning such migrations from the often considerable vagaries of the housing market.

The example concerns a migration from the north west to the south east of England:

I was promoted again to the head office, which was in London. ... And that was a great problem because house prices had escalated and there was very little for sale in the South East. It was an enormous area that we could choose from anyway ... and it was terribly difficult because we made many, many trips down there to try and find something and it was a case of estate agents 'phoning us up in the North, saying 'There's a house just up for sale, can you get down here today if possible'. It took us several months to find something and in the end we picked something in desperation and it was a bungalow which we said we would never have again. ... But we did pick another one and it was right out in a very rural part of West Sussex, right out in a pine forest. We saw it on a lovely hot summer's day in July and we didn't see it again until we moved in in the middle of November and we were devastated. Very remote and ....

... I would travel into Waterloo [London] on the train, so I then got British Rail maps and we hatched off areas round the main railway lines into Waterloo Station, which were either in walking distance of the station or driving distance....

It was a Scandinavian house in a wood. But we also didn't expect to be there very long. We were told when I was going there that it would only be for a couple of years, but in fact it was less than that, it was only twenty months.

We sold that house in Nantwich [their previous property] for £43 000 and we bought the bungalow for £56 000. ... The lady who owned it was away on holiday so the agent took us round and we couldn't put an offer in. We then went away on holiday ... and I rang the estate agent from a call box to find out if we had got the bungalow because we had put an offer in of £56 000 with the estate agents. ... When I rang up the guy said somebody had offered £56 500 and I said I would match that offer – the company sponsors the move. In other words it is a guaranteed purchase, because she wouldn't get involved in a chain with the changing price situation, and it was accepted. ...

... from then on when we were in Sussex, the children were in an awful school. It turned out to be a dreadful school, so from then on we have picked schools and then houses afterwards.

I then moved to a management post, near Liverpool,

(quoted in Forrest and Murie 1987: 345–6)

Source: Forrest and Murie 1987.

be explained by the concentration of employment opportunities in this region, including work in London for service class women (Boyle and Halfacree 1995). In the south east, service class people are trained and promoted, thereby ascending the escalator of occupational success. In middle age, many then step off the escalator through a posting out of the region or through becoming self-employed, with many of the remainder leaving the region on retirement. Key regions that attract these migrants are the south west, the East Midlands and East Anglia. These exchanges have promoted the embourgeoisement of the south east, although, as Fielding (1992c: 15) observed, it makes the region appear almost as a 'non-place urban realm ... a spatially extended zone of almost formless urban development'. South east England has considerable prosperity but it is also characterised by very privatised lifestyles, making it, for Fielding, more banal than the north or the west, with their stronger local identities.

The practice of spiralism and the status of south eastern England as an escalator region suggest that place-based identities and feelings of belonging can be the high prices paid for the economic rewards that a service class career can bring. For service class members who have become more settled, an attempt to regain a stake in the community is thus understandable and is often reflected in active membership of conservation and other pressure groups (Savage 1988; Short, Fleming and Witt 1986). For the rest of the service class, a local political apathy and a lack of interest in the immediate environment – as noted many years ago (Bell 1969) – is likely to remain a key feature of the spiralist culture of migration. This is reflected in the seeming 'placelessness' (Relph 1981) of many suburban landscapes, which are perceived as having been stripped of any inherent sense of meaning and identity. Thus, we may wish to argue that migration can serve to undermine the place-based aspect of individual or group identity.

Migration as escape and resistance

In contrast to spiralism, there are cultures of migration associated with opting out of a society. Migration may carry with it the hope for a new life freed from the shackles of the old. For example, we saw in Chapter 6 that an important current underpinning independent urbanward migration of women in developing countries involves attempts to escape traditional restrictions and norms. In developed nations, too, there is a feeling that the individual will be able to obtain more freedom in the city than in the countryside. The anonymity of the city is seen as benign: one can lose oneself there and create one's own life. Thus, Raban's (1974) 'soft city' is a space to be moulded to the wishes of the individual. Similarly, selective migration to subcultural areas of the city, such as homosexuals to parts of San Francisco, occurs in part because migrants are better able to live their chosen lifestyle and maintain their identity in such a context (Fischer 1982).

We also saw in Chapter 6 how quality of life factors are important in explaining counterurbanisation. Migration to very remote and isolated areas is typically driven by a desire to escape from the stresses and strains of the modern world. In this respect the rural environment represents a refuge from modernity, a place where an alternative cultural experience can be appropriated or created by the migrant (Halfacree 1997). An interesting example is the flow of 'counter-cultural individuals' to western Ireland (Kockel 1991). From the 1960s onwards, there developed a perception of the west of Ireland as a place with an unlimited potential for people to pursue 'alternative' lifestyles. This resulted in a number of waves of counter-cultural migrants (Table 9.1), such as 'polit-tourists', concerned with trying to build a post-colonial Irish lifestyle; 'Celtic twilighters', whose romanticism blended into the

Table 9.1 Characteristics of different waves of counter-cultural migrants in the west of Ireland.

	Galway Bay	Lough Key
Period of migration >	From *c.* 1971	From *c.* 1975
Peak years >	*c.* 1981–1987	Mid-1980s
Original initiative >	Artists/artisans	Business people
In-migration		
First-wave settlers	*c.* 1971–1979	c.1975–1982
	Internal migrants	Crafts people
	Hippy farmers	Hippy farmers
Second-wave settlers	*c.* 1975–1981	*c.* 1982–1986
	Returning migrants	Returning migrants
	Hippy farmers and artisans	Hippy farmers and artisans
	Polit-tourists	Polit-tourists
	Folk-freaks	Folk-freaks
	New Agers	New Agers
	Celtic twilighters	Celtic twilighters
Third-wave drifters	From *c.* 1981	From *c.* 1986
	Sea Gypsies	Eco-freaks
	Wandering poets	Bards
Out-migration		
Drifters	From *c.* 1987	[Negligible]
	Wandering poets	
	Migrant labourers	

Source: Kockel 1991: 76.

New Age movement; and 'Drifters' engaged in casual labour or eking a living out of self-employment. Generally, unlike examples considered later, the migrants experienced little conflict with the established population, since they tended to be quite quiet, respected local traditions and often provided much needed employment. The process by which the counter-cultural persons became established in the west of Ireland, in towns such as Westport in Mayo, was relatively *ad hoc*. Often, young foreigners with counter-cultural leanings came over on holiday and then decided to stay, either using contacts to gain land or settling in a caravan or abandoned building. Finally, the mobility of the migrants became increasingly important over time, enabling Kockel to draw up a table of 'drifter culture' (Table 9.2).

As suggested in Kockel's study, semi-nomadic lifestyles are often attempts at escaping and resisting the norms and expectations of conventional society. There have been throughout history the 'masterless men' who have roamed throughout the land seeking casual labour and alms but remaining beholden to no-one (Crowther 1992). While these tramps often used migration to reject societal norms in a rather passive way, more political forms of vagrancy have also emerged periodically. In the United States, for example, the constant travelling of the folk singer Woody Guthrie in the inter-war years of the Great Depression was strongly linked to his left-wing and pro-trade union political beliefs and activism. Through travelling around in the same way as the American hobos (tramps) and migrant workers, such as the Okies who fled the dustbowl states for California in the 1930s (Mitchell 1996; Steinbeck 1939), Guthrie got to understand and appreciate the struggles of the migrant workers for a decent living and basic human rights, an appreciation that informed his songwriting and radio presentations (Klein 1981). Guthrie's migratory life also reflected his discomfort with the settled confines of the nuclear family and other conservative elements of the American Dream, as expressed by Bob Dylan in his 1962 *Song to Woody*:

I'm out here a thousand miles from my home,
Walkin' a road other men have gone down.
I'm seein' your world of people and things,
Your paupers and peasants and princes and kings.
Hey, hey, Woody Guthrie, I wrote you a song
'Bout a funny ol' world that's a-comin' along.
Seems sick an' it's hungry, it's tired an' it's torn,
It looks like it's a-dyin' an' it's hardly been born.

(Dylan 1988: 14).

This cultural–political status for migration went on to become expressed most fully in the constant travelling across America of members of the Beat Generation in the 1940s and 1950s and, later still, in the travels of hippie groups, such as Ken Kesey's Merry Pranksters (Wolfe 1989). Indeed, this nomadic rejection of a settled home (see below) seems very much a geographical expression of the late Timothy Leary's counter-culture dictum to 'Turn on, tune in, drop out'. Mobility has long been a key theme for the counter-culture, especially in the United States, from Jack London, through Bob Dylan, *Bonnie and Clyde* and *Easy Rider*, to *Thelma and Louise* (Cresswell 1993).

Migration as nomadism

Nomadic cultures

There are those for whom migration is probably *the* key defining feature of their way of life; for these *no-*

Table 9.2 Classification of drifter culture.

	Vagrant	Hobo	Pilgrim
Economic basis	Begging and petty crime	Casual work	Own work (artists, artisans, professionals)
	Welfare benefits	Welfare benefits	Welfare benefits
Range of movement	Not defined	Clearly defined	Vaguely defined
Temporary group settlements	None (sometimes institutionalised; occasionally use open colonies for short stay)	Open colonies with a high turnover but sometimes a stable core of settlers	Open colonies with a high turnover but sometimes a stable core of settlers

Source: Kockel 1991: 78.

Box 9.3 Mobility, the 'American Dream' and the Beat Generation

An interesting way in which the adoption of a semi-nomadic life can be associated with cultural resistance comes in the lifestyles of the Beat Generation, a group who emerged in the two decades after the end of the Second World War. These years saw, on the one hand, the shadow of the threat of the atom bomb and the fears raised by activities such as the anti-Communist witch-hunts of Senator Joseph McCarthy and, on the other hand, the promise provided by greater personal freedoms, higher incomes and the lure of technological progress. In this era of rapid change and often sweeping expectations, people increasingly began to question many of the contours and contents of the American Dream (Ehrenreich 1983). One such group was the Beats, a name adopted by one of their leading writers, Jack Kerouac, to mean everything from dragged-down and exhausted (dead-beat) to blessed (beatific) to following the rhythms of jazz. The term was used to describe Kerouac, his literary compatriots and their associates.

Many of the Beats adopted a semi-nomadic lifestyle, none more so than Kerouac. Much of his writing, notably the novel *On the Road* (1957), is dominated by the travels of Kerouac and his friend Neal Cassady. Indeed, *On the Road* can be read in a way which suggests that the Beats used mobility as an expression of their resistance to the established norms of dominant United States culture in the 1950s. Kerouac constantly criss-crossed the United States and Mexico during this time, mostly by car (although he could not drive) and bus. The key aspect in this restlessness was the ability to be 'just going', since an excited anticipation of a destination led quickly to disillusion on arriving, only to be replaced almost immediately by a renewed anticipation of the next destination; the elusive search for 'IT' (in many respects an internal spiritual journey) was never successful. Something of the spirit of this sense of movement is expressed in these three extracts from *On the Road*:

It was drizzling and mysterious at the beginning of our journey. I could see that it was all going to be one big saga of the mist. 'Whooee!' yelled Dean. 'Here we go!' And he hunched over the wheel and gunned her; he was back in his element, everybody could see that. We were all delighted, we all realized we were leaving confusion and nonsense behind and performing our one and noble function of the time, *move*.

What is that feeling when you are driving away from people and they recede on the plain until you see their specks dispersing? – it's the too-huge world vaulting us, and it's goodbye. But we lean forward to the next crazy venture beneath the skies.

So in America when the sun goes down and I sit on the old broken-down river pier watching the long, long skies over New Jersey and sense all that raw land that rolls in one unbelievable huge bulge over to the West Coast...

(Kerouac 1972 [1957]: 127, 148, 291).

In adopting such a shiftless lifestyle, Kerouac rejected the normative pattern of falling into a relatively stable and settled nuclear family structure. In this respect, the Beats' mobility was also linked to a rejection of conventional sexual relations, with any journey typically involving a variety of casual sexual encounters. For Kerouac, the road was 'a symbol of holiness and purity inhabited by mad angels, hungry for experience' (Cresswell 1993: 255), a manic intensity that he tried to get across in his jazz-inflected writing style and in his practice of writing novels in one huge surge over a number of days on a continuous piece of paper, being kept awake by benzedrine, coffee and other stimulants.

Looking further at Kerouac's and the Beats' emphasis on 'just going', however, immediately illustrates the highly selective freedom granted by this lifestyle. Most significantly, it is a highly gender-specific experience, with most of the major figures of the Beat Generation, and virtually all of those with a mobile lifestyle, being men. Thus a clear sense of exclusion and marginalisation comes across in the account of the period given by Neal Cassady's wife and Kerouac's lover, Carolyn. In this respect, the 'resistance' of the Beats shows many parallels with the myth of the mobile male outlaw – itself part of the American mythology that the Beats in other ways were trying to resist. Only one side of the picture was presented in books such as *On the Road*, the neglected side being the home environment. The male Beats' double life was never transcended and their 'masculine selfishness' (McDowell 1996: 418) meant that they failed to problematise the vital connection between their (male) freedom and the (female) refuges/homes to which they periodically retreated. Moreover, the celebrated semi-nomadic lifestyle was much less open to women, especially where young children were also involved. In short, in emphasising the freedom of the open road, the Beats ignored both the experience of women and the scope for the home, too, to be a site of resistance. Thus was one culture of migration highly restrictive in its scope.

Sources: Cassady 1991; Clark 1984; Cresswell 1993; McDowell 1996.

madic people migration/mobility *is* their culture. Such groups and individuals have received relatively little attention from migration researchers, who have perhaps been too eager to focus on the origin and destination areas of the migration, rather than the action of migrating itself. Indeed:

Migration studies purport to be about movement but use the push and pull factors of points A and B as explanations. People leave point A because point B appears to be favourable. It is never the case that both point A and B are unbearable and that the motion in between is the 'pull' factor.

(Cresswell 1993: 259).

Or, coming from a literary angle, Milan Kundera observed:

Road: a strip of ground over which one walks. ... A road is a tribute to space. Every stretch of road has meaning in itself and invites us to stop. ... Before roads and paths disappeared from the landscape, they had disappeared from the human soul: man stopped wanting to walk, to walk on his own feet and to enjoy it.

(Kundera 1992: 249).

Although our experience within the developed world today is primarily one of a settled existence, as Kundera suggests, there are still many peoples in the world who either live a nomadic existence or who have a culture that remains rooted in such an existence. As Figure 9.1 demonstrates, within Africa and Asia many areas are inhabited by such cultures, while the New World also has many examples of indigenous nomadic peoples, such as the Aborigines of Australia. Although nomadic peoples have conventionally been regarded as backward and in a transitional stage to a settled life, today they are more intrinsically valued (Monbiot 1994). Nonetheless, even maps such as Figure 9.1 are in many ways a reflection of our ignorance since, for example, the use of the term 'Bedouin' to cover nomadic peoples in

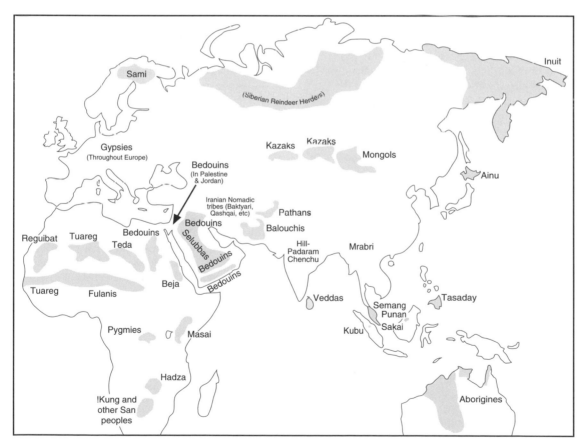

Figure 9.1 Nomads of Africa, Asia and Australia (source: Evans 1991: 23).

Arabia disguises the fact that there are very many different nomadic tribes in this region alone.

Again simplifying, Evans (1991) distinguishes three types of nomads. First, there are gatherer nomads, typified by constant movement in groups of around 20–60 persons engaged in hunting and gathering. Examples include the Pygmies of Zaire's Ituri Forest or the Bushmen of the Kalahari desert. Second, there are the pastoral nomads, probably the best-known group, which includes such diverse peoples as the Maasai of East Africa and the Sami of Finland. Third, there are trader nomads, who tend to live by their wits. The Gypsies (see later) are the best known, but the category also includes people such as the Sulubba tribe of Saudi Arabia. As we go down these three groups we find the interests of the nomads increasingly interwoven with those of the settled society.

Of the gatherer nomads, Taylor and Bell (1994, 1996) noted that the mobility of indigenous people of the New World, such the Australian Aborigines, is poorly understood. Aborigines tend to be highly mobile over relatively short distances, reflecting a high incidence of circular mobility. Circuits of migration between places combine to form networks, whose boundaries and divisions are defined by location of kin, traditional associations with the land, seasonal or short-term employment opportunities, and the location of public and community services (Altman 1987; Young and Doohan 1989). These networks have been maintained even with the erosion of the clan system since the nineteenth century – itself assisted by state policies such as New South Wales's Family Resettlement Scheme from 1969, which prompted a drift to the cities – and have been re-consolidated with the recent trend of a return to remote rural areas and clan-based settlements known as outstations or homeland centres.

The largest numbers of pastoral nomads are to be found around the edge of the Sahara Desert in Africa, such as in Sudan (Davies 1988; El-Arifi 1975), and in the deserts of southern Asia. As Stock (1995) notes, Africans have historically been very mobile, with the circulation pattern of pastoralism linked to the rhythm of the seasons (Figure 9.2). In particular, there is the movement of people with their animals from dry lands to wetter areas in the dry season and then back again in the rainy season. The latter is to prevent the overuse of the land and to avoid diseases. Detailed knowledge of the pastures is handed down through the family, reflecting the centrality of mobility to the pastoralist culture,

an importance also represented in animals providing a source of wealth and prestige. Finally, most pastoralists also have permanent settlements (*ibid.*), usually maintained throughout the year by women, children and some men, while herders take the animals to seasonally utilised pastures (Figure 9.3) (page 218).

Typically, pastoralists roamed over land too dry, rocky or steep for sedentary farming, thus complementing rather than competing with agriculture. However, in almost all studies of nomads, the perilousness of the nomadic existence is stressed, with migratory lifestyles threatened from a number of angles. First, there has been the impact of wars and political upheaval. For example, in Western Sahara the lives of the native Sahrawi nomads have been disrupted by the wall that has been built across the country to prevent attacks on the occupying Moroccan forces by the Polisario independence fighters (Arkell 1991). Second, drought has been a scourge of the nomads from time immemorial (Stock 1995) but its effects have been exacerbated in recent years by political factors. The tent-dwelling Beja nomads of Sudan's Red Sea Hills have been devastated by drought (Asher 1991), but this situation has been made worse by their geographical marginalisation by the Sudanese state (Egemi 1994). Third, direct and indirect political pressures have threatened the nomad. In particular, governments have often encouraged nomads to settle. Thus, in northern Côte d'Ivoire, Fulani herders arrived in the 1970s from Mali and Burkina Faso, fleeing drought, but came into conflict with local Senufo farmers, especially as the government had encouraged settlement (Bassett 1988). Protection of land for conservation or tourism purposes has seen the Maasai being excluded from Kenyan National Parks, undermining both their very livelihoods and the objectives of the conservationists (Monbiot 1994). Fourth, loss of land is blighting the future of nomadism. In Kenya, the costs of resisting the pressures of sedentary society can be extremely high:

the Barabaig [Maasai], following their rotation, tried to return to their pastures with their animals ... [but] they were told they were trespassing and were forcibly evicted. When they tried again, they were beaten up, fined and imprisoned. Homes were burnt down, dams were destroyed, and the mounds of the ancestor spirits were dragged under the plough.

(*ibid.*: 136).

In spite of such costs, nomads do not just accept

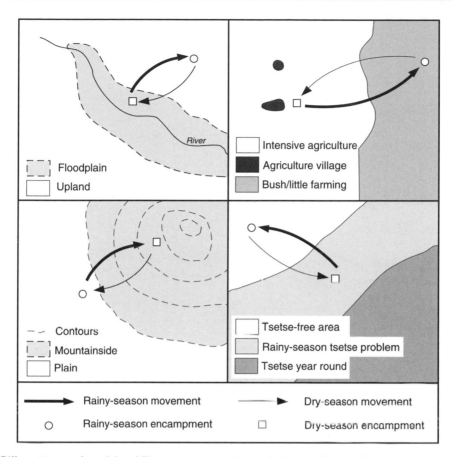

Figure 9.2 Different types of spatial mobility among pastoral nomads (source: Stock 1995: 156).

threats to their lifestyles and cultures. Resistance is most clearly demonstrated in the revolt of the 1.3 million Tuareg of Mali and Niger. This armed rebellion resulted from their political marginalisation in these post-colonial countries, the loss of their livelihoods due to state intervention and development projects, and from their experience as refugees (Krings 1995). Other forms of resistance include a revaluation and restatement of the importance of the nomadic lifestyle. This approach has probably been most successfully pursued by indigenous peoples in the New World.

It is important to stress the significance of the nomadic cultures to their members. Besides the value of the animals, mobile peoples demonstrate an immensely strong attachment to their territories. Even with the loss of tribal authority, drought and geographical marginalisation in the Sudanese state, the Hadendowa Beja of Sudan retained this attachment (Egemi 1994), eschewing migration to the Gulf states for work, for example. The nomadic culture of the Hadendowa is also reflected in their greeting, the *sakanab*, which involves a ritual verbal expression of news on the environment, rainfall amount and distribution, floods and the condition of the grazing resources (Morton 1988). For indigenous Australians, too, the land is in many respects venerated, and residential spaces are often conceived in terms of regions rather than single places (Taylor and Bell 1996). Lastly, nomadic cultures also stress reciprocity, since sharing resources and knowledge is crucial for minimising risks amongst such fragile and threatened cultures.

Gypsies

Gypsies arrived in Europe through the Balkans in Mediaeval times and, from around the 1430s, spread into the rest of Europe and beyond. Although their name originates from 'Egyptian', the most likely ori-

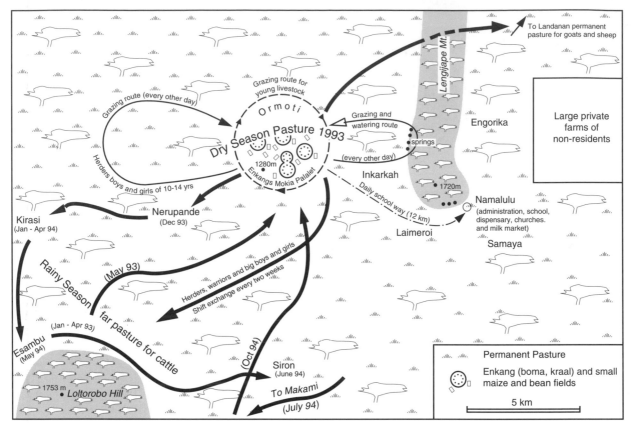

Figure 9.3 Annual movements of Maasai nomads in Lengijape, Tanzania (source: *Geojournal*, **36**, Ibrahim and Ibrahim, Pastoralists in transition – a case study from Lengijape, Massai Steppe,1995: 30, reprinted by kind permission of Kluwer Academic Publishers).

gin of the Gypsies is India. The Gypsy language (Romani) shows clear similarities with the Indian languages of Sanskrit and Hindi. However, the whole issue of Gypsy definition is fraught with difficulty and controversy, apparent from just looking at the names Gypsies call themselves. While in England, Gypsy (men) are often known as *Romanichal*, as they are in Australia, Canada and the United States; in Spain and southern France we have the *Calé* (Blacks); in Finland there are *Kaale*; and in Germany there are the *Sinti* (Fraser 1992).

Throughout their existence the Gypsies have met with distrust and resistance from the settled population, known as the *gadźo*. These experiences have been central to forging the contemporary Gypsy identity (Liegeois and Gheorghe 1995; Mayall 1992). In part, anti-Gypsy feelings and actions reflect the stereotypical identities with which they have been burdened, although more generally it also reflects an ages-old resentment of nomadic peoples in general.

With their migration into Europe, public attitudes towards the Gypsies deteriorated. Legislation resulted in beating, branding, execution and, in the Netherlands, organised *heidenjachten* (Gypsy hunts) in the eighteenth century (Fraser 1992). Anti-Gypsy repression revived following westward Gypsy migrations in the nineteenth century (Lucassen 1991). It was especially noteworthy in Germany (Fraser 1992). Early measures included the Prussian directive of 1906 on *Bekämpfung des Zigeunerunwesens* (combating the Gypsy nuisance), which promoted ways of impeding the Gypsy way of life. However, oppression peaked with the Nazi's genocidal policies of the 1930s and 1940s, where the Gypsies were singled out with the Jews for annihilation. Gypsies were classified by blood types, given specific identity papers and deported, with as many as 500,000 massacred in the Final Solution.

Since 1945, Europe's Gypsies have still been subject to persecution, such as that stimulated by the

Box 9.4 The songlines of Australian Aborigines

The British writer Bruce Chatwin's semi-autobiographical novel *The Songlines* gives us an insight into the culture of Australian Aborigines. Chatwin himself was favourably disposed towards nomadism, as reflected in his own widespread wanderings across the globe (Chatwin 1990), and this comes across strongly in the book. He describes his efforts to discover and understand the secret of the songlines. These are invisible pathways across the land that are 'sung' by indigenous Australians. When they are sung they tell the story of the creation of the world – the Dreaming – and it is a duty of the Aborigine to travel the lines – to go walkabout – singing at key points along the journey the stories which go with that place. In this way, the rich oral tradition of the Aborigines is maintained, a tradition also reflected in their works of art, which represent the landscape and, to the initiated, tell of history.

Witness a discussion between Chatwin and Arkady, a Russian–Australian who was mapping the Aborigines' sacred sites:

The Aboriginals had an earthbound philosophy. The earth gave life to a man; gave him his food, language and intelligence; and the earth took him back when he died. A man's 'own country', even an empty stretch of spinifex, was itself a sacred ikon that must remain unscarred.

'Unscarred, you mean, by roads or mines or railways?'

'To wound the earth', he answered earnestly, 'is to wound yourself, and if others wound the earth, they are wounding you. The land should be left untouched: as it was in the Dreamtime when the Ancestors sang the world into existence'.

(p. 13).

From this quote we can appreciate the centrality of the Aborigines' nomadic culture to their idea of the land, and their resistance to wounding the Earth. However, the songlines are by no means clearly defined or easy to understand and exhibit an extremely complex order. As Arkady put it: 'the whole of bloody Australia's a sacred site' (p. 5). This is particularly the case since individual Aborigines see themselves as descendants of 'totemic species', each with its own songline:

Every Wallaby Man believed he was descended from a universal Wallaby Father, who was the ancestor of all other Wallaby Men and all living wallabies. Wallabies, therefore, were his brothers. To kill one for food was both fratricide and cannibalism.

He [Arkady] went on to explain how each totemic ancestor, while travelling through the country, was thought to have scattered a trail of words and musical notes along the line of his footprints, and how these Dreaming-tracks lay over the land as 'ways' of communication between the most far-flung tribes.

(p. 15).

While Chatwin learned a great deal about the indigenous Australians and their way of life on his visit to Australia, he never succeeded in getting *that* close emotionally or spiritually to the songlines themselves. In part this reflects the secrecy with which the stories are guarded. Additionally, the fact that Chatwin came from a settled society very alien to that of the Aborigines meant that he never really had the cultural competence necessary to comprehend the alternative world view that was being espoused. Nonetheless, near the end of the book Chatwin speculates on the extension of the songlines to other continents:

I have a vision of the Songlines stretching across the continents and ages; that wherever men have trodden they have left a trail of song (of which we may, now and then, catch an echo).

(p. 314).

Source: Chatwin 1988.

---

collapse of the old order in Eastern Europe at the end of the 1980s (Crowe 1995; Liegeois and Gheorghe 1995). Usually, the persecution has taken the form of attempts to undermine the nomadic existence and promote settlement. Indeed, in Eastern and southern Europe, the majority of Gypsies are now permanently settled. Where nomadism remains commonplace, efforts to promote settlement often involve the provision of official sites, as in the British Caravan Sites Act of 1968 (Hawes and Perez 1995) or the Caravan Act of 1968 in the Netherlands (Cottaar and Willems 1992). At present, there are around 8–15 million Gypsies worldwide, with at least 5–6 million resident in Europe (Table 9.3) (page 220). There has been a growth of national and international Gypsy organisations since the 1960s, such as the *Comité International Tsigane* (CIT) (International Gypsy Committee). The CIT organised the First World Romany Congress in London in 1971, where delegates from fourteen countries adopted the name *Rom*, a flag and a slogan, *Opré Rom!* (Gypsies arise!). The influence of Gypsies in domestic politics has also been growing (Liegeois and Gheorghe 1995).

When considering the importance of nomadism to the Gypsies' cultural identity, we must note the two contrasting stereotypes which have plagued them throughout their history (Okely 1983; Sibley 1981, 1995). On the one hand, there is the Romantic image

Table 9.3 Classification of European countries by estimated number of Gypsies in each, late 1980s.

Number	Countries
500,000+	Former Yugoslavia, Romania
250,000+	Hungary, Spain, Bulgaria, former Union of Soviet Socialist Republics, former Czechoslovakia
100,000+	France
50,000+	Italy, Germany, United Kingdom, Greece
25,000+	Poland, Albania, Portugal
10,000+	Austria
1,000+	Sweden, Finland, Netherlands, Belgium, Switzerland, Denmark
<1,000	Ireland, Cyprus, Norway

Source: Fraser 1992: 299; Kalibová 1993.

of the Gypsy as being free, passionate and beholden to no one. On the other hand, there is the image of the Gypsy as a criminal; in the United States, there are so-called 'Gypsy experts' who lecture on why Gypsies are criminal by nature (Hancock 1992). While the latter stereotype is clearly more dangerous for the Gypsy, neither help in the better understanding of their way of life. We must, therefore, be very careful in generalising about Gypsy culture, as we must with all cultures. The creation of such stereotypes facilitates the 'geographies of exclusion' (Sibley 1995), whereby groups such as the Gypsies are defined as a polluting 'Other', a threat to the supposed purity of the (settled) established group.

The economic activities with which Gypsies engage constantly change with the changing demands of the settled population amongst whom they live (Fraser 1992; Sibley 1981; Sutherland 1975a). Occupational tribal names, such as *Lovara* (horse-dealers), often retain little specific meaning. Nonetheless, in spite of this constant flux, Gypsies show a sustained association with working on their own account and maintaining an ability to engage in a variety of activities (Liegeois and Gheorghe 1995). They tend to look for customers to whom they can offer a service but refrain from getting tied up with contracts or regular employment that links them too closely to settled society. Specifically, Gypsies often engage in salvage work, construction, entertainment, agriculture and horticulture (Table 9.4).

The Gypsies' communal structure (Figure 9.4) is vital to their identity, with a strong emphasis given to family and other ties from a very young age (Cottaar and Willems 1992; Fraser 1992; Okely 1983). Members of the same *vitsa* are expected to aid one another in times of hardship, while the *kumpánia* is typically an alliance formed through economic necessity, headed by the *rom baró* (big man). It forms the basic political unit, usually managing to make decisions through *diváno* (discussion). Where this fails, in many Gypsy societies a *kris Romani* (Romany trial) takes place.

A final very distinctive feature of Gypsy society is the purity code, where a series of strict taboos delimit a dread of contamination (Miller 1975; Okely 1983; Rao 1975). To be *marimé* (unclean) is to be the opposite of *romania* (the social order) (Sutherland 1975b). A taboo code, relating to persons, objects, parts of the body, foods and even topics of conversation, is relatively consistent across the range of Gypsy groups. It informs the conduct of male–female and Gypsy–*gadžo* interaction and to be declared polluted is 'social death' (Fraser 1992: 245), since communal life will be lost. Amongst the *Rom*, the only way to revoke *marimé* status is through a *kris*.

Table 9.4 The varied economic activities of British Gypsies.

Sales	Services	Seasonal labour
Hawking: bulk purchases of manufactured goods: blankets, linen, carpets, household wares	*Clearance*: scrap metal, rags, old cars, demolished material	*Agricultural work*: fruit and potato harvesting, beet hoeing, hop tying
Secondhand goods: cars, caravans, household appliances, clothes, furniture, antiques	*Building and gardening*: tarmac, crazy paving, kerbing, logging, tree pruning, gardening, roofing, painting	
Hawking 'Gypsy' goods: clothes pegs, white heather, hand-made flowers	*Casual entertainment*: fairground booths, singing, guitar playing, fortune telling	
Other goods: fruit and vegetables, dogs, horses, Christmas decorations	*Other services*: knife-grinding, cart construction	

Based on Okely 1983: 51.

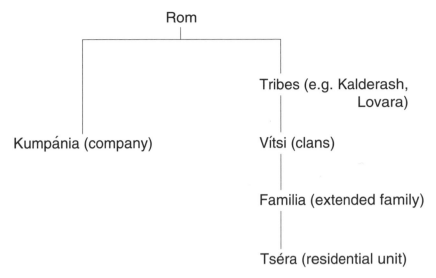

Figure 9.4 Community structure of Gypsy society (source: Fraser 1992: 238–40).

The various features of Gypsy culture fit well with the adoption of a nomadic existence: occupational flexibility allows the Gypsy to respond to the immediate demand in the area moved into; a strong kin structure enables a degree of social cohesion and support to be maintained throughout the rigours of travel; and the purity rules help to maintain the nomad's health and to control tensions arising from sexual rivalry while on the road. These features fit well with the need for the nomad to be versatile, flexible and able to make quick decisions (Okely 1983). Even for Gypsies who are no longer nomadic, the legacy of nomadism remains. Typically, 'Gypsy patterns' are still apparent, as they

carry over a style close to that of the encampment, uneasy with solitude, seeking company, and spending a good deal of time outside the house, even when at leisure.

(*ibid.*: 309).

Finally, there are in Europe and elsewhere people who have taken to the semi-nomadic life who are not from Gypsy backgrounds. In the Netherlands, there are the *Woonwagenbewoners*, caravan dwellers dating back to the middle of the last century (Cottaar, Lucassen and Willems 1992; Willems and Lucassen 1992). Irish travellers pre-date the arrival of the Gypsies in the British Isles and were traditionally tinsmiths, pedlars and horse-dealers. Originally rural tent and horse-drawn waggon dwellers, with the obsolescence of their livelihoods they have become more urban, although they often

also travel to England and Wales to earn a living (Adams *et al.* 1975; Barnes 1975; Fraser 1992). There are also the British New Age travellers, who combine a semi-nomadic existence with social protest. The appeal of a nomadic lifestyle can still be strong, even in the most 'advanced' of societies.

Migration as national diaspora

The diasporic experience

A nation can be defined as a 'group of persons who believe that they consist of a single "people" based upon historical and cultural criteria' (Taylor 1993: 332). The idea that nations have their own geographical political structure in the form of a state tends to be relatively taken for granted. However, this assumption is an over-generalisation, as there are nations that do not have their own state, such as the Kurds, whose Kurdistan is currently dispersed between Turkey, Iraq, Iran, Syria and Armenia. States themselves also never contain members of just one nation, a point brought across most brutally in the recent ethnic conflicts within Rwanda and former Yugoslavia. Moreover, members of specific nations are often found in a wide range of states, with the experience of migration forming a central plank in their cultural definition and identity. For example, the Western Samoan author Albert Wendt expressed in his novels the way in which his home country be-

---

Box 9.5 New Age Travellers in contemporary Britain

Britain in the 1980s and 1990s has seen the government and some members of the public expressing concern over a group dubbed New Age travellers (NATs), due to their apparent hippie lifestyles and their trespassory activities. Without stereotyping a diverse group, NATs have sought to combine a nomadic lifestyle with the idea of escaping the city (they are often unemployed) and engaging in forms of environmental and other protest. These travellers emerged first in the late 1970s in the wake of the free festival (music) scene. However, media prominence reached a peak in 1985, with the events surrounding the infamous Battle of the Beanfield in Savernake Forest, Wiltshire, England. This was a violent confrontation between travellers (then known popularly as the Peace Convoy, due to their connection with anti-nuclear protests) and the Wiltshire police, who were attempting to prevent the annual but illegal summer solstice gathering and music festival at Stonehenge. After this confrontation and related clashes the following year, the travellers declined somewhat in media prominence.

NATs reappeared in the media in May 1992, as the result of a gathering of around 50,000 people for another illegal music festival, this time at Castlemorton Common in Hereford and Worcester, England. This event became symbolic of changes in traveller culture as, in crude terms, it brought together travellers proper with weekend ravers (enthusiasts of high-intensity rave music). A key aspect of this often tension-filled intermingling was that it introduced the more numerous but largely apolitical ravers to some of the more environmental politics of the travellers. Consequently, ravers became political for the first time, a politicisation that, paradoxically, was reinforced considerably by the British government's attempts to extinguish the traveller lifestyle through measures in the 1994 Criminal Justice and Public Order Act. Whether the NATs are able to withstand the government's assault on their way of life that this legislation represents remains to be seen. However, radical direct action environmental protest, which can be linked in some ways to the travellers' lifestyle and beliefs, is currently on an upward trajectory. For many travellers, roads, which were once seen as a source of liberation from conventional (settled) society, came to be seen as part of the problem. In this context, the last (ambitious) words will be left to Alexandra Plows of the so-called Donga Tribe, a traveller group particularly active in anti-road campaigns such as the one waged against the M3 motorway extension in southern England:

We are a tribe in far more than name. We have a collective purpose and a cultural identity as the nomadic indigenous peoples of Britain – we have formulated our own customs, mythology, style of dress, beliefs and are evolving our own language.

(cited in McKay 1996: 137).

Sources: Cresswell 1996; Earle *et al.* 1994; Halfacree 1996b; McKay 1996.

---

comes 'constructed and reconstructed' (Connell 1995: 270) amongst Samoans living in New Zealand:

And so she continued throughout the years, until a new mythology woven out of her romantic memories, her legends, her illusions and her prejudices, was born in her sons: a new fabulous Samoa to be attained by her sons when they returned home after surviving the winters of a pagan country.

(Wendt 1973: 76, quoted in Connell 1995: 271).

These geographically scattered nations are known as *diasporas*, a term originally referring to the scattering of the Jews but that has been applied subsequently to other nations. Indeed, the Gypsies discussed in the previous section are representative of a diaspora (Liegeois and Gheorghe 1995).

Besides novels such as those by Wendt, one aspect of the contemporary cultural significance of the diaspora has been demonstrated in Gilroy's (1993) exploration of the African scattering that occurred initially as a result of the slave trade. The black cultures that emerged from these forced migrations – symbolised by the slave ships crossing the Atlantic – have now achieved a position of considerable authority within contemporary society. Through their strength and versatility alone, they critique the notion that 'cultures' should be ethnically homogenous and bounded by the nation-state. Instead of being absolute and fixed, the Black Atlantic as described by Gilroy is a restless, ever-evolving – through absorption and transformation – and subversive culture, disrupting the supposed certainties of ethnic and national cultures from a position on the margins of these cultures. Moreover, current trends towards a globalisation of daily life may put these diasporic groups, with their strong sense of connections and networks, in an enviable position on the world stage. The diaspora may provide an alternative focus of loyalty to the nation-state (Cohen 1995). This appears to be particularly true for some very successful diaspora populations, such as the Lebanese (Hourani and Shehadi 1993), in contrast to 'victim diasporas' (Cohen 1995), such as the Armenians. Finally, the concept of diaspora should not always be associated with forced exile, hardship and loneliness (*ibid.*). The Greek origin of the word refers to expansion and settler colonisation, and hence to a degree of choice on the migrant's part.

The Jews

Although they now have their own state in Israel, the Jews remain the group most strongly associated with the experience of diaspora. Migration has long been linked with stereotypes of Jewish identity, such as that of the wandering Jew, as coined in the eighteenth century (Plaut 1996). Today, Jews are still regarded as the archetypal 'global tribe' (Cohen 1995: 12) and their homelessness is a key *leitmotif* in their literature and art (Ages 1973). Historically, the exile of the Jews from the Promised Land of *Eretz Israel* is associated with the destruction of the Temple in Jerusalem in 586 BC and AD 70 (Goldberg and Raynor 1989). From that time, Jews became strangers and wanderers for nearly 2000 years.

It is partly to this long experience of exile that a Jewish tradition of friendship and welcome to strangers, migrants and refugees can be traced. This tradition was demonstrated by the support North American Jewry gave to the Sanctuary Movement, a group that harboured illegal immigrants in churches and other locations, in an effort to avoid their deportation, in the 1980s and 1990s (Plaut 1996). In contrast, much less welcome and acceptance was often afforded the Jews themselves, especially by supposedly Christian peoples. Resultant oppression reached a horrifying climax with the Nazis' Final Solution in the Second World War (Chapter 7), in the aftermath of which the British withdrew their mandate from Palestine and the state of Israel was set up in 1948. Since then, Jews have been 'returning' to a land most of them had never been to before, exercising their right under the Law of Return.

A recent survey located Jews in nearly 100 countries in the world (Table 9.5). The United States has nearly half of the world's Jews and provides the 'cultural leadership of Diaspora Jewry' (Lerman *et al.* 1989: 183). Jews in the United States are a highly educated and affluent minority group, a self-confident population with a high profile in the economic, cultural, political, legal and academic leadership of the

Table 9.5 Major world Jewish populations, *c*. 1989.

Country	Number
United States	5,835,000
Israel	3,590,000
Union of Soviet Socialist Republics	1,810,000
France	535,000
United Kingdom	330,000
Canada	325,000
Argentina	228,000
Brazil	150,000
South Africa	120,000
Australia	90,000

Source: Lerman *et al.* 1989.

country (Vital 1990). In most countries with a sizeable Jewish population, synagogues are joined by kosher butchers and restaurants, schools, museums, libraries, cultural celebrations of the arts, research centres, a vocal press, welfare organisations, youth groups and other expressions of a communal identity that has not been assimilated into that of the host country (Lerman *et al.* 1989). Thus is a migrant identity stamped on to the landscape. Indeed, the theme of community/culture remains fundamental (Waterman and Kosmin 1987; Waterman and Schmool 1995), although an individual freedom to leave that community is also apparent, in contrast to the experience of living within the Jewish state itself (Vital 1990).

The culture of migration for the Jews is reflected in the concept of *Eretz Israel*, an object or symbol that is a central feature of Jewish identity. For example, the Jewish legal code contains many references to territory, but this only applies to the land of *Eretz Israel*. Such specificity helped tie the Jews of the diaspora to their Promised Land, no matter whence they scattered. Moreover, the Old Testament promised redemption through land for those sticking to its strictures (*Exodus* 34: 24) and called for the repossession of the Promised Land (*Deuteronomy* 1) (Rowley 1991). Consequently, *Eretz Israel* 'is part of the Jewish religious and cultural essence throughout the world' (Shilhav 1993: 274), internalising the geography of the homeland within a (religious) heritage. Hence, new Jewish settlements on the West Bank of the Jordan River and elsewhere (Chapter 7) can be seen as a reassertion of the settlers' Jewishness (Rowley 1991).

With the creation of the Israeli state in the shadow of Zionism in 1948, the diaspora concept of *Eretz Israel* began both to become a reality and to fragment. This has led to strains within the Jewish world, which illustrates the central importance of a migratory/exile identity to definitions of Jewishness. Vital (1990) outlined tensions and divisions between the Jews of Israel and those still living in the diaspora. He argued that 'the rise of an independent Jewish state has both revolutionized and destabilized the Jewish world' (p. 147). There is a 'waning of the Jewish nation' (p. 148) as a single – if scattered and far from monolithic – nation is replaced by a series of islands of Jewishness. Middle class, 'part-time', Euro-American Jews have become increasingly distinct from the more religious, working class, Euro-Mediterranean Jewry. Moreover, with the foundation of the Israeli state, diaspora Jews have had to deal with split national loyalties more directly than in the past. Without the experience of diaspora and, hence, migration, the Jewish identity established for over 2000 years is changing fundamentally.

The Irish and exile

The Irish are another nation – this time less clearly unified by religion – for whom the experience of migration forms a central component of their national cultural identity (O'Sullivan 1992). Indeed, they are a group where the strongly normative status of (e)migration conveys a very distinctive culture of migration (Ni Laoire 1997). Although the 26 counties of the Republic of Ireland have been independent from the United Kingdom since 1922, the influence and importance of migration remains. The 1991 census put the population of the Republic at just 3.52 million, but it has been estimated that 60 million people worldwide can trace Irish descent (Duffy 1995). This is because emigration has been a central experience for the Irish for around 300 years:

The sustained nature of emigration over such a long time span, combined with its high level relative to the population of the island, have conferred on Ireland a unique status in the modern history of emigration.

(Breathnach and Jackson 1991: 1).

Emigration was boosted historically by such human tragedies as the Great Famine of the 1840s and long-standing political oppression. However, as time has gone on, Irish emigration has taken on something of a logic of its own, perceived as 'an inevitable natural path' (MacÉinrí 1991: 38). From a structuralist perspective (Breathnach 1988; Breathnach and Jackson 1991; MacLaughlin 1994), Ireland has become a peripheral country linked to a more developed set of core countries – notably Britain – through migration.

Box 9.6 Algeria, France, Israel: identity in an Algerian Jewish community

After the Algerian Revolution saw the end of France's colonial rule of Algeria in 1962, most Jews resident in the country emigrated to France. This was because they had remained loyal to France during the war and so left with the remainder of the European population. One city in France where many of these Jews ended up was Aix-en-Provence, in the south, where typically they originate from the town of Batna, at the foot of Algeria's Aures Mountains. Some of these extended families were the subject of a piece of research in the mid-1970s. When the author revisited the community in 1983, she talked again to some of the people she had met eight years before. One of these respondents was 'Richard G.', who expresses in the extract below his difficulty in finding a clear sense of identity as someone for whom both diaspora and migration have played key roles in defining his life:

I asked him my old question from eight years ago: 'Do you feel French?' No, he doesn't feel French. 'Feeling really French is to feel united with the soil, to have a sense of territory here in France. ... You really need to have a home village to be from. ...' With that he turned up the music with its Hebrew then Arabic tunes, and, turning to me, said a '*francais profonde*' (a true Frenchman) wouldn't be listening to *Radio Juive* [Jewish radio station].

'For my generation, our sentiment of exile lies at our very center. ... That's why my generation is drawn to writing novels and creating movies and plays – it's our only way of finding ourselves. In the novels and films there's two things combined – one is a form of nostalgia ... *la nostalgérie*, he jokingly punned. 'The other is the hope of finding what was lost, that sense of belonging, of community, of family.'

'Another profound sentiment for my generation is that France is a country of refuge: we're well off in France because it's worse elsewhere.' And he dreamily talks about Israel while admitting that when he was there he felt 'rootless', like a 'foreigner'. 'Israel', he says, 'is my mythic world'.

(Friedman 1988: x–xi).

Source: Friedman 1988.

Having entrenched this peripheral status over the past 300 years, it is difficult for Ireland to break free and build a diversified economy; hence, emigration continues for sound economic reasons.

Until recently, Irish emigration was dominated by people from rural backgrounds, often with little in the way of formal skills and qualifications. Prior to the First World War, migrants were particularly attracted to the New World, especially the United States (McCaffrey 1976) but also Canada and Australia. After 1918, as immigration controls became tighter in these countries, Britain became the key destination. A 'new emigration' of around 200,000 people emerged in the 1980s, distinguished by migrants with higher levels of formal education and more urban backgrounds (King *et al.* 1990; Walsh 1991). Their destinations became more varied (MacÉinrí 1991), with King, Shuttleworth and Walsh (1996: 207) describing Ireland as the 'human resource warehouse of Europe'. As Shuttleworth (1991) demonstrated in a survey of undergraduates, when compared with English students the Irish were more highly internationalised, in terms of having parents who had worked abroad, having themselves had holiday jobs abroad, having close family abroad and/or being well aware of overseas job opportunities.

The social history of Ireland has been dominated by emigration, with numerous *Exiles of Erin* (Lees 1979) describing a culture of (e)migration. The themes of exile and isolation particularly expressed this experience, encompassing a fatalistic view of migration as being undesired but inevitable (Miller 1985). With this feeling about migration came an attachment to an often idealised homeland and a desire to return, which could hinder assimilation in the destination country (Duffy 1995). For example, in a study of the Irish in the suburbs of Sydney, Australia, Grimes (1992) showed the continued importance of ethnic networks, especially for more recent arrivals and those from rural backgrounds.

Cutting across feelings of loss is the sense that emigration can also provide an escape, especially for women (Walter 1991), from the constraints of a (typically) rural life. Indeed, in contrast to structuralist accounts of emigration, King, Shuttleworth and Walsh (1996) stressed how lifestyle, family and social network considerations overlay more strictly economic reasons for moving. In particular, the maintenance of social networks facilitates the acquisition of both work and accommodation for the new emigrant (Grimes 1992; MacLaughlin 1991). Similarly, describing new Irish arrivals in New York, Almeida (1992) emphasised the importance attached to Irish friends and socialising, especially in bars. However, she also noted how many of the newcomers were living and working illegally in the United States, thus necessitating a degree of anonymity from the wider public.

The Irish people's culture of migration is expressed eloquently in literary products. While some writers, such as James Joyce, reflected on their origin society from exile, others have concentrated more directly on the migration process. Going into exile has been most prominently featured (Duffy 1995), being played down in euphemisms such as 'taking the boat' or 'crossing the water', or dealt with directly in discussions of the American Wake, a party held for the migrant prior to departure. For example:

As night fell, most of the people of the district would gather into the 'convoy-house' – the old people in the first instance up to ten o'clock and then the young people. There'd be drinking and dancing then until morning. The person leaving would be keened [lamented] three times altogether during the night ... and the whole gathering would accompany him three or four miles along the beginning of the journey. Then they'd stand until the emigrant was well out of sight. It wasn't to be wondered at, of course, as often enough that would be the last sight of him a lot of them would ever have.

(MacGabhann 1973: 49).

Other literary accounts have concentrated on the escape from the boredom and oppressiveness of rural society, or on the experiences in the New World. In the latter, 'great reserves of nostalgia' (Duffy 1995: 31) are often apparent:

the things he saw most clearly were the green hillside, and the bog lake and the rushes about it, and the greater lake in the distance, and behind it the blue line of wandering hills.

(Moore 1914: 49).

Similar themes also emerge in Irish songs and ballads.

Migration as conflict

Migration and the experience of racism

Migration frequently involves moving to a country, region or even local area characterised by very different people from those at the origin, with their own relatively distinct customs, traditions and ways of life. Consequently, migration can be associated with a degree of cultural disjuncture, alienation and even conflict, experienced especially by the migrants but also by the established residents of the destination area. In general, 'cultural' conflict can be represented either by racist prejudice of a majority population towards immigrants or by more strictly cultural mismatches between the two groups.

Miles (1989: 11) expressed the consequences of migration thus:

Box 9.7 Emigration and exile in the songs of the Pogues

Irish folk songs and ballads are replete with reference to exile and emigration, a theme coming second only to the independence struggle. We have ballads on the emigration theme (*The Emigrant's Farewell*, *The Exiles of Erin*, *Goodbye Johnny Dear*), while other compositions (*Will You Come to the Bower?*, *Óró 'sé do Bheatha 'Bhaile*) combine the theme of exile with calls to return home to struggle for freedom (Duffy 1995; Loesberg 1979). Contemporary songs, influenced by rock as well as traditional music, also concern themselves with emigration. Good examples can be found in the songs of the Pogues, an Anglo-Irish group formed in London in 1982. All of their albums, plus the solo work of their principal singer–songwriter, Shane MacGowan, contain references to the experience of exile. Specific early examples include *Transmetropolitan*, *The Dark Streets of London* and *The Old Main Drag*, which peer into the often bleak experience of the Irish immigrant to London, and *Sally MacLennane*, with the following chorus:

We walked in to the station in the rain
We kissed him as we put him on the train
And we sang him a song of times long gone
Though we knew that we'd be seeing him again
(Far away) sad to say I must be on my way
So buy me beer and whiskey 'cos I'm going far away (far away)
I'd like to think of me returning when I can
To the greatest little boozer and to Sally MacLennane.

The band also covered the traditional song *The Irish Rover* with the Dubliners, sang of an Irish-American being returned home for burial in *The Body of an American* and described the experience of travelling from Ireland to London on *The Boat Train*. MacGowan's solo work covers exile, history and nationalism in *The Snake with the Eyes of Garnet*. However, it was on 1988's *If I Should Fall from Grace with God* album that the emigration theme was expressed most strongly. Two songs stand out. First, *Fairytale of New York* tells of a drunken Christmas Eve spent reminiscing about a failed relationship and shattered dreams, with a chorus:

The boys of the NYPD [New York Police Department] choir
Were singing *Galway Bay*
And the bells were ringing
Out for Christmas Day.

Second, *Thousands are Sailing*, a contribution from Philip Chevron, draws upon the now 'silent' Ellis Island (the key point of entry of the Irish to the United States in the nineteenth century) to ruminate on the migration of the Irish across the Atlantic in the infamous 'coffin ships':

Where e'er we go, we celebrate
The land that makes us refugees
From fear of Priests with empty plates
From guilt and weeping effigies
And we dance....

A more telling expression of the ambiguity of the Irish culture of migration is hard to find.

Source: The works of the Pogues and Shane MacGowan are available through WEA Records and ZTT Records, respectively.

Migration, determined by the interrelation of production, trade and warfare, has been a precondition for the meeting of human individuals and groups over thousands of centuries. In the course of this interaction, imagery, beliefs and evaluations about the Other have been generated and reproduced amongst all the participants in the process in order to explain the appearance and behaviour of those with whom contact has been established and in order to formulate a strategy for interaction and reaction.

Some of the most acute cultural tensions arising from such interactions express themselves in forms of racism. Indeed, migration has become fundamentally entwined with the politics of race (Solomos and Back 1996). While the institutional dimension of racism and migration, as expressed through government policy, has been considered in Chapter 7, the experience of racism frequently accompanies and is associated with migration. In this context, cultures of migration may become imbued with the experience of prejudice, with the original hopes of economic betterment and/or political freedom becoming lost through the intolerance of the destination population.

'Cultural differences' have increasingly been the key way in which racist attitudes are expressed. Miles (1993) noted how the process of *racialising* groups attaches importance to the perceived failure of the stigmatised group to assimilate to the way of life of the majority population, while for Solomos and Back (1996: 27) there is a 'new racism', which centres on the defence of a mythic way of life 'coded in a cultural logic'. In general, racism depends upon 'making (and acting upon) predictions about people's character, abilities or behaviour on the basis of socially constructed markers of difference' (Castles and Miller 1993: 30).

Although cultural differences are used by racists

to justify their prejudices, culture is also used by minority groups to express and reinforce their own identity, often against the stereotypes placed upon them by the majority population. Migrant cultures become political cultures of resistance, often expressed in seemingly quite minor ways, such as the insistence of Muslim schoolgirls in France in the late 1980s on wearing *foulards* (Islamic headscarves) in defiance of the secular framework of the French education system. Alternatively, cultural resistance can be expressed violently in urban uprisings, such as those that have erupted in Britain, France and the United States in recent decades. Culture is not easily surrendered:

Processes of marginalisation and isolation of ethnic groups have gone so far in many countries that culture has become a marker for exclusion on the part of some sections of the majority population, and a mechanism of resistance by the minorities.

(*ibid.*: 273).

Given this background, we can expect the cultures of migration for ethnic minorities to be both prominent in their everyday lives and sharply defined. The migration experience is part of their cultural biography, a biography assailed by the forces of racism but providing a bulwark of identity and resistance for the migrant.

Turkish *Gastarbeiter* (guest worker) migration to Germany involves a situation where racism is typically most acute, namely when a 'non-white' minority population is recruited to work in a rich 'white' country. As Miles (1982: 159) notes, labour migration 'created the terrain for racial categorisation', whether in the colonial era or through present day 'interior racisms', such as those within Europe (Miles 1993). The migration of Turks to Germany first took place during a long period of economic prosperity but continued, in spite of official efforts to the contrary, through a period of socio-economic restructuring, when the majority population itself felt most threatened by job losses and other changes and looked to scapegoat groups such as immigrants.

Germany has a long history of recruiting foreign labour but, building on the efforts of the *Bundesanstalt für Arbeit* (Federal Labour Office) in the 1950s, between 1960 and the 1973 oil shock the country made a special effort to obtain workers to fill the vacancies created by a booming economy (Rist 1978; Wilpert 1992). In adopting this strategy, Germany was not unusual within Western Europe (Miles 1987), although it did not have the (former)

colonial sources of recruitment of many other nations (for example, Britain or France). As sources of recruitment such as former Yugoslavia dried up, bilateral contracts were concluded with Turkey in 1961 and 1964, and numbers of Turkish *Gastarbeiter* rose from about 2500 in 1960 to 605,000 in 1973. By then, an economic downturn had encouraged the German government to rein in this recruitment but numbers of Turks in Germany continued to rise, reaching around 1.9 million of the 6.8 million foreign nationals resident in the country in 1993 (Fassman and Münz 1994; Rudolph 1994). This continued rise reflected the increased importance of immigration linked to family reunion, especially since the migrants – notably those from more rural areas – tended to maintain strong social networks facilitating chain migration (Chapter 2) (Wilpert 1992). From 1973 onwards, Turks have outnumbered all other foreign nationals in Germany.

A variety of meanings can be given to this migration stream. For Germany, the key issue was the means of obtaining relatively cheap and flexible labour at a time of labour market tightness. For the Turkish government, the economic planning of the 1960s and 1970s had concentrated on capital-intensive industrialisation (Collinson 1993), such that economic expansion was unlikely to be sufficient to cope with the high rates of unemployment and underemployment in the country. Therefore, emigration was encouraged. Furthermore, the country was desperate for the foreign exchange the *Gastarbeiter* could provide through remittances to their relatives in Turkey. For the Turkish workers, migration to Germany suggested the possibility of an escape from poverty, unemployment and their dependency upon semi-feudal landowners (Castles and Miller 1993; Martin 1991). However, migration also meant separation from spouses, children and other close kin, since only the migration of individual workers was allowed initially by Germany, with very strict rules governing family reunion.

The work obtained was very arduous, dangerous and poorly paid, concentrated in the manufacturing sector (Miles 1987; Rist 1978). Initial settlement was often in hostels and camps near the work site, and subsequent housing was typically poor quality and cramped, reflecting the workers' low wages and lack of savings, with local authorities often refusing to house them in public projects. However, the key experience for many Turks, which helps to explain these marginal material positions, was that of exclusion, rejection and racism (Martin 1991). Turks have borne the brunt of anti-foreigner discrimination in Germany in recent years (Cohen 1987; Fischer and McGowan 1995; Thränhardt 1995). Attacks increased, the tabloid press presented sensationalist anti-Turkish reports, and crude jokes were commonplace. Even Chancellor Kohl argued in 1982 that the issue of foreigners was a key problem facing the government. Although this tense situation subsided during the 1980s, it revived in 1990–91 with German reunification prompting renewed migration into the country of a 'third wave' (Jones 1994) of *Aussiedler* (Chapter 7) and asylum seekers. Former East Germany witnessed a wave of racist violence, often orchestrated by neo-Nazi groups. Only recently has public revulsion at these attacks appeared to dampen the activities of the neo-Nazis and the more general anti-foreigner sympathies that were aroused (Thränhardt 1995).

Unsurprisingly, the culture of migration for the Turks in Germany has been moulded and overshadowed by the experience of discrimination. The economic dreams that many brought with them to Germany have often been nullified by the strains of everyday survival. Although the one-millionth arrival in 1969 was greeted personally by the president of the Federal German Institute of Labour (Ashkenasi 1990), this welcome does not reflect the reality of Turkish lives in Germany. Turks are generally regarded as a group that cannot be assimilated within German society, a rejection officially reflected by Germany's position of not being a country of immigration. For example, the 1965 *Ausländergesetz* (Foreigners Law) had instructions which read:

Foreigners enjoy all basic rights, except the basic rights of freedom of assembly, freedom of association, freedom of movement and free choice of occupation, place of work and place of education and protection from extradition abroad.

(quoted in Cohen 1987: 157).

Under this legislative framework migrants cannot vote and even the second generation find it hard to obtain citizenship. Although this situation has been getting more liberal, in part due to the *de facto* permanent immigration of the migrants, even the 1991 *Ausländergesetz* demands assimilation to the 'legal, social and economic order of the Federal Republic, its cultural and political values' (Fischer and McGowan 1995: 41). For German policy, a putative common culture is placed at the heart of nation building. For non-Germans, even the various terms

used to describe them, such as *Gastarbeiter* and *Ausländer*, stress what is seen as their otherness. Clearly, the consciousness of the migrants will be influenced by this position, undermining their ability to take a long-term perspective and to plan and build a future within German society.

Islam is a key cultural badge that renders the Turkish minority highly visible within Germany. Combined with the strong ethnocentric cultures apparent in both countries, Islam draws attention to the difficulties of both assimilation and mutual tolerance and respect. Tensions between the two communities are often exacerbated by being refracted through frictions due to school overcrowding, high levels of unemployment and housing shortages (Ashkenasi 1990). The net result is that – within their cultures of migration – Turks in Germany see themselves positioned within German society relative to other ethnic groups. The Italian immigrant writer Franco Biondi expressed it thus:

It's obvious, German: biggest fish. Italian, big fish. Turk, little fish. You [Pakistani], even smaller fish. African: all the worst jobs.

(quoted in Fischer and McGowan 1995: 43).

Migrant writing in Germany, much of it by Turks, expresses well their cultures of migration (Fischer and McGowan 1995). This literature focuses on the migrants' immediate experiences; it contrasts original naïve perceptions of Germany as a 'promised land' with the reality of heavy, dirty and unhealthy work in often appalling conditions; it articulates the prejudice, indifference and rejection shown to the immigrants; it dwells on their resultant feelings of homesickness and their dreams of return; and it presents the dilemma of a people torn between two worlds and two languages. For example, Güney Dal's (1979) novel *Wenn Ali die Glocken Läuten Hört* (When Ali Hears the Bells Ringing), set amongst Turkish industrial workers in Cologne and Berlin, covers the dreams of wealth that motivated the migrations, hostel life, feelings of isolation and puzzlement at an alien culture and language, mutual suspicion, everyday experiences of racism, stress-related illness and the corruption of traditional values. Reflecting the diversity of Turkish society, the book shows a people divided by politics, apathy and religious fundamentalism – multiple voices that were developed in a later novel, *Europastraße 5* (Dal 1990 [1981]).

Recognising diversity is crucial, as it is unhelpful to suggest that there is *one* culture of migration for

Box 9.8 Narratives of identity and residential history of Turks in Germany

The value of adopting a biographical analysis of migration can be illustrated with respect to the residential experiences of a Turkish couple, Mr and Mrs 'Keyser', resident in the German city of Munich. Three key narratives or 'presiding fictions' through which an overall identity was sustained were apparent, although the relative importance of each of these narratives varied over time and with circumstance, emphasising their highly contextual character. The narratives were:

- *Return* – the ever-present desire by the Keysers to return to Turkey. Mrs Keyser illustrated how this narrative came into play when the family had thought of buying a new flat:

  If we had bought it, we would have had to pay back 2,500 marks per month. And then the stress, always in debt. We wouldn't have been able to afford anything else. ... And if we had wanted to go back to Turkey, maybe after a few years or so, we would still have been in debt. And then it's impossible to sell straight away, and then we wouldn't have been able to go back.
  (quoted in Gutting 1996: 487).

- *Proper life* – the desire to live normal, rounded lives in Germany. This was expressed by Mrs Keyser when explaining why she had moved to a different but more expensive hostel before she married:

  I felt much better because we lived like *real* people, not like animals anymore. ... And although I was single, I didn't live like other people, I always bought proper things, cooked and ate properly. ... I really tried to live properly.
  (quoted in Gutting 1996: 485).

- *Family* – the considerable importance attached to the nuclear family. This was expressed thus by Mr Keyser:

  And the family is always the most important thing. In many families, the children only see their mum and dad on Saturdays and Sundays. We're not like that. For example, for many guestworkers ... work is everything. They work a lot. All the time.
  (quoted in Gutting 1996: 485–486).

Together, these three narratives can be used to help the researcher interrogate and understand the migration decisions made by the Keysers, since these decisions are 'constituted through biography' (*ibid.*: 484).

Source: Gutting 1996.

the Turkish *Gastarbeiter*. The actual experience of an individual Turk in Germany has become increasingly diverse. It is very dependent upon previous experiences, socialisation, opportunities, motivations and perspectives on life (Fischer and McGowan 1995; Gitmez and Wilpert 1987). A good reflection of this diversity is the way in which the extremism of Turkish politics has spilled over into conflicts within the German Turkish community. Both left-wing groups and neo-Fascists are politically active, as are the half a million or so Kurds (Ashkenasi 1990; Castles and Miller 1993). A gender perspective on Turkish immigrant literature also demonstrates the non-uniformity of the *Gastarbeiter* experience (Fischer and McGowan 1995). Although Turkish women often preceded men to Germany, being recruited explicitly into manufacturing jobs after the 1966–67 recession (Miles 1987; Rist 1978), their literary output remained the 'minority in the minority' until the 1980s. Subsequent output raises issues of sexual as well as ethnic discrimination, as well as the tensions between liberal Christian and conservative Islamic values.

As with other immigrant groups, especially those which have failed to assimilate, Turks' cultures of migration are represented on the ground by ethnic concentrations in the large cities, notably Kreuzberg in Berlin (Ashkenasi 1990; Castles and Miller 1993; Gitmez and Wilpert 1987). Shops, cafés, professional practices, mosques, newspapers, welfare organisations and other services have sprung up to meet their needs. Also associated with these spaces are political groupings demanding rights such as dual citizenship. Often reluctantly, the city administrations have reacted to these developments by setting up bodies such as Commissions for Foreigners, which seek cooperation between the two groups. In this respect, cultures of migration are being transformed into *de facto* cultures of settlement.

Migration and the experience of cultural conflict

Moving away from the extreme and violent confrontations that characterise experiences of racism, migration can also result in conflict at a more narrowly cultural level. English migration into rural Wales is an historically well-established phenomenon but one that has been noted as being numerically significant only in recent decades. The classic village-based community studies of the 1950s and 1960s showed little awareness of the in-migration of the English (Halfacree and Boyle 1992), although where such migrants could be found they were often seen as a basis for social conflict (G. Lewis 1986). Instead, the studies reflected concerns about rural depopulation – between 1871 and 1951 Mid Wales lost around 25 percent of its population (Williams 1985) – a problem still apparent from census data for the 1961–71 period for most rural areas in Wales (Halfacree and Boyle 1992). By the 1970s, this situation had changed dramatically and unexpectedly (Day 1989), with rural population growth as a result of net migration gains becoming widespread throughout rural Wales. Survey work in Ceredigion, a district in the south west, and on the island of Ynys Mon (Anglesey) in the north west showed the inmigrants to be mainly of working age, typically in professional and/or self-employed groups, and frequently to be English (Ceredigion District Council 1981; Hedger 1981; Morris 1989). While employment factors were important in explaining these migration flows, the significance of quality of life factors was clear. This was true for the 1980s, when the net migration gains were sustained, with an Institute of Welsh Affairs (1988) survey demonstrating the influx of the English and the way the migrants linked living in rural Wales with health, high quality of life, beautiful scenery and low crime rates.

The areas to which the English migrants moved were not of course 'empty', in spite of the ravages of depopulation, but were populated by the Welsh, with their own culture(s). While the internal diversity of Welsh culture must be stressed in contrast to strong external stereotypes (Cloke and Davies 1992; Cloke, Goodwin and Milbourne forthcoming; Cloke and Milbourne 1992), it can be summarised by Balsom's (1986) 'three Wales model' (Figure 9.5). First, in rural north west and west central Wales there is *Y Fro Gymraeg*, the Welsh-speaking, Welsh-identifying heartland. Second, there is industrial Welsh Wales, Welsh-identifying but generally not Welsh-speaking. Finally, there is British Wales, where a lack of Welsh speaking and identifying is evident. Migration has had the most impact and generated the most controversy in *Y Fro Gymraeg*, where it has been regarded as both a cause and a symbol of the erosion and decline of Welsh culture. In this area, the problem of stemming rural depopulation was increasingly overshadowed by the issue of community change. As one newcomer to rural Wales noted:

The problem of rural deprivation gave way to the problem of creating a new kind of community, where incomers respect and contribute to the local culture, and locals can have a continuing sense of identity, robust enough to be open to new influences without feeling overwhelmed.

(Jones 1993: 10).

There are many ways in which the English in-migrants have been accused of undermining Welsh culture (Day 1989; Williams 1985). First, there is the issue of language, frequently central to Welsh cultural identity. The English can almost never speak Welsh when they move to these areas and, although some of them subsequently make the effort, most do not learn. Thus, the linguistic cohesiveness of the area is undermined by in-migration. Second, although studies (for example, Cloke *et al.* 1995) have shown that not all English migrants are economically wealthy, there is resentment over their socio-economic status, especially the way in which their higher spending power serves to inflate house prices in the area, restricting access to accommodation for working class 'local' people. This housing question is linked with resentment over the high numbers of second or holiday homes in these areas. Third, the social and cultural norms and behavioural patterns of the migrants are often distinct from those of the established village residents. There is resentment over those in-migrants who try to hijack local cultural institutions *and* over those who appear to be aloof from such institutions, perhaps because they socialise elsewhere (Cloke, Goodwin and Milbourne forthcoming).

A 'loss of community' is perceived, as the English in-migrants are seen as lacking the necessary 'cultural competence' to live in the villages of *Y Fro Gymraeg* (Cloke, Goodwin and Milbourne 1997, forthcoming). As one in-migrant put it: 'It's a shock to find that even when the Welsh and the English speak the same language, we do not mean the same thing' (quoted in Jones 1993: 325). In this respect, the term 'English' takes on something of a catch-all category, with national identities assuming symbolic roles as boundaries of cultural distinctiveness:

What we see is a clash of values between different communities, languages and lifestyles. The categories 'Welsh' and 'English' are merely a telegrammatic means of expressing these tensions.

(Bowie 1993: 174).

Thus, in spite of many English in-migrants actively and often quite vigorously rejecting an association with the south east English stereotype of the ruling class, they become subsumed under such an identity

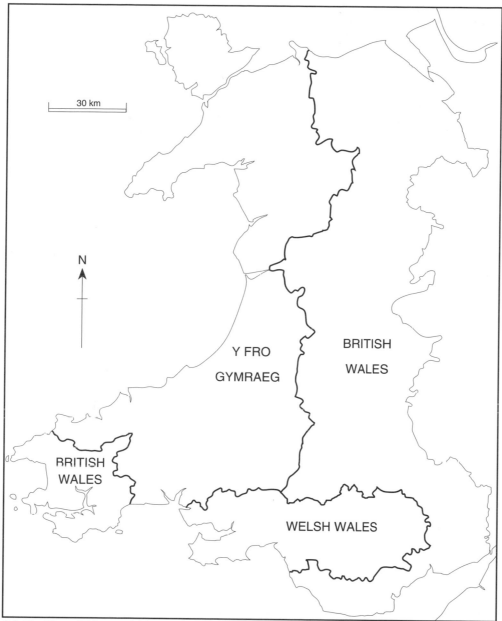

Figure 9.5 The Three Wales model (source: Balsom 1985: 5 in *The National Question Again*, John Osmond (ed), reprinted by kind permission of the Gomer Press).

(Cloke, Goodwin and Milbourne forthcoming). The dilemma that this raises was expressed by an in-migrant from the north of England:

When we first came we were ostracised – literally. Taboo. They wanted us out, full stop. They didn't tell us to our face. That's not their way. [...] We're not going to change the valley, I want it to stay the way it is. [...] If I'd ever had the chance to sit down and talk to the locals about how I feel about the valley and keeping it the way it is, they might have less animosity towards me. But I haven't had the chance. The way I was brought up it's natural when somebody passes to ask them in for a brew and a natter. But here you never get asked inside their houses. [...] I'm not saying we should get together every Saturday night, because I agree with having your own privacy as well. The

Box 9.9 In-migration and the Welsh language

The issue of the Welsh language and its association with migration must be seen in the context of the considerable standardising force of a metropolitan, mid-Atlantic culture based on the English language. In order for any culture or language to resist this force, considerable initiative and effort must be employed. This has been the case within Wales, where the Welsh language remains very much a living language – Wales is a linguistically plural society – albeit one whose future 'is clearly poised on a knife edge' (Aitchison and Carter 1991: 78).

At the turn of the century, half of the population of Wales spoke Welsh but by 1961 this had fallen to little over a quarter; by 1981 about half a million people spoke this Celtic tongue, a figure falling slightly but stabilising by 1991. Prior to the 1960s there was a relatively clear distinction between the predominantly Welsh-speaking and English-speaking areas of Wales. The former was focused on *Y Fro Gymraeg* and change over time was characterised by a gradual westward movement of the boundary. However, the core area has now fragmented, with distinct fracture zones opening up.

Migration has played a key role in this changing geography of the Welsh language. First, there was the development of out-migration from and consequent depopulation of the rural core areas of the language, typically involving the younger, more dynamic members of the population. Numbers of Welsh speakers declined and, although the proportion of the population in these areas speaking Welsh remained steady, the ageing of the rural population saw the language losing its vitality. Out-migration has continued in recent decades but now it is not just a case of a loss of absolute numbers of Welsh speakers from *Y Fro Gymraeg* but also a decline in the proportion of the population speaking Welsh. This latter trend is related directly to in-migration, especially of English-speaking people, and associated effects such as suburbanisation and tourism growth.

Since the 1960s, considerable concern has been expressed over the decline of the Welsh language. Its cause has been taken up, albeit rather grudgingly at times, by central government. There is now no longer a state-sponsored effort to eliminate the language. Its marginalisation since the Act of Union of 1536, which joined Wales politically to England, has been reversed. Welsh-language education is compulsory in the national curriculum and, from 1982, one of the four television stations, S4C (*Sianel Pedwar Cymraeg*, Welsh Channel Four), broadcasts weekly at least 25 hours of peak-time Welsh-language material. There is also *Radio Cymraeg*, plus numerous initiatives, movements (for example, *Cymdeithas Yr Iaith Cymraeg*, the Welsh Language Society) and other legislation. Finally, there has been increasing professional interest in the language, especially around the national capital, Cardiff. Welsh has become fashionable and a social cachet, not least because it can be the key to high-paying jobs and careers.

Efforts have also been made to encourage English speakers to become involved with the Welsh language. The nationwide group *Pont*, formed in 1988, aims to build bridges between locals and incomers and explain the Welsh culture, while *Cyd* also helps learners cross the bridge to Welsh-speaking culture. However, there has been resistance to the language from many English-speaking in-migrants, especially in the field of education. The group Education First, consisting mainly of such migrants, argues that parents should have the final choice of the language in which their children are educated. Particular concern was expressed over initiatives by some Welsh local authorities that children should be taught exclusively in Welsh until seven years of age. The Welsh language remains a hot political topic, enhanced by the effects of migration.

Sources: Aitchison and Carter 1990, 1991, 1993; Cloke, Goodwin and Milbourne forthcoming; Giggs and Pattie 1992; Jones 1993.

---

trouble here is that they're out there looking at you through the window, watching what you're doing, but they won't come up and talk to you straight. [...] You can't say there's one community here can you? It's a lot of different groups like anywhere else.

(quoted in Jones 1993: 25-28).

For the English migrants to rural Wales, their culture of migration may be very much one of obtaining a better quality of life in an area perceived to conform to the rural idyll (Chapter 6). A higher quality of life is often seen as more than adequate compensation for lower wages and poorer amenities (Cloke *et al.* 1995). This is not the thinking of the established residents. In-migrants frequently fail to appreciate that the areas to which they move have their own priorities for jobs and services, and their own cultures of Welshness, into which the in-migrants often fail to fit. At a time when the Welsh culture of *Y Fro Gymraeg* is perceived as being under threat from all sides, with Wales a 'nation in retreat', questions of identity loom large (Osmond 1987, 1988). An increased cultural diversification of rural Wales, largely through in-migration, has prompted heightened levels of disagreement (Day 1989), the significance of which is deeper than that generated by similar counterurbanisation moves within England:

With an indigenous and living language and associated social, cultural and political organisations, as well as the deeply rooted community ethic based on like-speaking kith and kin, rural Wales has a distinctive set of social relations which cannot easily be conflated with ideas of 'locals' and 'newcomers'.

(Cloke and Davies 1992: 352).

Moreover, the experience of the English migrants to rural Wales is thus often far from the expected idyll (Cloke, Goodwin and Milbourne 1997, forthcoming; Cloke and Milbourne 1992). While some locals welcome the changes and the new people and ideas coming into their villages, with advantages including greater levels of money circulating in the local economy and a possible revival of rural Wales's dispersed settlement pattern (Day 1989), other established residents are much less happy. Usually, this unhappiness is expressed by feelings of regret over the inevitability of change and a grudging acceptance of the migrants. However, in other cases, in-migration stimulates stronger anti-English feelings, which can verge on racism. The late 1980s saw a peak of actions by the nationalist extremist group *Meibion Glyndwr* (Sons of Glyndwr), the successor of the Free Wales Army of the 1960s, against holiday homes within Wales and estate agents' premises in both Wales and England that offered Welsh properties for sale. The burning of a holiday house, this key symbol of the changes affecting rural Wales, became a key 'icon of resistance' (Cloke, Goodwin and Milbourne forthcoming). Threats and attacks have also been made on individuals, mostly within *Y Fro Gymraeg* (Aitchison and Carter 1990; Croon 1995).

A contrasting example of the cultural impacts and experiences of counterurbanisation concerns the migration of English people to rural France. This 'international counterurbanisation' (Buller and Hoggart 1994a) blossomed in the 1980s. It was reflected in popular books such as Peter Mayle's *A Year in Provence* (1989) and *Toujours Provence* (1990) and popularised by the media and property agents (Hoggart and Buller 1994). By 1991, evidence from *cartes de séjour* (residence permits) indicated the presence of 762,000 British residents in France. In contrast to the established destinations of Paris or the Côte d'Azur, the more recent migrants, most of whom were from southern England (Hoggart and

Buller 1995a), were settling in the rural *départements* of the west and south. A key motivation behind this migration stream was the feeling that a more 'genuine' rural experience and lifestyle could be obtained in rural France. Many migrants felt that British rurality had been lost or was too expensive to buy into and France became Britain's lost rurality. Migrants typically sought out isolated housing in a peaceful rural location with scenic views, although some moves symbolised a search for escape. As one migrant put it:

We all came here chasing some kind of dream. For some, that is fulfilled by spending a few weeks of each year in their French home, cooking cassoulets, drinking 'cahors' on their vine covered terrace as the sun goes down. ... So far, so good [but]. ... A large proportion seem to come to France to escape difficult work, family or financial situations in Britain. They come *from* rather than *to* and the move is seldom a complete solution.

(quoted in Buller and Hoggart 1994a: 86).

In rural France, unlike in Wales, there is less evidence of resentment against the English in-migrants. Indeed:

while... in-migrants might always be *étrangers* [strangers, foreigners, outsiders] to local residents, many of them appear to be no more so than non-local French nationals and, arguably, are far less so than many middle class in-migrants have been in rural Britain.

(Buller and Hoggart 1994b: 209).

Consequently, English residents showed considerable satisfaction with their migration. They had found somewhere ideal and realised a dream. Furthermore, nearly 98 percent of those surveyed regarded their French neighbours as 'very friendly':

English friends ... are absolutely amazed when they stay with us at the friendliness of the French. I can honestly say not one French person has been rude to us in our own time here.

(quoted in Buller and Hoggart 1994a: 105).

For many respondents, the biggest resentment was saved for fellow British migrants, especially those who had not tried to integrate with the local population (Buller and Hoggart 1994b).

This seemingly different experience of migration to that experienced in rural Wales reflects a number of factors (Buller and Hoggart 1994a, b). First, France has experienced a severe population decline in recent years, leaving many vacant and decaying properties, many of which the English purchased without having the effect of raising property prices locally (Hoggart and Buller 1995b). Second, there is little demand from French migrants for these properties, as French counterurbanisation has largely been concentrated in peri-urban zones (Winchester and Ogden 1989). This reflects the less romantic view that the French have of the countryside. Third, the French recognise the money that the English migrants inject into the local economy, the jobs they create in renovating and servicing the properties, and how the migrants often sympathise with their environmental concerns. Finally, the numbers of migrants are generally too small to 'capture' French rurality, and to come into major conflict with local people over issues such as hunting. In this respect, the enthusiasm of the migrants for assimilation and their efforts to achieve this must be emphasised. Yet there are still some barriers to assimilation, such as the French people's perceived greater family orientation and religious fervour, their nationalism and an aloofness. English migrants must make a deliberate effort to integrate completely.

Conclusion

This book has shown that, within every society, migration tends to be associated with certain life events and experiences, resulting in distinctive patterns and characteristics. Consequently, migration takes on specific cultures of migration for those involved. Given the huge variety of cultures, peoples and experiences in the world today, it has been impossible in this chapter to cover every aspect of migration's intersection with culture. Nonetheless, a range of examples have illustrated a few of the many fascinating ways in which migration can form a crucial component of people's maps of meaning. In this respect, therefore, we can now readily appreciate Chambers's (1994: 2) recent observation – expressed in Chapter 1 – that 'migration ... is ... deeply inscribed in the itineraries of much contemporary reasoning'. Whether we are migrants ourselves – and we are *all* likely to acquire that status several times – it is clearly hard to escape from or avoid the issue of migration in our everyday lives. Indeed, Chambers has suggested that we now live in age of *migrancy*, where 'the promise of homecoming ... becomes an impossibility' (p. 5). This experience imprints itself heavily upon our identities, with the migrant's transgression of borders – of countries, nations, regions, local places – opening up new ways of seeing the world and one's place within it. Migration is lodged firmly within our contemporary maps of meaning. Hence, the crucial importance of studying and understanding this phenomenon in all of its many guises.

Bibliography

Abu-Lughod, J. (1961) 'Migrant adjustment to city life: the Egyptian case', *American Journal of Sociology* 67: 22–32.

Abu-Lughod, J. (1995) 'The displacement of the Palestinians', in R. Cohen (ed.) *The Cambridge survey of world migration*, Cambridge: Cambridge University Press, pp. 410–14.

Abu-Sahlieh, S. (1996) 'The Islamic conception of migration', *International Migration Review* 30: 37–57.

Adams, B., Okely, J., Morgan, D. and Smith, D. (1975) *Gypsies and government policy in England*, London: Heinemann.

Adler, P. and Adler, P. (1987) *Membership roles in field research*, Newbury Park, California. Sage.

Ages, A. (1973) *The diaspora dimension*, The Hague: Martinus Nijhoff.

Agyeman, J. (1990) 'Black people in a white landscape. Social and environmental justice', *Built Environment* 16: 231–6.

Ahlburg, D. (1991) 'Remittances and their impact: a study of Tonga and Western Samoa', *Pacific Policy Papers* 7, National Centre for Development Studies, Australian National University, Canberra.

Aitchison, J. and Carter, H. (1990) 'Battle for a language', *Geographical Magazine* 62(3): 44–6.

Aitchison, J. and Carter, H. (1991) 'Rural Wales and the Welsh language', *Rural History* 2: 61–79.

Aitchison, J. and Carter, H. (1993) 'The Welsh language in 1991 – a broken heartland and a new beginning?', *Planet* 97: 3–10.

Ali, S., Habibullah, A., Hossain, A., Islam, R., Mahmud, W., Osmani, S., Rahman, Q. and Siddiqui, A. (1981) *Labor migration from Bangladesh to the Middle East*, Washington: The World Bank.

Allon-Smith, R. (1982) 'The evolving geography of the elderly in England and Wales', in A. Warnes (ed.) *Geographical perspectives on the elderly*, Chichester: Wiley, pp. 35–52.

Almeida, L. (1992) '"And they still haven't found what they're looking for". A survey of the New Irish in New York city', in P. O'Sullivan (ed.) *Patterns of migration*, Leicester: Leicester University Press, pp. 196–221.

Altman, J. (1987) *Hunter-gatherers today: an Aboriginal economy in north Australia*, Canberra: Australian Institute of Aboriginal Studies.

Altsimadja, F. (1992) 'The changing geography of Central Africa', in G. Chapman and K. Baker (eds) *The changing geography of Africa and the Middle East*, London: Routledge, pp. 52–79.

Ambrose, P. (1974) *The quiet revolution*, London: Chatto & Windus/University of Sussex Press.

Ambrose, P. (1992) 'The rural/urban fringe as battleground', in B. Short (ed.) *The English rural community*, Cambridge: Cambridge University Press, pp. 175–94.

Amin, S. (1974) 'Modern migrations in Western Africa', in S. Amin (ed.) *Modern migrations in Western Africa*, London: Oxford University Press, pp. 65–124.

Amrhein, C. and Flowerdew, R. (1992) 'The effect of data aggregation on a Poisson regression model of Canadian migration', *Environment and Planning A* 24: 1381–91.

Anderson, M. (1982) 'Indicators of population stability and change in nineteenth-century cities: some sceptical comments', in C. Pooley and J. Johnson (eds) *The structure of nineteenth-century cities*, London: Croom Helm, pp. 283–98

Anderson, M. (1985) 'The emergence of the modern life cycle in Britain', *Social History* 10: 69–87.

Ansari, S. (1994) 'The movement of Indian Muslims to West Pakistan after 1947', in J. Brown and

R. Foot (eds) *Migration: the Asian experience*, Basingstoke: Macmillan, pp. 149–69.

Anthony, C. (1991) 'Africa's refugee crisis: state building in historical perspective', *International Migration Review* 25: 574–91.

Appleyard, R. (1985) 'Processes and determinants of international migration', paper presented to the IUSSP seminar on Emerging Issues in International Migration, Bellagio.

Appleyard, R. (1989) 'International migration and developing countries', in R. Appleyard (ed.) *The impact of international migration on developing countries*, Paris: OECD, pp. 19–45.

Arkell, T. (1991) 'The decline of pastoral nomadism in the Western Sahara', *Geography* 76: 162–6.

Asher, M. (1991) 'How the mighty are fallen', *Geographical Magazine* 64(3): 24–6.

Ashkenasi, A. (1990) 'The Turkish minority in Germany and Berlin', *Immigrants and Minorities* 9: 303–16.

Atkinson, J. (1987) 'Relocating managers and professional staff', *IMS Report* 39, Institute of Manpower Studies, University of Sussex.

Atkinson, P. and Hammersley, M. (1994) 'Ethnography and participant observation', in N. Denzin and Y. Lincoln (eds) *Handbook of qualitative research*, Thousand Oaks: Sage, pp. 248–61.

Auriat, N. (1991) 'Who forgets? An analysis of memory effects in a retrospective survey on migration history', *European Journal of Population* 7: 311–42.

Auriat, N. (1993) ' "My wife knows best". A comparison of event dating accuracy between the wife, the husband, the couple, and the Belgium population register', *Public Opinion Quarterly* 57: 165–90.

Bach, R. (1978) 'Mexican immigration and the American state', *International Migration Review* 12: 536–8.

Bach, R. and Schraml, L. (1982) 'Migration, crisis and theoretical conflict', *International Migration Review* 16: 320–41.

Badcock, B. (1995) 'Building upon the foundations of gentrification: inner-city housing development in Australia in the 1990s', *Urban Geography* 16: 70–90.

Baker, K. (1992) 'The changing geography of West Africa', in G. Chapman and K. Baker (eds) *The changing geography of Africa and the Middle East*, London: Routledge, pp. 80–113.

Balán, J. (1992) 'The role of migration policies and social networks in the development of a migration system in the Southern Cone', in M. Kritz, L. Lim and H. Zlotnik (eds) *International migration systems. A global approach*, Oxford: Clarendon Press, pp. 115–30.

Baldassare, M. (1986) *Trouble in paradise: the suburban transformation in America*, New York: Columbia University Press.

Baldassare, M. (1992) 'Suburban communities', *Annual Review of Sociology* 18: 475–94.

Balsom, D. (1985) 'The three Wales model', in J. Osmond (ed.) *The national question again*, Llandysul: Gomer, pp. 1–17.

Barff, R. (1990) 'The migration response to the economic turnaround in New England', *Environment and Planning A* 22: 1497–516.

Barlow, J. (1992) 'Planning practice, housing supply and migration', in A. Champion and A. Fielding (eds) *Migration processes and patterns. Volume 2. Research progress and prospects*, London: Belhaven Press, pp. 65–76.

Barnes, B. (1975) 'Irish travelling people', in F. Rehfisch (ed.) *Gypsies, tinkers and other travellers*, London: Academic Press, pp. 231–56.

Baron, J. (1988) *Thinking and deciding*, Cambridge: Cambridge University Press.

Bartholomew, K. (1991) 'Women migrants in mind: leaving Wales in the nineteenth and twentieth centuries', in C. Pooley and I. Whyte (eds) *Migrants, emigrants and immigrants: a social history of migration*, London: Routledge, pp. 174–87.

Bascom, J. (1994) 'The dynamics of refugee repatriation: the case of the Eritreans in eastern Sudan', in W. Gould and A. Findlay (eds) *Population migration and the changing world order*, Chichester: John Wiley & Sons, pp. 225–48.

Basok, T. and Benifand, A. (1995) 'Soviet Jewish emigration', in R. Cohen (ed.) *The Cambridge survey of world migration*, Cambridge: Cambridge University Press, pp. 502–7.

Bassett, T. (1988) 'The political ecology of peasant–herder conflicts in the northern Ivory Coast', *Annals of the Association of American Geographers* 78: 453–72.

Batra, R. and Scully, G. (1972) 'Technical progress, economic growth and the north–south wage difference', *Journal of Regional Science* 12: 375–86.

Batty, M. and Sikdar, P. (1982a) 'Spatial aggregation in gravity models: 1. An information–theoretic framework', *Environment and Planning A* 14: 377–405.

Batty, M. and Sikdar, P. (1982b) 'Spatial aggregation

in gravity models: 2. One-dimensional population density models', *Environment and Planning A* 14: 525–53.

Batty, M. and Sikdar, P. (1982c) 'Spatial aggregation in gravity models: 3. Two-dimensional trip distribution and location models', *Environment and Planning A* 14: 629–58.

Batty, M. and Sikdar, P. (1982d) 'Spatial aggregation in gravity models: 4. Generalisations and large scale applications', *Environment and Planning A* 14: 795–822.

Beale, C. (1975) *The revival of population growth in non-metropolitan America*, United States Department of Agriculture, Economic Research Service, ERS 605.

Beals, R., Levy, M. and Moses, L. (1967) 'Rationality and migration in Ghana', *Review of Economics and Statistics* 49: 480–6.

Bean, F., King, A. and Passel, J. (1983) 'The number of illegal migrants of Mexican origin in the United States: sex-ratio based estimates for 1980', *Demography* 20: 99–109.

Bean, P. and Melville, J. (1990) *Lost children of the Empire: the untold story of Britain's child migrants*, London: Unwin Hyman.

Beauregard, R. (1986) 'The chaos and complexity of gentrification', in N. Smith and P. Williams (eds) *Gentrification of the city*, Boston: Allen & Unwin, pp. 35–55.

Behr, M. and Gober, P. (1982) 'When a residence is not a house: examining residence-based migration definitions', *Professional Geographer* 34: 178–84.

Beiser, M. (1993) 'After the door has been opened: the mental health of immigrants and refugees in Canada', in V. Robinson (ed.) *The international refugee crisis: British and Canadian responses*, Basingstoke: Macmillan, pp. 213–29.

Bell, C. (1969) *Middle class families*, London: Routledge & Kegan Paul.

Bennett, R. and Chorley, R. (1978) *Environmental systems: philosophy, analysis and control*, London: Methuen.

Berelson, B. (1979) 'Romania's 1966 anti-abortion decree: the demographic experience of the first decade', *Population Studies* 33: 209–22.

Berger, J. and Mohr, J. (1975) *A seventh man*, Harmondsworth: Penguin.

Berry, B. (ed.) (1976a) *Urbanization and counterurbanization*, Beverly Hills, California: Sage Publications.

Berry, B. (1976b) 'The counterurbanization process: urban America since 1970', in B. Berry (ed.) *Urbanization and counter-urbanization*, Beverly Hills, California: Sage Publications, pp. 17–30.

Berry, B. (1988) 'Migration reversals in perspective: the long-wave evidence', *International Regional Science Review* 11: 245–51.

Bertram, G. and Watters, R. (1986) 'The MIRAB process: earlier analysis in context', *Pacific Viewpoint* 27: 47–59.

Bielby, W. and Baron, J. (1986) 'Men and women at work: sex segregation and statistical discrimination', *American Journal of Sociology* 91: 759–99.

Bielby, W. and Bielby, D. (1992) 'I will follow him: family ties, gender role beliefs, and reluctance to relocate for a better job', *American Journal of Sociology* 97: 1241–67.

Birks, S., Seccombe, I. and Sinclair, C. (1986) 'Migrant workers in the Arab Gulf: the impact of declining oil revenues', *International Migration Review* 20: 799–814.

Birman, I. (1979) 'Jewish emigration from the USSR', *Soviet Jewish Affairs* 9: 46–63.

Black, J. and Stafford, D. (1988) *Housing Policy and Finance*, London: Routledge.

Black, R. (1993) 'Refugees and asylum seekers in Western Europe: new challenges', in R. Black and V. Robinson (eds) *Geography and refugees: patterns and processes of change*, London: Belhaven, pp. 87–104.

Black, R. (1994a) 'Environmental change in refugee-affected areas of the Third World', *Disasters* 18: 107–16.

Black, R. (1994b) 'Refugee migration and local economic development in Eastern Zambia', *Tijdschrift voor Economische en Sociale Geografie* 85: 249–62.

Blackwood, L. and Carpenter, E. (1978) 'The importance of anti-urbanism in determining residential preferences and migration patterns', *Rural Sociology* 43: 31–47.

Blainey, G. (1984) *All for Australia*, North Ryde: Methuen.

Bogue, D. (1959) 'Internal migration', in P. Hauser and O. Duncan (eds) *The study of population: an inventory and appraisal*, Chicago: University of Chicago Press, pp. 486–509.

Bogue, D. (1969) *Principles of demography*, New York: Wiley.

Bolton, N. and Chalkley, B. (1989) 'Counter-urbanisation – disposing of the myths', *Town and Country Planning* 58: 249–50.

Bonaguidi, A. (1990) 'Italy', in C. Nam, W. Serow

and D. Sly (eds) *International handbook on internal migration*, New York: Greenwood Press, pp. 239–55.

Bonaguidi, A. and Terra Abrami, V. (1992) 'The metropolitan aging transition and metropolitan redistribution of the elderly in Italy', in A. Rogers (ed.) *Elderly migration and population redistribution*, London: Belhaven, pp. 143–63.

Bonaguidi, A. and Terra Abrami, V. (1996) 'The pattern of internal migration: the Italian case', in P. Rees, J. Stillwell, A. Convey and M. Kupiszewski (eds) *Population migration in the European Union*, Chichester: John Wiley & Sons, pp. 231–45.

Bondi, L. (1991) 'Gender divisions and gentrification: a critique', *Transactions of the Institute of British Geographers* 16: 190–8.

Bone, M. (1984) *Registration with general medical practitioners in inner London*, London: HMSO.

Bonney, N. (1988) 'Dual earning couples: trends of change in Britain', *Work, Employment and Society* 2: 89–102.

Bonney, N. and Love, J. (1991) 'Gender and migration: geographical mobility and the wife's sacrifice', *Sociological Review* 39: 335–48.

Borgegärd, L.-E. and Murdie, R. (1993) 'Socio-demographic impacts of economic restructuring on Stockholm's inner city', *Tijdschrift voor Economische en Sociale Geografie* 84: 269–80.

Borjas, G. (1989) 'Economic theory and international migration', *International Migration Review* 23: 457–85.

Borjas, G. (1990) *Friends or strangers*, New York: Basic Books.

Bottomley, G. (1992) *From another place: migration and the politics of culture*, Cambridge: Cambridge University Press.

Boudoul, J. and Faur, J. (1982) 'Renaissance des communes rurales ou nouvelle forme d'urbanisation?', *Economie et Statistique* 149: I–XVI.

Bover, O., Muellbauer, J. and Murphy, A. (1988) 'Housing, wages and the UK labour markets', *Centre for Economic Policy Research Discussion Paper* 268.

Bowie, F. (1993) 'Wales from within: conflicting interpretations of Welsh identity', in S. MacDonald (ed.) *Inside European identities*, Oxford: Berg Publishers, pp. 167–93.

Bowlby, S. and Silk, J. (1982) 'Analysis of qualitative data using GLIM: two examples based on shopping survey data', *Professional Geographer* 34: 80–90.

Boyd, M. (1989) 'Family and personal networks in international migration: recent developments and new agendas', *International Migration Review* 23: 638–64.

Boyer, R. and Savageau, D. (1989) *Places rated almanac*, Englewood Cliffs, New Jersey: Prentice-Hall.

Boyle, M. and Hughes, G. (1991) 'The politics of the representation of 'the real': discourses from the Left on Glasgow's role as European City of Culture, 1990', *Area* 23: 217–28.

Boyle, P. (1995a) 'Rural in-migration in England and Wales 1980–1981', *Journal of Rural Studies* 11: 65–78.

Boyle, P. (1995b) 'Modelling population movement into the Scottish highlands and islands from the remainder of Britain, 1990–91', *Scottish Geographical Magazine* 111: 5–12.

Boyle, P. (1995c) 'Public housing as a barrier to long-distance migration', *International Journal of Population Geography* 1: 147–64.

Boyle, P. and Halfacree, K. (1995) 'Service class migration in England and Wales, 1980–81: identifying gender-specific mobility patterns', *Regional Studies* 29: 43–57.

Boyle, P. and Halfacree, K. (1997, forthcoming) 'Migration into rural areas and collective behaviour', in P. Boyle and K. Halfacree (eds) *Migration into rural areas: theories and issues*, Chichester: Wiley.

Bradley, H. (1989) *Men's work, women's work*, Cambridge: Polity.

Brandes, S. (1975) *Migration, kinship and community: tradition and transition in a Spanish village*, New York: Academic Press.

Breathnach, P. (1988) 'Uneven development and capitalist peripheralisation: the case of Ireland', *Antipode* 20: 122–41.

Breathnach, P. and Jackson, J. (1991) 'Ireland, emigration and the New International Division of Labour', in R. King (ed.) *Contemporary Irish Migration*, Dublin: Geographical Society of Ireland, pp. 1–10.

Bright, M. and Thomas, D. (1941) 'Interstate migration and intervening opportunities', *American Sociological Review* 6: 773–83.

Bristow, M. (1976) 'Britain's response to the Ugandan Asian crisis: government myths versus political and resettlement realities', *New Community* 5: 265–79.

Brown, A. (1976) 'Towards a world census', *Population Trends* 14: 17–19.

Brown, L. and Moore, E. (1970) 'The intra-urban migration process: a perspective', *Geografiska Annaler* 52B: 1–13.

Brown, L. and Sanders, R. (1981) 'Toward a development paradigm of migration, with particular reference to third world settings', in G. De Jong and R. Gardner (eds) *Migration decision making: multidisciplinary approaches to microlevel studies in developed and developing countries*, New York: Pergamon Press, pp. 149–85.

Brown, R. (1982) 'Work histories, career strategies and class structure', in A. Giddens and G. Mackenzie (eds) *Social class and the division of labour*, Cambridge: Cambridge University Press, pp. 119–36.

Brown, R. and Connell, J. (1993) 'The global flea market: migration, remittances and the informal economy in Tonga', *Development and Change* 24: 611–47.

Brydon, L. (1992) 'Ghanaian women in the migration process', in S. Chant (ed.) *Gender and migration in developing countries*, London: Belhaven, pp. 91–108.

Bryman, A., Bytheway, B., Allatt, P. and Keil, T. (1987) 'Introduction', in A. Bryman, B. Bytheway, P. Allatt and T. Keil (eds) *Rethinking the life cycle*, Basingstoke: Macmillan, pp. 1–16.

Buller, H. and Hoggart, K. (1994a) *International counterurbanization*, Aldershot: Avebury.

Buller, H. and Hoggart, K. (1994b) 'The social integration of British home owners into French rural communities', *Journal of Rural Studies* 10: 197–210.

Bulusu, L. (1991) *A review of migration data sources*, London: OPCS Occasional Paper 39.

Bunce, M. (1994) *The countryside ideal. Anglo-American images of landscape*, London: Routledge.

Bunge, W. (1962) *Theoretical geography*, Lund: Gleerup.

Burgers, J. and Engbersen, G. (1996) 'Globalisation, migration and undocumented immigrants', *New Community* 22: 619–35.

Burnley, I (1989) 'Settlement dimensions of the Vietnam-born population in metropolitan Sydney', *Australian Geographical Studies* 27: 129–54.

Buttimer, A. (1976) 'Grasping the dynamism of the lifeworld', *Annals of the Association of American Geographers* 66: 277–92.

Cadwallader, M. (1986) 'Migration and intra-urban mobility', in M. Pacione (ed.) *Population geography: progress and prospect*, London: Croom Helm, pp. 257–83.

Cadwallader, M. (1989) 'A conceptual framework for analysing migration behaviour in the developed world', *Progress in Human Geography* 13: 494–511.

Cadwallader, M. (1991) 'Metropolitan growth and decline in the United States: an empirical analysis', *Growth and Change* 22: 1–16.

Cadwallader, M. (1993) 'Commentary on Zelinsky's model', *Progress in Human Geography* 17: 215–17.

Calavita, K. (1995) 'Mexican immigration to the USA: the contradictions of border control', in R. Cohen (ed.) *The Cambridge survey of world migration*, Cambridge: Cambridge University Press, pp. 236–45.

Cameron, S. (1992) 'Housing, gentrification and urban regeneration policies', *Urban Studies* 29: 3–14.

Campani, G. (1995) 'Women migrants: from marginal subjects to social actors', in R. Cohen (ed.) *The Cambridge survey of world migration*, Cambridge: Cambridge University Press, pp. 546–50.

Campbell, I., Fincher, R. and Webber, M. (1991) 'Occupational mobility in segmented labour markets: the experience of immigrant workers in Melbourne', *Australian and New Zealand Journal of Sociology* 27: 172–94.

Campbell, R. and Garkovich, L. (1984) 'Turnaround migration as an episode of collective behavior', *Rural Sociology* 49: 89–105.

Carpenter, J. and Lees, L. (1995) 'Gentrification in New York, London and Paris: an international comparison', *International Journal of Urban and Regional Research* 19: 286–303.

Carrothers, G. (1956) 'An historical review of the gravity model and potential concepts of human interaction', *Journal of the American Institute of Planners* 22: 94–102.

Cassady, C. (1991) *Off the road*, London: Flamingo.

Castles, L. (1991) 'Jakarta: the growing centre', in H. Hill (ed.) *Unity and diversity: regional economic development in Indonesia since 1970*, Singapore: Oxford University Press, pp. 232–54.

Castles, S. (1995) 'Contract labour migration', in R. Cohen (ed.) *The Cambridge survey of world migration*, Cambridge: Cambridge University Press, pp. 510–514.

Castles, S., Booth, H. and Wallace, T. (1984) *Here for good: Western Europe's new ethnic minorities*, London: Pluto.

Castles, S. and Kosack, G. (1973) *Immigrant workers*

and class structure in Western Europe, London: Oxford University Press.

Castles, S. and Miller, M. (1993) *The age of migration: international population movements in the modern world*, Basingstoke: Macmillan.

Caulfield, J. (1993) 'Responses to growth in the Sun-Belt state – planning and coordinating policy initiatives in Queensland', *Australian Journal of Public Administration* 52: 431–42.

Cebula, R. (1979) *The determinants of human migration*, Lexington, Massachusetts: D.C. Heath and Company.

Cebula, R. and Vedder, R. (1973) 'A note on migration, economic opportunity, and the quality of life', *Journal of Regional Science* 13: 205–11.

Ceredigion District Council (1981) *Ceredigion migration and labour mobility survey*, Aberaeron, Dyfed: Ceredigion District Council.

Cernea, M. (1990) 'Internal refugee flows and development-induced population displacement', *Journal of Refugee Studies* 3: 320–39.

Cernea, M. (1995) 'Understanding and preventing impoverishment from displacement: reflections on the state of knowledge', *Journal of Refugee Studies* 8: 245–64.

Chambers, I. (1994) *Migrancy, culture, identity*, London: Routledge.

Champion, A. (1987) 'Recent changes in the pace of population deconcentration in Britain', *Geoforum* 18: 379–401.

Champion, A. (ed.) (1989) *Counterurbanization: the changing pace and nature of population deconcentration*, London: Edward Arnold.

Champion, A. (1992) 'Urban and regional demographic trends in the developed world', *Urban Studies* 29: 461–82.

Champion, A., Green, A. and Owen, D. (1988) 'House prices and local labour market performance: an analysis of building society data', *Area* 20: 253–63.

Chan, K. (1994) 'Urbanization and rural–urban migration in China since 1982: a new base line', *Modern China* 20: 243–81.

Chan, T. (1981) 'A review of micro migration research in the third world context', in G. De Jong and R. Gardner (eds) *Migration decision making: multidisciplinary approaches to microlevel studies in developed and developing countries*, New York: Pergamon Press, pp. 303–27.

Chandler, M. (1989) 'Voices crying in the suburbs', in B. Kelly (ed.) *Suburbia re-examined*, New York: Greenwood Press, pp. 215–22.

Chant, S. (1991a) 'Gender, migration and urban development in Costa Rica: the case of Guanacaste', *Geoforum* 22: 237–53.

Chant, S. (1991b) *Women and survival in Mexican cities: perspectives on gender, labour markets and low income households*, Manchester: Manchester University Press.

Chant, S. (1992a) 'Migration at the margins: gender, poverty and population movement on the Costa Rican periphery', in S. Chant (ed.) *Gender and migration in developing countries*, London: Belhaven, pp. 49–72.

Chant, S. (1992b) 'Conclusion: towards a framework for the analysis of gender-selective migration', in S. Chant (ed.) *Gender and migration in developing countries*, London: Belhaven, pp. 197–206.

Chant, S. and Radcliffe, S. (1992) 'Migration and development: the importance of gender', in S. Chant (ed.) *Gender and migration in developing countries*, London: Belhaven, pp. 1–29.

Chapman, G. (1992) 'Change in the South Asian core: patterns of stagnation and growth in India', in G. Chapman and K. Baker (eds) *The changing geography of Asia*, London: Routledge, pp. 10–43.

Chapman, G. and Baker, K. (eds) (1992a) *The changing geography of Africa and the Middle East*, London: Routledge.

Chapman, G. and Baker, K. (eds) (1992b) *The changing geography of Asia*, London: Routledge.

Chapman, L. (1976) 'Illegal aliens: time to call a halt!', *Readers Digest* 109: 188–92.

Chapman, M. (1985) 'Me go walkabout: you too?', in M. Chapman and R. Prothero (eds) *Circulation in population movement: substance and concepts from the Melanesian case*, London: Routledge & Kegan Paul, pp. 429–43.

Chapman, M. (1991) 'Pacific island movement and socioeconomic change: metaphors of misunderstanding', *Population and Development Review* 17: 263–92.

Chapman, M. and Prothero, R. (1985) 'Preface', in M. Chapman and R. Prothero (eds) *Circulation in population movement: substance and concepts from the Melanesian case*, London: Routledge & Kegan Paul, pp. xix–xxiii.

Charney, A. (1993) 'Migration and the public sector: a survey', *Regional Studies* 27: 313–26.

Chatwin, B. (1988) *The Songlines*, London: Picador.

Chatwin, B. (1990) *What am I doing here?*, London: Picador.

Checkoway, B. (1984) 'Large builders, federal housing programs, and postwar suburbanisation', in

W. Tabb and L. Sawers (eds) *Marxism and the metropolis: new perspectives in urban political economy*, Oxford: Oxford University Press, pp. 152–73.

Cherry, G. (1988) *Cities and plans: the shaping of urban Britain in the nineteenth and twentieth centuries*, London: Edward Arnold.

Christenson, J. (1979) 'Value orientations of potential migrants and nonmigrants', *Rural Sociology* 44: 331–44.

Christopher, A. (1987) 'Apartheid planning in South Africa: the case of Port Elizabeth', *Geographical Journal* 153: 195–204.

Christopher, A. (1994) *The atlas of apartheid*, London: Routledge.

Clark, D. (1982) *Urban geography*, London: Croom Helm.

Clark, D. and Cosgrove, J. (1991) 'Amenities versus labor market opportunities: choosing the optimal distance to move', *Journal of Regional Science* 31: 311–28.

Clark, D. and Hunter, W. (1992) 'The impact of economic opportunity, amenities and fiscal factors on age-specific migration rates', *Journal of Regional Science* 32: 349–65.

Clark, G. (1987) 'Job search theory and indeterminate information', in M. Fischer and P. Nijkamp (eds) *Regional labour markets*, Amsterdam: North Holland, pp. 169–88.

Clark, R. and Wolf, W. (1992) 'Proximity of children and elderly migration', in A.Rogers (ed.) *Elderly migration and population redistribution*, London: Belhaven, pp. 77–97.

Clark, T. (1984) *Jack Kerouac. A biography*, New York: Marlowe and Company.

Clark, W. (1982) 'Recent research on migration and mobility: a review and interpretation', *Progress in Planning* 18: 1–56.

Clark, W. (1986) *Human migration*, Beverly Hills, California: Sage.

Clark, W., Duerloo, M. and Dielemen, F. (1984) 'Housing consumption and residential mobility', *Annals of the Association of American Geographers* 74: 29–43.

Clark, W. and Huff, J. (1977) 'Some empirical tests of duration-of-stay effects in intra-urban migration', *Environment and Planning A* 9: 1357–74.

Clarke, C., Peach, C. and Vertovec, S. (1990) *South Asian communities overseas*, Cambridge: Cambridge University Press.

Clarke, G. (1994) 'The movement of population to the west of China: Tibet and Qinghai', in J. Brown and R. Foot (eds) *Migration: the Asian experience*, Basingstoke: Macmillan, pp. 221–58.

Clay, J. (1986) 'Refugees flee Ethiopian collectivization', *Cultural Survival Quarterly* 10: 80–5.

Cliff, A., Martin, R. and Ord, J. (1974) 'Evaluating the friction of distance parameters in gravity models', *Regional Studies* 8: 281–6.

Cliff, A., Martin, R. and Ord, J. (1975) 'Map pattern and friction of distance parameters: reply to comments by RJ Johnston and L Curry, DA Griffith and ES Sheppard', *Regional Studies* 9: 285–8.

Cliff, A., Martin, R. and Ord, J. (1976) 'A reply to the final comment', *Regional Studies* 10: 341–2.

Cloke, P. and Davies, L. (1992) 'Deprivation and lifestyles in rural Wales. I. Towards a cultural dimension', *Journal of Rural Studies* 8: 349–58.

Cloke, P., Goodwin, M., Milbourne, P. and Thomas, C. (1995) 'Deprivation, poverty and marginalization in rural lifestyles in England and Wales', *Journal of Rural Studies* 11: 351–65.

Cloke, P., Goodwin, M. and Milbourne, P. (1997, forthcoming) 'Inside looking out; outside looking in. Different experiences of cultural competence in rural lifestyles', in P. Boyle and K. Halfacree (eds) *Migration into rural areas: theories and issues*, Chichester: Wiley.

Cloke, P., Goodwin, M. and Milbourne, P. (forthcoming) 'Cultural change and conflict in rural Wales: competing constructs of identity', *Environment and Planning A*.

Cloke, P. and Milbourne, P. (1992) 'Deprivation and lifestyles in rural Wales. II. Rurality and the cultural dimension', *Journal of Rural Studies* 8: 359–71.

Cloke, P., Phillips, M. and Thrift, N. (1997, forthcoming) 'Class, colonisation and lifestyle strategies in Gower', in P. Boyle and K. Halfacree (eds) *Migration into rural areas: theories and issues*, Chichester: Wiley.

Cloke, P., Philo, C. and Sadler, D. (1991) *Approaching human geography*, London: Paul Chapman Publishing.

Cohen, B. (1992) 'Israel's expansion through immigration', *Middle East Policy* 1: 120–35.

Cohen, R. (1987) *The new helots: migrants in the international division of labour*, Aldershot: Gower.

Cohen, R. (1995) 'Rethinking "Babylon": iconoclastic conceptions of the diasporic experience', *New Community* 21: 5–18.

Cohen, Y. and Tyree, A. (1994) 'Palestinian and Jewish Israeli-born immigrants in the United

States', *International Migration Review* 28: 243–55.

Coleman, D. and Haskey, J. (1986) 'Marital distance and its geographical orientation', *Transactions of the Institute of British Geographers* 2: 337–55.

Coleman, D. and Salt, J. (1992) *The British population: patterns, trends, and processes*, Oxford: Oxford University Press.

Collins, J. (1991) *Migrant hands in a distant land: Australia's post-war immigration*, Leichhardt: Pluto Press.

Collinson, S. (1993) *Europe and international migration*, London: Pinter.

Collver, A. (1963) 'The family life cycle in India and the United States', *American Sociological Review* 28: 86–96.

Colpi, T. (1991) *The Italian factor: the Italian community in Great Britain*, Edinburgh: Mainstream Publishing.

Condon, R. (1987) *Inuit youth: growth and change in the Canadian Arctic*, New Brunswick: Rutgers University Press.

Connell, J. (1978) *The end of tradition. Country life in central Surrey*, London: Routledge & Kegan Paul.

Connell, J. (1992) 'Far beyond the Gulf: the implications of warfare for Asian labour migration', *Australian Geographer* 23: 44–50.

Connell, J. (1995) 'In Samoan worlds. Culture, migration, identity and Albert Wendt', in R. King, J. Connell and P. White (eds) *Writing across worlds. Literature and migration*, London: Routledge, pp. 263–79.

Connell, J. and Brown, R. (1995) 'Migration and remittances in the South Pacific: towards new perspectives', *Asian and Pacific Migration Journal* 4: 1–33.

Cook, I. and Crang, M. (1995) *Doing ethnography*, Norwich: School of Environmental Science, CATMOG 58.

Coombes, M. and Charlton, M. (1992) 'Flows to and from London. A decade of change?', in J. Stillwell, P. Rees and P. Boden (eds) *Migration patterns and processes: population redistribution in the United Kingdom*, London: Belhaven, pp. 56–81.

Cordey-Hayes, M. (1975) 'Migration and the dynamics of multiregional population system', *Environment and Planning A* 7: 793–814, London: Pion Ltd.

Cornelius, W. (1991) 'Los migrantes de la crisis: the changing profile of Mexican migration to the United States', in M. González de la Rocha and A. Escobar Latapí (eds) *Social responses to Mexico's economic crisis of the 1980s*, San Diego: Center for US-Mexican Studies, pp. 155–93.

Cottaar, A. and Willems, W. (1992) 'The image of Holland: caravan dwellers and other minorities on Dutch society', *Immigrants and Minorities* 11: 67–80.

Cottaar, A., Lucassen, L. and Willems, W. (1992) 'Justice or injustice? A survey of government policy towards Gypsies and caravan dwellers in Western Europe in the nineteenth and twentieth centuries', *Immigrants and Minorities* 11: 42–66.

Coughlan, J. (1989a) 'The spatial distribution and concentration of Australia's three Indo-Chinese born communities', Griffith University CSAAR, Nathan, New South Wales.

Coughlan, J. (1989b) 'A comparative study of the labour force performance of Indo-Chinese born immigrants in Australia', Griffith University CSAAR, Nathan, New South Wales.

Courchene, T. and Melvin, J. (1986) 'Canadian regional policy: lessons from the past and prospects for the future', *Canadian Journal of Regional Science* 9: 49–67.

Courgeau, D. (1978) 'Les migrations internes en France de 1954 à 1975, I. Vue d'ensemble', *Population* 35: 525–45.

Courgeau, D. (1984) 'Relations entre cycle de vie et migrations', *Population* 39: 483–514.

Courgeau, D. (1985) 'Interaction between spatial mobility, family and career life-cycle: a French survey', *European Sociological Review* 1: 139–62.

Courgeau, D. (1989) 'Recent conceptual advances in the study of migration in France', in P. Ogden and P. White (eds) *Migrants in modern France*, London: Unwin Hyman, pp. 60–73.

Court, Y. (1989) 'Denmark: towards a more deconcentrated settlement system', in A. Champion (ed.) *Counterurbanization*, London: Edward Arnold, pp. 121–40.

Cousens, S. (1960) 'The regional pattern of emigration during the Great Irish Famine 1846–51', *Transactions of the Institute of British Geographers* 28: 119–34.

Cox, D. (1972) 'Regression models and life tables', *Journal of the Royal Statistical Society B* 34: 187–202.

Cresswell, T. (1993) 'Mobility as resistance: a geographical reading of Kerouac's "On the road"', *Transactions of the Institute of British Geographers* 18: 249–62.

Cresswell, T. (1996) *In place/Out of place*, Minneapolis: University of Minnesota Press.

Crompton, R. and Sanderson, K. (1986) 'Credentials and careers: some implications of the increase in professional qualifications amongst women', Sociology 20: 25–42.

Croon, M. (1995) Wales, an extremely regional geography, Utrecht: Utrecht University, Department of Geography.

Crossman, R. (1975) The diaries of a cabinet minister. Volume 1, London: Hamish Hamilton.

Crowe, D. (1995) A history of the Gypsies of Eastern Europe and Russia, London: IB Tauris.

Crowther, M. (1992) 'The tramp', in R. Porter (ed.) Myths of the English, Cambridge: Polity, pp. 91–113.

Crush, J. (1995) 'Vulcan's brood. Spatial narratives of migration in Southern Africa', in R. King, J. Connell and P. White (eds) Writing across worlds. Literature and migration, London: Routledge, pp. 229–47.

Curry, L. (1972) 'A spatial analysis of gravity flows', Regional Studies 6: 131–47.

Curry, L., Griffith, D. and Sheppard, E. (1975) 'Those gravity parameters again', Regional Studies 9: 289–96.

Curtin, P. (1969) The Atlantic slave trade: a census, Madison: University of Wisconsin Press.

Dahal, D., Manzardo, A. and Rai, N. (1977) Land and migration in far western Nepal, Kathmandu: Institute of Nepal and Asian Studies, Tribhuvan University.

Dail, P. (1993) 'Homelessness in America: involuntary family migration', Marriage and Family Review 19: 55–75.

Dal, G. (1979) Wenn Ali die Glocken läuten hört, Berlin: Edition der 2.

Dal, G. (1990) [1981] Europastraße 5, Munich: Piper.

Dale, A. (1993) 'Whither the census', in A. Dale and C. Marsh (eds) The 1991 census user's guide, London: HMSO, pp. 352–363.

Datta, K. (1996) 'Historical observations on the impact of gender on labour migration in Botswana', Migration Unit Research Paper 12,: University of Wales Swansea: Department of Geography.

Da Vanzo, J. (1980) Micro economic approaches to studying migration decisions, Santa Monica: Rand Corporation.

Davies, H. (1988) 'Population change in the Sudan since independence', Geography 73: 249–55.

Davies, R. (1981) 'The spatial formation of the South African city', Geojournal (Supplementary issue) 2: 59–72.

Davies, R. (1991) 'The analysis of housing and migration careers', in J. Stillwell and P. Congdon (eds) Migration models: macro and micro approaches, London: Belhaven, pp. 207–27.

Davies, R. and Flowerdew, R. (1992) 'Modelling migration careers, using data from a British survey', Geographical Analysis 24: 35–57.

Davies, R. and Pickles, A. (1983) 'The estimation of duration-of-residence effects: a stochastic modelling approach', Geographical Analysis 15: 305–17.

Davies, R. and Pickles, A. (1985a) 'The longitudinal analysis of housing careers', Journal of Regional Science 25: 85–101.

Davies, R. and Pickles, A. (1985b) 'A panel study of life-cycle effects in residential mobility', Geographical Analysis 17: 199–216.

Davis, V. (1945) 'Development of a scale to rate attitudes of community satisfaction', Rural Sociology 10: 246–55.

Day, G. (1989) 'A million on the move?: population change and rural Wales', Contemporary Wales 3: 137–59.

Dear, M. (1988) 'The postmodern challenge: reconstructing human geography', Transactions of Institute of British Geographers 13: 262–74.

Deedes, W. (1968) Race without rancour, London: Conservative Political Centre.

Delf, G. (1963) Asians in East Africa, London: Oxford University Press.

Denham, C. (1993) 'Census geography. 1 An overview', in A. Dale and C. Marsh (eds) The 1991 census user's guide, London: HMSO, pp. 52–69.

Denzin, N. (1989) Interpretive interactionism, Newbury Park, California: Sage.

De Oliveira, O. (1991) 'Migration of women, family organisation and labour markets in Mexico', in E. Jelin (ed.) Family, household and gender relations in Latin America, London: Kegan Paul International, pp. 101–18.

Desbarats, J. (1985) 'Indochinese resettlement in the United States', Annals of the Association of American Geographers 75: 522–38.

Dickenson, J. (1983) A geography of the Third World, London: Methuen.

Dickenson, J., Clarke, C., Gould, W., Hodgkiss, A., Prothero, R., Siddle, D., Smith, C. and Thomas-Hope, E. (1983) A geography of the Third World, London: Routledge.

Doorn, P. (1989) 'Selective migration in the Dutch labour force', in J. Stillwell and H. Scholten (eds)

Contemporary research in population geography, Dordrecht: Kluwer, pp. 102–15.

Douglas, J. (1985) *Creative interviewing*, Beverly Hills, California: Sage.

Dowty, A. (1987) *Closed borders: the contemporary assault on freedom of movement*, London: Yale University Press.

Drugge, S. (1985) 'Factor mobility, the elasticity of substitution and interregional wage differentials', *The Annals of Regional Science* 19: 34–9.

Drugge, S. (1987) 'A critique of the neoclassical migration model as a normative approach to Canadian regional policy – a comment', *Canadian Journal of Regional Science* 10: 91–5.

D'Souza, F. and Crisp, J. (1985) *The refugee dilemma*, London: Minority Rights Group.

Duffy, P. (1995) 'Literary reflections on Irish migration in the nineteenth and twentieth centuries', in R. King, J. Connell and P. White (eds) *Writing across worlds. Literature and migration*, London: Routledge, pp. 20–38.

Duncan, J. (1990) *The city as text: the politics of landscape interpretation in the Kandyan Kingdom*, Cambridge: Cambridge University Press.

Duncan, O. and Duncan, B. (1955) 'A methodological analysis of segregation indexes', *American Sociological Review* 20: 210–17.

Duncan, S. (1991) 'The geography of gender divisions of labour in Britain', *Transactions of the Institute of British Geographers* 16: 420–39.

Durand, J. and Massey, D. (1992) 'Mexican migration to the United States: a critical review', *Latin American Research Review* 27(2): 3–42.

Duvall, E. (1977) *Marriage and family development*, Philadelphia: Lippincott.

Dylan, B. (1988) *Lyrics, 1962–1985*, London: Paladin.

Earle, F., Dearling, A., Whittle, H., Glasse, R. and Gubby (1994) *A time to travel? An introduction to Britain's newer travellers*, Lyme Regis: Enabler Publications.

Egemi, O. (1994) 'The political ecology of subsistence crisis in the Red Sea Hills, Sudan', PhD (Dr.Polit) thesis, Department of Geography, University of Bergen.

Ehrenreich, B. (1983) *The hearts of men: American dreams and the flight from commitment*, London: Pluto Press.

El-Arifi, S. (1975) 'Pastoral nomadism in the Sudan', *East African Geographical Review* 13: 89–103.

El-Bushra, E. (1989) 'The urban crisis and rural–urban migration in Sudan', in R. Potter and T. Unwin (eds) *The geography of urban–rural interaction in developing countries*, London: Routledge, pp. 109–40.

Elder, G. (1978) 'Family history and the life-course', in T. Hareven (ed.) *The family and the life course in historical perpective*, New York: Academic Press, pp. 19–64.

Elkin, T., McLaren, D. and Hillman, M. (1991) *Reviving the city*, London: Friends of the Earth/Policy Studies Institute.

Elridge, H. (1965) 'Primary, secondary and return migration in the United States, 1955–60', *Demography* 2: 444–55.

Engels, R. and Healy, M. (1981) 'Measuring interstate migration flows: an origin–destination network based on internal revenue service records', *Environment and Planning A* 13: 1345–60.

England, K. (1991) 'Gender relations and the spatial structure of the city', *Geoforum* 22: 135–47.

Entrikin, N. (1976) 'Contemporary humanism in geography', *Annals of the Association of American Geographers* 66: 615–32.

Erikson, K. (1967) 'A comment on disguised observation in sociology', *Social Problems* 14: 366–73.

Evans, A. (1989) 'South East England in the eighties: explanations for a house price explosion', in M. Breheny and P. Congdon (eds) *Growth and change in a core region: the case of South East England*, London: Pion, pp. 130–49.

Evans, D. (1984) 'Demystifying suburban landscapes', in D. Herbert and R. Johnston (eds) *Geography and the Urban Environment. Volume VI*, Chichester: Wiley, pp. 321–48.

Evans, R. (1991) 'Born to roam free', *Geographical Magazine* 64(3): 22–3.

Evers, G. (1989) 'Simultaneous models for migration and commuting: macro and micro economic approaches', in J. van Dijk, H. Folmer, H. Herzog, and A. Schlottman, (eds) *Migration and labor market adjustment*, Dordrecht: Kluwer, pp. 177–97.

Fakiolas, R. (1995) 'Italy and Greece; from emigrants to immigrants', in R. Cohen (ed.) *The Cambridge survey of world migration*, Cambridge: Cambridge University Press, pp. 313–16.

Fassman, H. and Münz, R. (1994) 'Patterns and trends of international migration in Western Europe', in H. Fassman and R. Münz (eds) *European migration in the late twentieth century*, Aldershot: Edward Elgar, pp. 3–33.

Ferris, E (1985) 'Refugees and world politics', in E. Ferris (ed.) *Refugees and world politics*, New York: Praeger, pp. 1–25.

Fielding, A. (1982) 'Counterurbanization in Western Europe', *Progress in Planning* 17: 1–52.

Fielding, A. (1989) 'Inter-regional migration and social change: a study of South East England based upon data from the Longitudinal Study', *Transactions of the Institute of British Geographers* 14: 24–36.

Fielding, A. (1990) 'Counterurbanisation: threat or blessing?', in D. Pinder (ed.) *Western Europe: challenge or change*, London: Belhaven Press, pp. 226–39.

Fielding, A. (1992a) 'Migration and culture', in A. Champion and A. Fielding (eds) *Migration processes and patterns. Volume 1. Research progress and prospects*, London: Belhaven Press, pp. 201–12.

Fielding, A. (1992b) 'Migration and social change', in J. Stillwell, P. Rees and P. Boden (eds) *Migration processes and patterns. Volume 1. Population redistribution in the United Kingdom*, London: Belhaven Press, pp. 225–47.

Fielding, A. (1992c) 'Migration and social mobility: South East England as an escalator region', *Regional Studies* 26: 1–15.

Fielding, A. (1993a) 'Mass migration and economic restructuring', in R. King (ed.) *Mass migrations in Europe. The legacy and the future*, London: Belhaven Press, pp. 7–18.

Fielding, A. (1993b) 'Migration and the metropolis: an empirical and theoretical analysis of inter-regional migration to and from south east England', *Progress in Planning* 39: 71–166.

Findlay, A. (1988) 'From settlers to skilled transients: the changing structure of British international migration', *Geoforum* 19: 401–10.

Findlay, A. (1992) 'Population geography', *Progress in Human Geography* 16: 88–97.

Findlay, A. (1993) 'End of the Cold War: end of Afghan relief aid?', in R. Black and V. Robinson (eds) *Geography and refugees: patterns and processes of change*, London: Belhaven, pp. 185–98.

Findlay, A. (1994a) *The Arab world*, London: Routledge.

Findlay, A. (1994b) 'An economic audit of contemporary immigration', in S. Spencer (ed.) *Strangers and citizens: a positive approach to migrants and refugees*, London: Rivers Oram Press, pp. 159–202.

Findlay, A. and Findlay, A. (1987) *Population and development in the Third World*, London: Routledge.

Findlay, A. and Garrick, L. (1990) 'Scottish emigration in the 1980s: a migration channels approach to the study of skilled international migration', *Transactions of the Institute of British Geographers* 15: 177–92.

Findlay, A. and Graham, E. (1991) 'The challenge facing population geography', *Progress in Human Geography* 15: 149–62.

Findlay, A., Li, F., Jowett, A. and Skeldon, R. (1996) 'Skilled international migration and the global city: a study of expatriates in Hong Kong', *Transactions of the Institute of British Geographers* 21: 49–61.

Findlay, A. and Rogerson, R. (1993) 'Migration, places and quality of life: voting with their feet?', in T. Champion (ed.) *Population matters*, London: Paul Chapman Publishing, pp. 33–49.

Findlay, A. and Stewart, A. (1986) 'Manpower policies of British firms with offices in the Middle East', *Bulletin Committee for Middle East Trade* 18: 22–6.

Finnan, C. (1981) 'Occupational assimilation of refugees', *International Migration Review* 15: 292–309.

Fischer, C. (1982) *To dwell among friends: personal networks in town and city*, Chicago: University of Chicago Press.

Fischer, M. and Nijkamp, P. (1987) 'Labour market theories: perspectives, problems and policy implications', in M. Fischer and P. Nijkamp (eds) *Regional labour markets*, Amsterdam: North Holland, pp. 37–52.

Fischer, S. and McGowan, M. (1995) 'From *Pappkoffer* to pluralism. Migrant writing in the German Federal Republic', in R. King, J. Connell and P. White (eds) *Writing across worlds. Literature and migration*, London: Routledge, pp. 39–56.

Fitchen, J. (1995) 'Spatial redistribution of poverty through migration of poor people to depressed rural communities', *Rural Sociology* 60: 181–201.

Flett, H., Henderson, J. and Brown, B. (1979) 'The practice of racial dispersal in Birmingham 1969–75', *Journal of Social Policy* 8: 289–309.

Fliegel, F. and Sofranko, A. (1984) 'Non-metropolitan population increase, the attractiveness of rural living, and race', *Rural Sociology* 49: 298–308.

Flowerdew, R. (1991) 'Poisson regression models of migration', in J. Stillwell and P. Congdon (eds)

Migration models: macro and micro approaches, London: Belhaven Press, pp. 92–112.

Flowerdew, R. (1992) 'Labour market operation and geographical mobility', in A. Champion and A. Fielding (eds) *Migration processes and patterns. Volume 1. Research progress and prospects*, London: Belhaven Press, pp. 135–47.

Flowerdew, R. and Aitkin, M. (1982) 'A method of fitting the gravity model based on the Poisson distribution', *Journal of Regional Science* 22: 191–202.

Flowerdew, R. and Boyle, P. (1992) 'Migration trends for the West Midlands: suburbanisation, counterurbanisation or rural depopulation?', in J. Stillwell, P. Rees and P. Boden (eds) *Migration processes and patterns. Volume 2. Population redistribution in the United Kingdom*, London: Belhaven Press, pp. 144–61.

Flowerdew, R. and Halfacree, K. (1994) 'Logit modelling of migration propensity in 1980s Britain', *Migration Unit Research Paper* 7, University of Wales Swansea: Department of Geography.

Fontana, A. and Frey, J. (1993) 'Interviewing: the art of science', in N. Denzin and Y. Lincoln (eds) *Handbook of qualitative research*, Thousand Oaks, California: Sage, pp. 361–76.

Forbes, D. (1984) *The geography of underdevelopment*, London: Croom Helm.

Forrest, R. (1987) 'Spatial mobility, tenure mobility and emerging social divisions in the UK housing market', *Environment and Planning A* 19: 1611–30.

Forrest, R. and Murie, A. (1987) 'The affluent homeowner: labour-market position and the shaping of housing histories', in N. Thrift and P. Williams (eds) *Class and space*, London: Routledge & Kegan Paul, pp. 330–59.

Forrest, R. and Murie, A. (1988) *Selling the welfare state: the privatisation of public housing*, London: Routledge.

Forrest, R. and Murie, A. (1990) 'Moving strategies among home owners', in J. Johnson and J. Salt (eds) *Labour migration*, London: David Fulton Publishers, pp. 191–209.

Forrest, R. and Murie, A. (1991) 'Housing markets, labour markets and housing histories', in J. Allen and C. Hamnett (eds) *Housing and labour markets. Building the connections*, London: Unwin Hyman, pp. 63–93.

Forrest, R. and Murie, A. (1992) 'Housing as a barrier to the geographical mobility of labour', in A. Champion and A. Fielding (eds) *Migration processes and patterns. Volume 1. Research progress and prospects*, London: Belhaven Press, pp. 77–101.

Forsyth, D. (1992) *Migration and remittances in the South Pacific Forum island countries*, Suva: Department of Economics monograph, University of the South Pacific.

Fotheringham, A. (1983) 'A new set of spatial interaction models: the theory of competing destinations', *Environment and Planning A* 15: 15–36.

Fotheringham, A. (1984) 'Spatial flows and spatial patterns', *Environment and Planning A* 16: 529–43.

Fotheringham, A. (1991) 'Migration and spatial structure: the development of the competing destinations model', in J. Stillwell and P. Congdon (eds) *Migration models: macro and micro approaches*, London: Belhaven Press, pp. 57–72.

Fotheringham, A. and O'Kelly, M. (1989) *Spatial interaction models: formulations and applications*, Dordrecht: Kluwer.

Fotheringham, A. and Pellegrini, P. (1996) 'Microdata for migration analysis: a comparison of sources in the US, Britain and Canada', *Area* 28: 347–57.

Fotheringham, A. and Wong, D. (1991) 'The modifiable areal unit problem in multivariate statistical analysis', *Environment and Planning A* 23: 1025–44.

Frank, A. (1969) *Capitalism and underdevelopment in Latin America*, New York: Monthly Review Press.

Frankenburg, R. (1987) 'Life: cycle, trajectory or pilgrimage?', in A. Bryman, B. Bytheway, P. Allatt, and T. Keil (eds) *Rethinking the life cycle*, Basingstoke: Macmillan, pp. 122–38.

Fraser, A. (1992) *The Gypsies*, Oxford: Blackwell.

Frederick, M. (1993) 'Rural tourism and economic development', *Economic Development Quarterly* 7: 215–24.

Freke, D. (1990) 'History', in V. Robinson and D. McCarroll (eds) *The Isle of Man: celebrating a sense of place*, Liverpool: Liverpool University Press, pp. 103–23.

Frey, W. (1989) 'United States: counterurbanization and metropolis depopulation', in A. Champion (ed.) *Counterurbanization*, London: Edward Arnold, pp. 34–61.

Frey, W. (1992) 'Metropolitan redistribution of the US elderly, 1960–70, 1970–80, 1980–90', in A. Rogers (ed.) *Elderly migration and population redistribution*, London: Belhaven, pp. 123–43.

Frey, W. (1993) 'The new urban revival in the United States', *Urban Studies* 30: 741–74.

Frey, W. (1995a) 'Immigration and internal migration 'flight' from US metropolitan areas: toward a new demographic balkanisation', *Urban Studies* 32: 733–57.

Frey, W. (1995b) 'Immigration and internal migration 'flight': a California case study', *Population and Environment* 16: 353–75.

Frey, W. and Fielding, E. (1995) 'Changing urban populations: regional restructuring, racial polarization, and poverty concentration', *Cityscape: a Journal of Policy Development* 1(2): 1–66.

Frey, W. and Speare, A. (1991) 'U.S. metropolitan area growth 1960–1990: census trends and explanations', *Research Report* No. 91–212, Population Studies Center, University of Michigan.

Friedburg, R. and Hunt, J. (1995) 'The impact of immigrants on host country wages, employment and growth', *Journal of Economic Perspectives* 9: 23–44.

Friedman, E. (1988) *Colonialism and after. An Algerian Jewish community*, South Hadley, Massachusetts: Bergin and Garvey Publishers, Inc.

Frobel, F., Heinrichs, J. and Kreye, O. (1980) *The new international division of labour*, Cambridge: Cambridge University Press.

Frost, M. and Spence, N. (1993) 'Global city characteristics and central London's employment', *Urban Studies* 30: 547–58.

Fuguitt, G. and Zuiches, J. (1975) 'Residential preferences and population distribution', *Demography* 12: 491–504.

Gabriel, S., Shack-Marquez, J. and Wascher, W. (1993) 'Does migration arbitrage regional labor market differentials?', *Regional Science and Urban Economics* 23: 211–33.

Gallaway, L. and Cebula, R. (1972) 'The impact of property rights in human capital on regional factor proportions', *Zeitschrift fur Nationalökonomie* 32: 501–3.

Garside, J. (1993) 'Inner city gentrification in South Africa: the case of Woodstock, Cape Town', *GeoJournal* 30: 29–35.

Garson, J.-P. (1992) 'Migration and interdependence: the migration system between France and Africa', in M. Kritz, L. Lim and H. Zlotnik (eds) *International migration systems. A global approach*, Oxford: Clarendon Press, pp. 80–93.

Gasarasi, C. (1987) 'The tripartite approach to the resettlement and integration of rural refugees in Tanzania', in J. Rogge (ed.) *Refugees. A Third World dilemma*, Totowa: Rowman and Littlefield, pp. 99–115.

Geyer, H. and Kontuly, T. (1993) 'A theoretical foundation for the concept of differential urbanization' *International Regional Science Review* 15: 157–77.

Ghimire, K. (1994) 'Refugees and deforestation', *International Migration* 32: 561–70.

Giddens, A. (1984) *The constitution of society*, Cambridge: Cambridge University Press.

Giggs, J. and Pattie, C. (1992) 'Croeso i Gymru – welcome to Wales: but welcome to whose Wales?', *Area* 24: 268–82.

Gilani, I., Khan, M. and Iqbal, M. (1981) *Labour migration from Pakistan to the Middle East and its impact on the domestic economy*, final report, research project on export of manpower from Pakistan to the Middle East, Washington DC: The World Bank.

Gilbert, A. (1993) 'Third World cities: the changing national settlement system', *Urban Studies* 30: 721–40.

Gilbert, A. (1994) *The Latin American city*, London: Latin America Bureau.

Gilbert, A. and Gugler, J. (1982) *Cities, poverty and development. Urbanisation in the Third World*, Oxford: Oxford University Press.

Gilbert, A. and Gugler, J. (1992) *Cities, poverty and development. Urbanization in the Third World*, second edition, Oxford: Oxford University Press.

Gilbert, A. and Kleinpenning, J. (1986) 'Migration, regional inequality and development in the Third World', *Tijdschrift voor Economische en Sociale Geografie* 77: 2–6.

Gilg, A. (1983) 'Population and employment', in M. Pacione (ed.) *Progress in rural geography*, Beckenham: Croom Helm, pp. 74–105.

Gilroy, P. (1993) *The black Atlantic. Modernity and double consciousness*, London: Verso.

Girouard, M. (1985) *Cities and people*, London: Yale University Press.

Gitmez, A. and Wilpert, C. (1987) 'A micro-society or an ethnic community? Social organisation and ethnicity amongst Turkish migrants in Berlin', in J. Rex, D. Joly and C. Wilpert (eds) *Immigrant associations in Europe*, Aldershot: Gower, pp. 86–125.

Glass, D. (1967) *Population policies and movements in Europe*, London: Frank Cass.

Glass, D. (ed.) (1954) *Social mobility in Britain*, London: Routledge & Kegan Paul.

Glass, R. (1964) 'Introduction: aspects of change', in Centre for Urban Studies (eds) *London: aspects of change*, London: MacGibbon and Kee, pp. xiii–xlii.

Glick, P. (1947) 'The family life cycle', *American Sociological Review* 12: 164–74.

Golant, S. (1990) 'The metropolitanization and sub-urbanization of the US elderly population, 1970–88', *The Gerontologist* 30: 80–5.

Golant, S. (1992) 'The suburbanisation of the American elderly', in A. Rogers (ed.) *Elderly migration and population redistribution*, London: Belhaven, pp. 163–81.

Goldberg, D. and Raynor, J. (1989) *The Jewish people: their history and religion*, Harmondsworth: Penguin.

Goldstein, S. (1954) 'Repeated migration as a factor in high mobility rates', *American Sociological Review* 19: 536–41.

Goldthorpe, J. (with Llewellyn, C. and Payne, C.) (1980) *Social mobility and class structure in modern Britain*, Oxford: Oxford University Press.

Golledge, R. (1980) 'A behavioral view of mobility and migration research', *Professional Geographer* 32: 14–21.

González, J. and Puebla, J. (1996) 'Spain: return to the south, metropolitan deconcentration and new migration flows', in P. Rees, J. Stillwell, A. Convey and M. Kupiszewski (eds) *Population migration in the European Union*, Chichester: Wiley, pp. 175–89.

Goodman, J. (1981) 'Information, uncertainty, and the microeconomic model of migration decision making', in G. de Jong and R. Gardner (eds) *Migration decision making. Multidisciplinary approaches to microlevel studies in developed and developing countries*, New York: Pergamon, pp. 130–48.

Goodwin-Gill, G. (1982) 'The obligations of states and the protection function of the Office of the United Nations High Commissioner for Refugees', in *Michigan yearbook of international legal studies*, New York: Clark Boardman and Company, pp. 291–337.

Goodwin-Gill, G. (1986) 'Non-refoulement and the new asylum seekers', *Virginia Journal of International Law* 26: 897–915.

Gordon, I. (1991) 'Multi-stream migration modelling', in J. Stillwell and P. Congdon (eds) *Migration models: macro and micro approaches*, London: Belhaven, pp. 73–91.

Gordon, I. (1994) *Pay, conditions and segmentations in the London labour market*, mimeo, Reading: University of Reading.

Gordon, I. (1995) 'Migration in a segmented labour market', *Transactions of the Institute of British Geographers* 20: 139–55.

Gordon, R. (1977) *Mines, masters and migrants: life in a Namibian compound*, Johannesburg: Raven.

Goss, J. and Lindquist, B. (1995) 'Conceptualizing international labor migration: a structuration perspective', *International Migration Review* 29: 317–51.

Gould, P. and White, R. (1974) *Mental maps*, London: Penguin.

Gould, W. (1994) 'Population movements and the changing world order: an introduction', in W. Gould and A. Findlay (eds) *Population migration and the changing world order*, London: Belhaven, pp. 3–14.

Gould, W. and Findlay, A. (1994) 'Refugees and skilled transients', in W. Gould and A. Findlay (eds) *Population migration and the changing world order*, London: Belhaven, pp. 17–25.

Gould, W. and Prothero, R. (1975) 'Space and time in African population mobility', in L. Kosínski and R. Prothero (eds) *People on the move: studies on internal migration*, London: Methuen, pp. 39–49.

Grafton, D. and Bolton, N. (1987) 'Counter-urbanisation and the rural periphery: some evidence from north Devon', in B. Robson (ed.) *Managing the city*, London: Croom Helm, pp. 191–210.

Graham, E. (1995) 'Population geography and post-modernism', Paper presented at the International Conference on Population Geography, University of Dundee, September.

Graves, P. (1979a) 'Income and migration revisited', *Journal of Human Resources* 14: 112–21.

Graves, P. (1979b) 'A life-cycle empirical analysis of migration and climate, by race', *Journal of Urban Economics* 6: 135–47.

Graves, P. (1980) 'Migration and climate', *Journal of Regional Science* 20: 227–37.

Graves, P. (1983) 'Migration with a composite amenity: the role of rents', *Journal of Regional Science* 23: 541–6.

Graves, P. and Linneman, P. (1979) 'Household migration: theoretical and empirical results', *Journal of Urban Economics* 6: 383–404.

Graves, P. and Regulska, J. (1982) 'Amenities and migration over the life-cycle', in D. Diamond and G. Tolley (eds) *The economics of urban amenities*, New York: Academic Press, pp. 211–21.

Graves, P. and Waldman, D. (1991) 'Multimarket amenity compensation and the behavior of the elderly', *American Economic Review* 81: 1374–81.

Grayson, L. and Young, K. (1994) *Quality of life in cities*, London: British Library/London Research Centre.

Greenwood, M. (1969) 'The determinants of labor migration in Egypt', *Journal of Regional Science* 9: 283–90.

Greenwood, M. (1975) 'Research on internal migration in the United States: a survey', *Journal of Economic Literature* 13: 397–433.

Greenwood, M. (1985) 'Human migration: theory, models, and empirical studies', *Journal of Regional Science* 25: 521–44.

Greenwood, M. and Hunt, G. (1984) 'Migration and interregional employment redistribution in the United States', *American Economic Review* 74: 957–69.

Greenwood, M., Mueser, P., Plane, D. and Schlottmann, A. (1991) 'New directions in migration research: perspectives from some North American regional science disciplines', *Annals of Regional Science* 25: 237–70.

Gregory, D. (1978) *Ideology, science and human geography*, London: Hutchinson.

Gregory, J. and Piche, V. (1978) 'African migration and peripheral capitalism', *African Perspectives* 1.

Griffiths, G. (1992) 'Gerrymander: the place of suburbia in Australian fiction', in A. Rutherford (ed.) *Populous places: Australian cities and towns*, Sydney: Dangeroo Press, pp. 19–30.

Grigg, D. (1977) 'E.G. Ravenstein and the "laws of migration"', *Journal of Historical Geography* 3: 41–54.

Grigg, D. (1980) 'Migration and overpopulation', in P. White and R. Woods (eds) *The geographical impact of migration*, London: Longman, pp. 60–83.

Grimes, S. (1992) 'Friendship patterns and social networks amongst post-war Irish migrants in Sydney', in P. O'Sullivan (ed.) *Patterns of migration*, Leicester: Leicester University Press, pp. 164–82.

Grundy, E. (1985) 'Divorce, widowhood, remarriage and geographic mobility among women', *Journal of Biosocial Science* 17: 415–35.

Grundy, E. (1986) 'Migration and fertility behaviour in England and Wales: a record linkage study', *Journal of Biosocial Science* 18: 403–23.

Grundy, E. (1992) 'The household dimension in migration research', in A. Champion and A. Fielding (eds) *Migration processes and patterns. Volume 1. Research progress and prospects*, London: Belhaven, pp. 165–74.

Grundy, E. and Fox, A. (1984) 'Changes of address in the early years of marriage', *Population Trends* 38: 25–9.

Gugler, J. (1986) 'Internal migration in the third world', in M. Pacione (ed.) *Population geography: progress and prospects*, London: Croom Helm, pp. 194–223.

Gugler, J. (1988) 'Introduction', in J. Gugler (ed.) *The urbanization of the Third World*, Oxford: Oxford University Press, pp. 8–10.

Gugler, J. (1991) 'Life in a dual system revisited: urban rural ties in Enugu, Nigeria, 1961–87', *World Development* 19: 399–409.

Gulati, L. (1993) *In the absence of their men*, New Delhi: Sage.

Gurak, D. and Caces, F. (1992) 'Migration networks and the shaping of migration systems', in M. Kritz, L. Lim and H. Zlotnik (eds) *International migration systems. A global approach*, Oxford: Clarendon Press, pp. 150–76.

Gurtov, M. (1991) 'Open borders: a global humanist approach to the refugee crisis', *World Development* 19: 485–96.

Gutting, D. (1996) 'Narrative identity and residential history', *Area* 28: 482–90.

Gwynne, R. (1990) *New horizons? Third World industrialization in an international framework*, Harlow: Longman.

Haas, W. and Serow, W. (1993) 'Amenity retirement migration process', *The Gerontologist* 33: 212–20.

Hadden, J. and Barton, J. (1973) 'An image that will not die: thoughts on the history of anti-urban ideology', *Urban Affairs Annual Review* 7: 79–116.

Hainsworth, P. (1992a) 'The extreme Right in post-war France: the emergence and success of the Front National', in P. Hainsworth (ed.) *The extreme Right in Europe and the USA*, London: Pinter, pp. 29–60.

Hainsworth, P. (1992b) *The extreme Right in Europe and the USA*, London: Pinter.

Halfacree, K. (1994) 'The importance of 'the rural' in the constitution of counterurbanization: evidence from England in the 1980s', *Sociologia Ruralis* 34: 164–89.

Halfacree, K. (1995a) 'Household migration and the structuration of patriarchy: evidence from the U.S.A.', *Progress in Human Geography* 19: 159–82.

Halfacree, K. (1995b) 'Talking about rurality: social representations of the rural as expressed by residents of six English parishes', *Journal of Rural Studies* 11: 1–20.

Halfacree, K. (1996a) 'Ruralism and the postmodern experience: some evidence from England in the late 1980s', in M. Gentileschi and R. King (eds) *Questioni di popolazione in Europa: una prospettiva geografica*, Bologna: Pàtron Editore, pp. 163–78.

Halfacree, K. (1996b) 'Out of place in the country: travellers and the "rural idyll"', *Antipode* 28: 42–71.

Halfacree, K. (1997) 'Contrasting roles for the post-productivist countryside: a post-modern perspective on counterurbanisation', in P. Cloke and J. Little (eds) *Contested countryside cultures*, London: Routledge, pp. 70–93.

Halfacree, K. and Boyle, P. (1992) 'Population migration within, into and out of Wales in the late twentieth century. A general overview of the literature', *Migration Unit Research Paper* 1, University of Wales Swansea: Department of Geography.

Halfacree, K. and Boyle, P. (1993) 'The challenge facing migration research: the case for a biographical approach', *Progress in Human Geography* 17: 333–48.

Halfacree, K. and Boyle, P. (1995) ' "A little learning is a dangerous thing": a reply to Ron Skeldon', *Progress in Human Geography* 19: 97–9.

Halfacree, K., Flowerdew, R. and Johnson, J. (1992) 'The characteristics of British migrants in the 1990s: evidence from a new survey', *The Geographical Journal* 158: 157–69.

Hall, R. (1995) 'Households, families and fertility', in R. Hall and P. White (eds) *Europe's population: towards the next century*, London: UCL Press, pp. 34–50.

Hallos, M. (1991) 'Migration, education and the status of women in Southern Nigeria', *American Anthropologist* 93: 852–70.

Hamilton, L. and Seyfrit, C. (1993) 'Town–village contrasts in Alaskan Youth Aspirations', *Arctic* 46: 255–63.

Hammar, T. (1993) 'The Sweden-wide strategy of refugee dispersal', in R. Black and V. Robinson (eds) *Geography and refugees: patterns and processes of change*, London: Belhaven, pp. 104–18.

Hamnett, C. (1984) 'Gentrification and residential location theory: a review and assessment', in D. Herbert and R. Johnston (eds) *Geography and the urban environment. Volume 6*, Chichester: Wiley, pp. 283–319.

Hamnett, C. (1988) 'Regional variations in house prices and house price inflation in Britain 1969-1988', *Royal Bank of Scotland Review* 159: 28–40.

Hamnett, C. (1990) 'Back to the future', *Housing Review* 39: 5.

Hamnett, C. (1991) 'The blind men and the elephant: the explanation of gentrification', *Transactions of the Institute of British Geographers* 16: 173–89.

Hamnett, C. (1992) 'House-price differentials, housing wealth and migration', in A. Champion and A. Fielding (eds) *Migration processes and patterns. Volume 2. Research progress and prospects*, London: Belhaven Press, pp. 55–64.

Hancock, I. (1992) 'The roots of inequity: Romani cultural rights in their historical and social context', *Immigrants and Minorities* 11: 3–20.

Hansen, A. (1990) 'Long term consequences of two African refugee resettlement strategies', Paper presented at a meeting of the Society for Applied Anthropology, York University, UK.

Hansen, J. (1989) 'Norway: the turnaround which turned around', in A. Champion (ed.) *Counterurbanization*, London: Edward Arnold, pp. 103–20.

Hareven, T. (1982) *Family time and industrial time*, Cambridge: Cambridge University Press.

Harrigan, F. and McGregor, P. (1993) 'Equilibrium and disequilibrium perspectives on regional labor migration', *Journal of Regional Science* 33: 49–67.

Harrigan, F., Jenkins, J. and McGregor, P. (1986) 'A behavioural model of migration: an evaluation of neo-classical and Keynesian alternatives', *Working Paper* 42, University of Strathclyde: Frazer of Allander Institute.

Harris, C. (1987) 'The individual and society: a processual approach', in A. Bryman, B. Bytheway, P. Allatt and T. Keil (eds) *Rethinking the life cycle*, Basingstoke: Macmillan, pp. 17–29.

Harris, C. (1988) 'Images of blacks in Britain, 1930–60', in S. Allen and M. Macey (eds) *Race and social policy*, London: Economic and Social Research Council, pp. 15–32.

Harris, J. and Todaro, M. (1970) 'Migration, unemployment and development: a two sector analysis', *American Economic Review* 60: 139–49.

Hart, R. (1975) 'Interregional economic migration: some theoretical considerations (part II)', *Journal of Regional Science* 15: 289–305.

Hartshorn, T. (1992) *Interpreting the city: an urban geography*, second edition, New York: Wiley.

Harvey, D. (1978) 'Labor, capital, and class struggle around the built environment in advanced capitalist societies', in K. Cox (ed.) *Urbanization and conflict in market societies*, London: Methuen, pp. 9–37.

Harvey, D. (1982) *The limits to capital*, Oxford: Blackwell.

Haskey, J. (1989) 'Current prospects for the proportion of marriages ending in divorce', *Population Trends* 55: 34–7.

Haskey, J. (1996) 'The proportion of married couples who divorce: past patterns and current prospects', *Population Trends* 83: 25–36.

Hawes, D. and Perez, B. (1995) *The Gypsy and the state*, Bristol: School of Advanced Urban Studies Publications.

Hawkins, F. (1989) *Critical years in immigration: Canada and Australia compared*, Kensington, New South Wales: New South Wales University Press.

Hayes, L., Al-Hamad, A. and Geddes, A. (1995) 'Marriage, divorce and residential change: evidence from the houshold Sample of Anonymised Records', Paper presented at the British Society for Population Studies Annual Conference, Brighton.

Healey, N. (1987) 'Housing and the north–south divide', *Housing Review* 36: 189–90.

Heaton, T., Fredrickson, C., Fuguitt, G. and Zuiches, J. (1979) 'Residential preferences, community satisfaction, and the intention to move', *Demography* 16: 565–73.

Hedger, M. (1981) 'Reassessment in rural Wales', *Town and Country Planning* 50: 261–3.

Heller, W. and Hofmann, H.-J. (1992) '*Aussiedler* migration to Germany and the problems of economic and social integration', *Migration Unit Research Paper* 2, University of Wales Swansea: Department of Geography.

Helsinki Watch (1992) *Struggling for ethnic identity: Czechoslovakia's endangered Gypsies*, London: Human Rights Watch.

Herzog, H. and Schlottmann, A. (1984) 'Labor force mobility in the United States: migration, unemployment, and remigration', *International Regional Science Review* 9: 43–58.

Herzog, H., Schlottmann, A. and Boehm, T. (1993) 'Migration as spatial job-search: a survey of empirical findings', *Regional Studies* 27: 327–40.

Hicks, J. (1932) *A theory of wages*, London: Macmillan.

Hiltermann, J. (1991) 'Settling for war: Soviet emigration and Israel's settlement policy in East Jerusalem', *Journal of Palestine Studies* 20: 71–85.

Hiltner, J. and Smith, B. (1974) 'Intraurban residential location of the elderly', *Journal of Geography* 73: 23–33.

Hindson, D. (1987) 'Alternative urbanisation strategies in South Africa: a critical evaluation', *Third World Quarterly* 9: 583–600.

Hiro, D. (1991) *Black British, White British*, second edition, London: Grafton.

Hitch, P (1983) 'The mental health of refugees: a review of research', in R. Baker (ed.) *The psychosocial problems of refugees*, London: British Refugee Council, pp. 42–50.

Hochstadt, S. (1981) 'Migration and industrialisation in Germany', Social *Science History* 5: 445–68.

Hoggart, K. and Buller, H. (1994) 'Property agents as gatekeepers in British house purchases in rural France', *Geoforum* 25: 173–87.

Hoggart, K. and Buller, H. (1995a) 'Geographical differences in British property acquisitions in rural France', *Geographical Journal* 161: 69–78.

Hoggart, K. and Buller, H. (1995b) 'British home owners and housing change in rural France', *Housing Studies* 10: 179–98.

Hohn, C. (1987) 'The family life cycle: needed extensions of the concept', in J. Bongaarts, T. Burch and K. Wachter (eds) *Family demography: methods and their application*, Oxford: Clarendon Press, pp. 65–80.

Holmes, C. (1991) *A tolerant country? Immigrants, refugees and minorities in Britain*, London: Faber & Faber.

Home Office (1996) *Statistical Bulletin 9/96*, London: Government Statistical Service.

Hourani, A. and Shehadi, N. (eds) (1993) *The Lebanese in the world: a century of emigration*, London: IB Tauris.

Howard, E. (1946) [1898] *Garden Cities of tomorrow*, London: Faber.

Huff, J. and Clark, W. (1978) 'Cumulative stress and cumulative inertia: a behavioral model of the decision to move', *Environment and Planning A* 10: 1101–19.

Hughes, G. and McCormick, B. (1981) 'Do council housing policies reduce migration between regions?', *Economic Journal* 91: 919–37.

Hughes, G. and McCormick, B. (1985) 'Migration intentions in the UK. Which households want to migrate and which succeed?', *Economic Journal (Conference Supplement)* 95: 113–23.

Hughes, G. and McCormick, B. (1987) 'Housing markets, unemployment and labour market flexibility in the UK', *European Economic Review* 31: 615–45.

Hughes, G. and McCormick, B. (1989) 'Does migration reduce differentials in regional unemployment rates?', in J. van Dijk, H. Folmer, H. Herzog, and A. Schlottman (eds) *Migration and labor market adjustment*, Dordrecht: Kluwer, pp. 85–108.

Hughes, G. and McCormick, B. (1994) 'Did migration in the 1980s narrow the north–south divide?', *Economica* 61: 509–27.

Hugo, G. (1986) *Australia's changing population*, Oxford: Oxford University Press.

Hugo, G. (1989) 'Australia: the spatial concentration of the turnaround', in A. Champion (ed.) *Counterurbanization*, London: Edward Arnold, pp. 62–82.

Hugo, G. (1990a) 'Adaptation of Vietnamese in Australia', *South East Asian Journal of Social Science* 18: 182–210.

Hugo, G. (1990b) 'Demographic and spatial aspects of immigration', in M. Wooden, R. Holten, G. Hugo and J. Sloan (eds) *Australian immigration: a survey of the issues*, Canberra: Australian Government Publishing Service, pp. 30–111.

Hugo, G. (1992) 'Women on the move: changing patterns of population movement of women in Indonesia', in S. Chant (ed.) *Gender and migration in developing countries*, London: Belhaven, pp. 174–96.

Hugo, G. (1994a) 'The turnaround in Australia: some first observations from the 1991 census', *Australian Geographer* 25: 1–17.

Hugo, G. (1994b) 'Demographic and spatial aspects of immigration', in M. Wooden, R. Holton, G. Hugo and J. Sloan (eds) *Australian immigration*, Canberra: Australian Government Publishing Service, pp. 30–110.

Hugo, G. (1996) 'Research review 3: Asia on the move: research challenges for population geography', *International Journal of Population Geography* 2: 95–118.

Human Rights Watch (1995) *Global report on women's human rights*, London: Human Rights Watch.

Hunt, G. (1993) 'Equilibrium and disequilibrium in migration modelling', *Regional Studies* 27: 341–9.

Hutchings, A. and Bunker, R. (1986) *With conscious purpose: a history of town planning in South Australia*, Adelaide: Wakefield Press.

Ibrahim, B. and Ibrahim, F. (1995) 'Pastoralists in transition – a case study from Lengijape, Massai Steppe', *GeoJournal* 36: 27–48.

Illeris, S. (1991) 'Counterurbanization revisited: the new map of population in central and northwestern Europe', in M. Bannon, L. Bourne and R. Sinclair (eds) *Urbanization and urban development*, Dublin: Service Industries Research Centre, pp. 1–16.

Ilvento, T. and Luloff, A. (1982) 'Anti-urbanism and nonmetropolitan growth: a re-evaluation', *Rural Sociology* 47: 220–33.

Inagami, T. (1983) *Labour–management communications at the workshop level*, Tokyo: Japan Institute of Labour.

Indra, D. (1988) 'An analysis of the Canadian private sponsorship programme for South East Asian Refugees', *Ethnic Groups* 7: 153–72.

Indra, D. (1993) 'The spirit of the gift and the politics of resettlement: Canadian private sponsorship of South East Asians', in V. Robinson (ed.) *The international refugee crisis: British and Canadian responses*, Basingstoke: Macmillan, pp. 229–55.

Institute of Welsh Affairs (1988) *Rural Wales: population changes and current attitudes. Volume 2: Part A: survey results*, Cardiff: Institute of Welsh Affairs.

International Monetary Fund (1994) *International financial statistics yearbook*, New York: International Monetary Fund.

International Organisation for Migration (1995) 'Background document' presented at the IOM seminar, Geneva.

Isbell, E. (1944) 'Internal migration in Sweden and intervening opportunities', *American Sociological Review* 9: 627–39.

Ishikawa, Y. (1987) 'An empirical study of the competing destinations model using Japanese interaction data', *Environment and Planning A* 19: 1359–73.

Isserman, A., Plane, D., Rogerson, P. and Beaumont, P. (1985) 'Forecasting interstate migration with limited data: a demographic–economic approach', *Journal of the American Statistical Association* 80: 277–85.

Jackson, K. (1985) *Crabgrass frontier: the suburbanisation of the United States*, New York: Oxford University Press.

Jackson, P. (1989) *Maps of meaning*, London: Unwin Hyman.

Jackson, R. and Flores, E. (1992) 'Filipino migrants-for-marriage to Australia', Paper presented at the International Geographical Congress Commission on Population Geography, Los Angeles.

Jager, M. (1986) 'Class definition and the esthetics of gentrification: Victoriana in Melbourne', in N. Smith and P. Williams (eds) *Gentrification of the city*, Boston: Allen & Unwin, pp. 78–91.

James, S., Jordan, B. and Kays, H. (1991) 'Poor people, council housing and the right-to-buy', *Journal of Social Policy* 20: 27–40.

Jansen, C. and King, R. (1968) 'Migrations et "occasions intervenantes" en Belgique', *Recherches Économiques de Louvain* 4: 519–26.

Jarvie, W. (1985) 'Structural economic change, labour market segmentation and inter-regional migration', *Papers of the Regional Science Association* 56: 129–44.

Jenkins, J. (1978) 'The demand for immigrant workers: labour scarcity or social control?', *International Migration Review* 12: 514–36.

Johnson, J., Salt, J. and Wood, P. (1974) *Housing and the mobility of labour in England and Wales*, Farnborough: Saxon House.

Johnston, R. (1973) 'On frictions of distance and regression coefficients', *Area* 5: 187–91.

Johnston, R. (1975) 'Map pattern and friction of distance parameters: a comment', *Regional Studies* 9 281–3.

Johnson, R. (1979) *Peasant and proletarian: the working class of Moscow in the late nineteenth century*, Leicester: Leicester University Press.

Johnston, R. (1994) 'Quality of life', in R. Johnston, D. Gregory and D. Smith (eds) *The dictionary of human geography*, third edition, Oxford: Blackwell, p. 493.

Johnston, R., Gregory, D. and Smith, D. (eds) (1994) *The dictionary of human geography*, third edition, Oxford: Blackwell.

Jones, G. and Varley, A. (forthcoming) 'The reconquest of the historic centre: urban conservation and gentrification in Puebla, Mexico', *Transactions of the Institute of British Geographers*.

Jones, H. (1986) 'Evolution of Scottish migration patterns: a social-relations-of-production approach', *Scottish Geographical Magazine* 102: 151–64.

Jones, H. (1990) *Population geography*, second edition, London: Paul Chapman Publishing.

Jones, H., Caird, J., Berry, W. and Dewhurst, J. (1986) 'Peripheral counterurbanization: findings from an integration of census and survey data in northern Scotland', *Regional Studies* 20: 15–26.

Jones, N. (1993) *Living in rural Wales*, Llandysul: Gomer.

Jones, O. (1995) 'Lay discourses of the rural: developments and implications for rural studies', *Journal of Rural Studies* 11: 35–49.

Jones, P. (1990) 'Recent ethnic German migration from Eastern Europe to the Federal Republic', *Geography* 75: 249–52.

Jones, P. (1994) 'Destination Germany: the spatial distribution and impacts of the 'third wave' of post-war immigrants', in W. Gould and A. Findlay (eds) *Population migration and the changing world order*, Chichester: Wiley, pp. 27–46.

Jones, P. (1996) 'Immigrants, Germans and national identity in the new Germany: some policy issues', *International Journal of Population Geography* 2: 119–31.

Jones, R. (1989) 'Causes of Salvadoran migration to the United States', *Annals of the Association of American Geographers* 79: 183–94.

Jupp, J. (1995) 'From "White Australia" to "part of Asia": recent shifts in Australian immigration policy towards the region', *International Migration Review* 29: 207–28.

Kabeer, N. (1988) 'Subordination and struggle: women in Bangladesh', *New Left Review* 168: 95–121.

Kabera, J. (1987) 'The refugee problem in Uganda', in J. Rogge (ed.) *Refugees. A Third World dilemma*, Totowa: Rowman and Littlefield, pp. 72–80.

Kalibová, K. (1993) 'Gypsies in the Czech Republic and the Slovak Republic: geographic and demographic characteristics', *GeoJournal* 30: 255–8.

Kane, E. and Macaulay, L. (1993) 'Interviewer gender and gender attitudes', *Public Opinion Quarterly* 57: 1–28.

Katz, E. and Stark, O. (1984) 'Migration and asymmetric information: comment', *American Economic Review* 74: 533–4.

Katz, E. and Stark, O. (1986) 'Labor mobility under asymmetric information with moving and signalling costs', *Economics Letters* 21: 89–94.

Kearney, M. (1986) 'From the invisible hand to visible feet: anthropological studies of migration and development', *Annual Review of Anthropology* 15: 331–61.

Kellerman, A. (1993) 'Settlement frontiers revisited: the case of Israel and the West Bank', *Tijdschrift voor Economische en Sociale Geografie* 84: 27–39.

Kendig, H. (1984) 'Housing careers, life cycle and residential mobility', *Urban Studies* 21: 271–83.

Kerouac, J. (1957) *On the road*, New York: Viking Press.

Kerouac, J. (1972) [1957] *On the road*, Harmondsworth: Penguin.

Khoser, K. (1993) 'Repatriation and information: a theoretical model', in R. Black and V. Robinson (eds) *Geography and refugees: patterns and processes of change*, London: Belhaven, pp. 171–85.

Kiernan, K. (1986) 'Leaving home: living arrangements of young people in six West European countries', *European Journal of Population* 2: 177–84.

Kiernan, V. (1995) 'The separation of India and Pakistan', in R. Cohen (ed.) *The Cambridge survey of world migration*, Cambridge: Cambridge University Press, pp. 356–59.

King, R. (1986) 'Return migration and regional economic development: an overview', in R. King (ed.) *Return migration and regional economic problems*, London: Croom Helm, pp. 1–37.

King, R., Connell, J. and White, P. (eds) (1995) *Writing across worlds. Literature and migration*, London: Routledge.

King, R. and Knights, M. (1994) 'Bangladeshis in Rome: a case of migratory opportunism', in W. Gould and A. Findlay (eds) *Population migration and the changing world order*, Chichester: John Wiley & Sons, pp. 127–44.

King, R., McGrath, F., Shuttleworth, I. and Strachan, A. (1990) 'Irish on the move', *Geography Review* 3(3): 23–7.

King, R., Shuttleworth, I. and Walsh, J. (1996) 'Ireland: the human resource warehouse of Europe', in P. Rees, J. Stillwell, A. Convey and M. Kupiszewski (eds) *Population migration in the European Union*, Chichester: Wiley, pp. 207–29.

Kinsman, P. (1995) 'Landscape, race and national identity: the photography of Ingrid Pollard', *Area* 27: 300–10.

Kinvig, R. (1975) *The Isle of Man: a social, cultural and political history*, Liverpool: Liverpool University Press.

Kirkby, R. (1985) *Urbanisation in China: town and country in a developing economy 1949–2000 A.D*, London: Croom Helm.

Klein, J. (1981) *Woody Guthrie. A life*, London: Faber & Faber.

Kliot, N. (1987) 'The era of the homeless man', *Geography* 72: 109–21.

Knopp, L. (1990) 'Some theoretical implications of gay involvement in an urban land market', *Political Geography Quarterly* 9: 337–52.

Kockel, U. (1991) 'Countercultural migrants in the west of Ireland', in R. King (ed.) *Contemporary Irish Migration*, Dublin: Geographical Society of Ireland, pp. 70–82.

Kontuly, T. and Vogelsang, R. (1989) 'Federal Republic of Germany: the intensification of the migration turnaround', in A. Champion (ed.) *Counterurbanization*, London: Edward Arnold, pp. 141–61.

Koo, H. and Smith, P. (1983) 'Migration, the urban informal sector, and earnings in the Philippines', *The Sociological Quarterly* 24: 219–32.

Krings, T. (1995) 'Marginalisation and revolt among the Tuareg in Mali and Niger', *GeoJournal* 36: 57–63.

Kritz, M., Keely, C. and Tomasi, S. (eds) (1981) *Global trends in migration: theory and research on international population movements*, New York: Centre for Migration Studies.

Kritz, M., Lim, L. and Zlotnik, H. (eds) (1992) *International migration systems. A global approach*, Oxford: Clarendon Press.

Kritz, M. and Zlotnik, H. (1992) 'Global interactions: migration systems, processes, and policies', in M. Kritz, L. Lim and H. Zlotnik (eds) *International migration systems. A global approach*, Oxford: Clarendon Press, pp. 150–76.

Kundera, M. (1985) *The unbearable lightness of being*, London: Faber & Faber.

Kundera, M. (1992) *Immortality*, London: Faber & Faber.

Kunz, E. (1973) 'The refugee in flight: kinetic models and forms of displacement', *International Migration Review* 7: 125–46.

Kunz, E. (1981) 'Exile and resettlement: refugee theory', *International Migration Review* 15: 42–51.

Kwok, V. and Leland, H. (1982) 'An economic model of the brain drain', *American Economic Review* 72: 91–100.

Lambert, A. (1989) 'Return of the vampire', *Geographical Magazine* February: 16–20.

Landale, N. (1994) 'Migration and the Latino family: the union formation behaviour of Puerto Rican women', *Demography* 31: 133–57.

Langevin, B. and Begeot, F. (1992) 'Censuses in the European Community', *Population Trends* 68: 33–5.

Langton, J. and Hoppe, G. (1990) 'Urbanisation, social structure and population circulation in pre-industrial times: flows of people through Vadstena (Sweden) in the mid-nineteenth century', in P. Corfield and D. Keene (eds) *Work in towns 850–1850*, Leicester: Leicester University Press, pp. 138–63.

Lanphier, M. (1993) 'Host groups: public meets private', in V. Robinson (ed.) *The international refugee crisis: British and Canadian responses*, Basingstoke: Macmillan, pp. 255–73.

Lansing, J. and Mueller, E. (1967) *The geographic mobility of labour*, University of Michigan: Institute for Social Research.

Larrabee, E. (1948) 'The six thousand houses that Levitt built', *Harpers Magazine*, **197**: 79–88.

Lasslett, P. (1989) *A fresh map of life: the emergence of the Third Age*, London: Weidenfeld & Nicolson.

Law, C. and Warnes, A. (1976) 'The changing geography of the elderly in England and Wales', *Transactions of the Institute of British Geographers* 1: 453–71.

Law, C. and Warnes, A. (1982) 'The destination decision in retirement migration', in A. Warnes (ed.) *Geographical perspectives on the elderly*, Chichester: Wiley, pp. 53–80.

Lawless, P. and Brown, F. (1986) *Urban growth and change in Britain: an introduction*, London: Harper & Row.

Lawton, R. (1978) 'Regional population trends in England and Wales, 1750–1971', in J. Hobcraft and P. Rees (eds) *Regional demographic development*, London: Croom Helm, pp. 29–70.

Lee, B., Oropesa, R. and Kanan, J. (1994) 'Neighborhood context and residential mobility', *Demography* 31: 249–70.

Lee, E. (1966) 'A theory of migration', *Demography* 3: 47–57.

Lees, L. (1979) *Exiles of Erin*, Manchester: Manchester University Press.

Lees, L. (1994) 'Rethinking gentrification: beyond the position of economics or culture', *Progress in Human Geography* 18: 137–50.

LeGates, R. and Hartman, C. (1986) 'The anatomy of displacement in the United States', in N. Smith and P. Williams (eds) *Gentrification of the city*, Boston: Allen & Unwin, pp. 178–200.

Leinbach, T. (1989) 'The transmigration programme in Indonesian national development strategy', *Habitat International* 13: 81–93.

Lerman, A., Jacobs, D., Stanley-Clamp, L., Frankel, A. and Montague, A. (1989) *The Jewish communities of the world*, Basingstoke: Macmillan.

Lever, W. (1993) 'Reurbanisation – the policy implications', *Urban Studies* 30: 267–84.

Lever-Tracey, C. and Quinlan, M. (1988) *A divided working class: ethnic segmentation and industrial conflict in Australia*, London: Routledge & Kegan Paul.

Lewis, G. (1982) *Human migration: a geographical perspective*, London: Croom Helm.

Lewis, G. (1986) 'Welsh rural community studies: retrospect and prospect', *Cambria* 13: 27–40.

Lewis, G. (1989) 'Counterurbanization and social change in the rural South Midlands', *East Midlands Geographer* 11: 3–12.

Lewis, G. and Maund, D. (1976) 'The urbanization of the countryside: a framework for analysis', *Geografiska Annaler* 60B: 16–27.

Lewis, J. (1986) 'International labour migration and uneven regional development in labour exporting countries', *Tijdschrift voor Economische en Sociale Geografie* 77: 27–41.

Ley, D. (1980) 'Liberal ideology and the postindustrial city', *Annals of the Association of American Geographers* 70: 238–58.

Ley, D. (1981) 'Inner-city revitalization in Canada: a Vancouver case study', *Canadian Geographer* 25: 124–48.

Ley, D. (1986) 'Alternative explanations for inner city gentrification: a Canadian assessment', *Annals of the Association of American Geographers* 76: 521–35.

Ley, D. (1994) 'Postmodernism', in R. Johnston, D. Gregory and D. Smith (eds) *The Dictionary of Human Geography*, third edition, Oxford: Blackwell, pp. 466–8.

Li, F., Findlay, A., Jowett, A. and Skeldon, R. (1996) 'Migrating to learn and learning to migrate: a study of the experiences and intentions of international student migrants', *International Journal of Population Geography* 2: 51–67.

Li, F., Jowett, A., Findlay, A. and Skeldon, R. (1995) 'Discourse on migration and ethnic identity: interviews with professionals in Hong Kong', *Transactions of the Institute of British Geographers* 20: 342–56.

Li, W. and Li, Y. (1995) 'Special characteristics of China's interprovincial migration', *Geographical Analysis* 27: 137–51.

Lichter, D. and De Jong, G. (1990) 'The United States', in C. Nam, W. Serow and D. Sly (eds) *International handbook on internal migration*, New York: Greenwood Press, pp. 391–417.

Lichter, D., McLaughlin, D. and Cornwell, G. (1994) 'Migration and the loss of human resources in rural America', in L. Beaulieu and D. Mulkey (eds) *Investing in people: the human capital needs of rural America*, Boulder, Colorado: Westview Press.

Lieblich, A. (1993) 'Looking at change. Natasha, 21:

new immigrant from Russia to Israel', in R. Josselson and A. Lieblich (eds) *The narrative study of lives*, Newbury Park, California: Sage, pp. 92–129.

Liegeois, J.-P. and Gheorghe, N. (1995) *Roma/ Gypsies: a European minority*, London: Minority Rights Group.

Light, I. and Bonacich, E. (1988) *Immigrant entrepreneurs*, Berkeley: University of California Press.

Lin-Yuan, Y. and Kosínski, L. (1994) 'The model of place utility revisited', *International Migration* 32: 49–70.

Lo, C. (1989) 'Recent spatial restructuring in Zhujiang Delta, South China; a study of socialist regional development strategy', *Annals of the Association of American Geographers* 79: 293–308.

Loesberg, J. (1979) *Folksongs and ballads popular in Ireland. Volume 1*, Cork: Ossian Publications.

Loiskandl, H. (1995) 'Illegal migrant workers in Japan', in R. Cohen (ed.) *The Cambridge survey of world migration*, Cambridge: Cambridge University Press, pp. 371–6.

Long, L. (1988) *Migration and residential mobility in the United States*, New York: Sage.

Long, L. and DeAre, D. (1988) 'US population redistribution: a perspective on the non-metropolitan turnaround', *Population and Development Review* 14: 433–50.

Longino, C. (1990) 'Geographic distribution and migration', in R. H. Binstock and I. K. George (eds) *Handbook of aging and the social sciences*, San Diego: Academic Press, pp. 45–63.

Lovett, A. and Flowerdew, R. (1991) 'Analysis of count data using Poisson regression', *Professional Geographer* 41: 190–98.

Lovett, A., Whyte, I. and Whyte, K. (1985) 'Poisson regression analysis and migration fields: the example of the apprenticeship records of Edinburgh in the seventeenth and eighteenth centuries', *Transactions, Institute of British Geographers* 10: 317–32.

Lowder, S. (1978) 'The context of Latin American labor migration: a review of the literature post-1970', *Sage Race Relations Abstracts* 6: 1-49.

Lowe, M. and Gregson, N. (1989) 'Nannies, cooks, cleaners, au pairs ... new issues for feminist geography?', *Area* 21: 414–17.

Lowenthal, D. (1972) *West Indian societies*, London: Oxford University Press.

Lowry, I. (1966) *Migration and metropolitan growth: two analytical models*, San Francisco: Chandler.

Lucassen, L. (1991) 'The power of definition, stigmatisation, minoritisation and ethnicity illustrated by the history of Gypsies in the Netherlands', *Netherlands Journal of Social Sciences* 27: 80–91.

Luling, V. (1989) 'Wiping out a way of life', *Geographical Magazine* July: 34–7.

Lutz, E. (ed.) (1991) *Future demographic trends in Europe and North America: what can we assume today?*, London: Academic Press.

Mabin, A. (1992) 'Comprehensive segregation: the origins of the Group Areas Act and its planning apparatuses', *Journal of Southern African Studies* 18: 405–29.

Mabogunje, A. (1970) 'Systems approach to a theory of rural–urban migration', *Geographical Analysis* 2: 1–18.

Mabogunje, A. (1990) 'Urban planning and the post-colonial state in Africa: a research overview', *African Studies Review* 33: 121–203.

MacÉinrí, P. (1991) 'The Irish in Paris: an aberrant community?', in R. King (ed.) *Contemporary Irish Migration*, Dublin: Geographical Society of Ireland, pp. 32–41.

MacGabhann, M. (1973) *The hard road to Klondyke*, London: Routledge & Kegan Paul.

MacLaughlin, J. (1991) 'Social characteristics and destinations of recent emigrants from selected regions in the west of Ireland', *Geoforum* 22: 319–31.

MacLaughlin, J. (1994) *Ireland: the emigrant nursery and the world economy*, Cork: Cork University Press.

Mamdani, M. (1993) 'The Ugandan Asian expulsion: twenty years after', *Economic and Political Weekly* January 16: 93–6.

Mangin, W. (ed.) (1970) *Peasants in cities*, Boston: Houghton-Mifflin.

Marcus, G. and Fischer, M. (1986) *Anthropology as cultural critique: an experimental moment in the human sciences*, Chicago: University of Chicago Press.

Marett, V. (1989) *Immigrants settling in the city*, London: Leicester University Press.

Markham, W. and Pleck, J. (1986) 'Sex and willingness to move for occupational advancement: some national sample results', *Sociological Quaterly* 27: 121–43.

Marrus, M. (1985) *The unwanted: European refugees in the twentieth century*, Oxford: Oxford University Press.

Marshall, B. (1992) 'German migration policies', in G. Smith, W. Paterson, P. Merkl and S. Padgett (eds) *Developments in German politics*, Basingstoke: Macmillan, pp. 247–64.

Martin, P. (1991) *The unfinished story: Turkish labour migration to Western Europe*, Geneva: International Labour Office.

Martin, P. (1992) 'Trade, aid and migration', *International Migration Review* 26: 162–72.

Marx, L. (1964) *The machine in the garden*, New York: Oxford University Press.

Masser, I. and Brown, P. (1975) 'Hierarchical aggregation procedures for interaction data', *Environment and Planning A* 7: 509–23.

Masser, I. and Brown, P. (1977) 'Spatial representation and spatial interaction', *Environment and Planning A* 9: 71–92.

Masser, I. and Gould, W. (1975) *Interregional migration in tropical Africa*, London: Institute of British Geographers.

Massey, D. (1990) 'Social structure, household strategies, and the cumulative causation of migration', *Population Index* 56: 3–26.

Massey, D., Arango, J., Hugo, G., Kouaouci, A., Pellegrino, A. and Taylor, J. (1993) 'Theories of international migration: a review', *Population and Development Review* 19: 431–66.

Massey, D. and Denton, N. (1993) *American apartheid: segregation and the making of the underclass*, London: Harvard University Press.

Matsukawa, I. (1991) 'Interregional gross migration and structural changes in local industries', *Environment and Planning A* 23: 745–56.

Mayall, D. (1992) 'The making of British Gypsy identities, c. 1500–1980', *Immigrants and Minorities* 11: 21–41.

Mayle, P. (1989) *A year in Provence*, London: Hamish Hamilton.

Mayle, P. (1990) *Toujours Provence*, London: Hamish Hamilton.

McCaffrey, L. (1976) *The Irish diaspora in America*, Bloomington: Indiana University Press.

McCarthy, K. (1976) 'The household life cycle and housing choices', *Papers of the Regional Science Association* 37: 55–80.

McCarthy, K. and Morrison, P. (1978) *The changing demographic and economic structure of non-metropolitan areas in the 1970s*, Santa Monica: Rand Corporation, P6062.

McDowell, L. (1992) 'Multiple voices: speaking from inside and outside "the project"', *Antipode* 24: 56–72.

McDowell, L. (1996) 'Off the road: alternative views of rebellion, resistance and "the beats"', *Transactions of the Institute of British Geographers* 21: 412–19.

McGinnis, R. (1968) 'A stochastic model of social mobility', *American Sociological Review* 23: 712–22.

McGregor, A., Munro, M., Heafey, M. and Symon, P. (1992) 'Moving job, moving house: the impact of housing on long-distance labour mobility', *Centre for Housing Research Discussion Paper* 38, University of Glasgow, Glasgow.

McHugh, K. and Mings, R. (1991) 'On the road again: seasonal migration to a sunbelt metropolis', *Urban Geography* 12: 1–18.

McHugh, K., Hogan, T. and Happel, S. (1995) 'Multiple residence and cyclical migration: a life course perspective', *Professional Geographer* 47: 251–67.

McKay, G. (1996) *Senseless acts of beauty*, London: Verso.

McKay, J. and Whitelaw, J. (1977) 'The role of large private and government organisations in generating flows of inter-regional migrants: the case of Australia', *Economic Geography* 53: 28–44.

McNabb, R. (1979) 'A socio-economic model of migration', *Regional Studies* 13: 297–304.

McNabb, R. and Ryan, P. (1990) 'Segmented labour markets', in D. Sapsford and Z. Tzannatos (eds) *Current issues in labour economics*, Basingstoke: Macmillan, pp. 151–76.

Meillassoux, C. (1981) *Maidens, meal and money*, London: Edward Arnold.

Meinig, D. (1991) 'Spokane and the Inland Empire: historical geographical systems and a sense of place', in D. Stratton (ed.) *Spokane and the Inland Empire: an interior Pacific Northwest anthology*, Pullman, Washington: Washington State University Press, pp. 1–32.

Merrill Lynch Limited (1986) *Fourth annual study of employee mobility*, Swindon: PHH Homequity Limited.

Meyer, J. and Speare, A. (1985) 'Distinctly elderly mobility: types and determinants', *Economic Geography* 61: 79–88.

Miles, M. and Crush, J. (1993) 'Personal narratives as interactive texts: collecting and interpreting migrant life-histories', *Professional Geographer* 45: 84–94.

Miles, R. (1982) *Racism and migrant labour*, London: Routledge & Kegan Paul.

Miles, R. (1987) *Capitalism and unfree labour. Anomaly or necessity?*, London: Tavistock Publications.

Miles, R. (1989) *Racism*, London: Routledge.

Miles, R. (1993) *Racism after 'race relations'*, London: Routledge.

Miller, C. (1975) 'American Rom and the ideology of defilement', in F. Rehfisch (ed.) *Gypsies, tinkers and other travellers*, London: Academic Press, pp. 41–54.

Miller, K. (1985) *Emigrants and exiles: Ireland and the Irish exodus to North America*, Oxford: Oxford University Press.

Miller, M. (1995) 'Illegal migration', in R. Cohen (ed.) *The Cambridge survey of world migration*, Cambridge: Cambridge University Press, pp. 537–40.

Milne, W. (1991) 'The human capital model and its econometric estimation', in J. Stillwell and P. Congdon (eds) *Migration models. Macro and micro approaches*, London: Belhaven, pp. 137–51.

Mincer, J. (1978) 'Family migration decisions', *Journal of Political Economy* 86: 749–73.

Minford, P. (with Ashton, P., Peel, M., Davies, D. and Sprague, A.) (1983) *Unemployment: cause and cure*, Oxford: Basil Blackwell.

Minford, P., Peel, M. and Ashton, P. (1987) *The housing morass: regulation, immobility and unemployment*, London: The Institute of Economic Affairs.

Mings, R. (1984) 'Recreational nomads in the southwestern Sunbelt', *Journal of Cultural Geography* 4: 86–99.

Mitchell, D. (1996) *The lie of the land*, Minneapolis: University of Minnesota Press.

Mitchell, J. (1975) *Psychoanalysis and feminism*, Harmondsworth: Penguin.

Mitchell, P. (1988) 'Modelling migration to and from London using the NHSCR', Paper presented to the Regional Science Association, April.

Moch, L. (1989) 'The importance of mundane movements: small towns, nearby places and individual itineraries in the history of migration', in P. Ogden and P. White (eds) *Migrants in modern France: population mobility in the later 19th and 20th Centuries*, London: Unwin Hyman, pp. 97–117.

Molho, I. (1984) 'A dynamic model of inter-regional migration flows in Great Britain', *Journal of Regional Science* 24: 317–27.

Molho, I. (1986) 'Theories of migration: a review', *Scottish Journal of Political Economy* 33: 396–419.

Momsen, J. (1991) *Women and development in the Third World*, London: Routledge.

Monbiot, G. (1994) *No man's land*, London: Macmillan.

Monzel, K. (1993) 'Only the women know', in R. Black and V. Robinson (eds) *Geography and refugees: patterns and processes of change*, London: Belhaven, pp. 118–34.

Moon, B. (1995) 'Paradigms in migration research: exploring 'moorings' as a schema', *Progress in Human Geography* 19: 504–24.

Moore, G. (1914) *The untilled field*, London: Heinemann.

Morokvasic, M. (1984) 'Birds of Passage are also women ...', *International Migration Review* 18: 886–907.

Morris, D. (1989) 'A study of language contact and social networks in Ynys Mon', *Contemporary Wales* 3: 99–117.

Morris, E. (1977) 'Mobility, fertility and residential crowding', *Sociological and Social Research* 61: 363–75.

Morrison, P. (1973) 'Theoretical issues in the design of population mobility models', *Environment and Planning* 5: 125–34.

Morrison, P. and Wheeler, J. (1976) 'Rural renaissance in America?', *Population Bulletin* 31(3): 1–27.

Morton, J. (1988) 'Sakanab: greetings and information among the Northern Beja', *Africa* 58: 423–36.

Moseley, M. (1984) 'The revival of rural areas in advanced economies: a review of some causes and consequences', *Geoforum* 15: 447–56.

Mueller, C. (1982) *The economics of labor migration, a behavioural analysis*, New York: Academic Press.

Mulder, C. and Hooimeijer, P. (1994) 'Moving into owner occupation: compositional and contextual effects on the propensity to become a home-owner', Paper presented at the British–Swedish–Dutch Conference on Population Planning and Policies, Gävle, Sweden, September.

Mulgan, G. (1989) 'A tale of two cities', *Marxism Today* March: 18–25.

Muller, C. (1981) *Contemporary suburban America*, New Jersey: Prentice-Hall.

Mullins, L. and Tucker, R. (eds) (1988) *Snowbirds in the Sun Belt: older Canadians in Florida*, Tampa, Florida: University of South Florida, International Exchange Center on Gerontology.

Mumford, L. (1946) 'The Garden City idea and modern planning', in E. Howard (ed.) *Garden Cities of tomorrow*, London: Faber, pp. 29–40.

Munro, A. (1968) 'The shining houses', in A. Munro

(ed.) *Dance of the happy shades*, New York: Viking Penguin, pp. 19–29.

Murdoch, J. and Marsden, T. (1994) *Reconstituting rurality*, London: UCL Press.

Murphy, H. (1955) *Flight and resettlement*, Paris: UNESCO.

Murphy, M. (1987) 'Measuring the family life cycle: concepts, data and methods', in A. Bryman, B. Bytheway, P. Allatt, and T. Keil (eds) *Rethinking the life cycle*, Basingstoke: Macmillan, pp. 30–50.

Murphy, P. and Zehner, R. (1988) 'Satisfaction and sunbelt migration', *Australian Geographical Studies* 26: 320–34.

Muth, R. (1971) 'Migration: chicken or egg?', *Southern Economic Journal* 37: 295–306.

Nair, S. (1985) 'Fijians and indo-Fijians in Suva: rural–urban movements and linkages', in M. Chapman and R. Prothero (eds) *Circulation in population movement: substance and concepts from the Melanesian case*, London: Routledge & Kegan Paul, pp. 306–29.

Nash, A. (1994) 'Population geography', *Progress in Human Geography* 18: 385–95.

Neuwirth, G. (1984) 'The private sponsorship programme: some reflections on its effectiveness', in R. Nann, P. Johnson and M. Beiser (eds) *Refugee resettlement: South East Asians in transition*, Vancouver: University of British Columbia Press.

Neuwirth, G. (1988) 'Refugee resettlement', *Current Sociology* 36: 27–41.

Neuwirth, G. (1993) 'The marginalisation of South East Asian refugees', in V. Robinson (ed.) *The international refugee crisis: British and Canadian responses*, Basingstoke: Macmillan, pp. 295–319.

Neuwirth, G. and Clark, L. (1981) 'Indochinese refugees in Canada: sponsorship and adjustment', *International Migration Review* 15: 131–40.

Newman, D. (1985) 'The evolution of a political landscape: geographical and territorial implications of Jewish colonization in the West Bank', *Middle Eastern Studies* 21: 192–205.

Ni Laoire, C. (1997) 'Migration, power and identity: life-path formation among Irish rural youth', PhD thesis, Department of Geography, University of Liverpool.

Nobel, P. (1982) 'Refugees, law and development in Africa', in *Michigan yearbook of international legal studies*, New York: Clark Boardman and Co., pp. 255–87.

Noin, D. and Chauviré, Y. (1987) *La population de la France,* Paris: Masson.

Nord, M. (1994) *Keeping the poor in their place: the proximate processes that maintain social concentration of poverty in the United States*, PhD dissertation, Pennsylvania State University, University Park, Pennsylvania.

Oakley, A. (1974) *The sociology of housework*, Oxford: Martin Robertson.

Oberai, A. and Singh, H. (1980) 'Migration, remittances and rural development: findings of a case study in the Indian Punjab', *International Labour Review* 119: 229–41.

Oberai, A., Prasad, P. and Sardana, M. (1989) *Determinants and consequences of internal migration in India: studies in Bihar, Kerala, and Uttar Pradesh*, New Delhi: Oxford University Press.

O'Connor, A. (1983) *The African city*, London: Hutchinson.

Office of Population Censuses and Surveys (1983) *Recently moving households*, London: OPCS.

Okely, J. (1983) *The traveller-Gypsies*, Cambridge: Cambridge University Press.

Oliver-Smith, A. (1991) 'Involuntary resettlement, resistance and political empowerment', *Journal of Refugee Studies* 4: 132–49.

Oliver-Smith, A. and Hansen, A. (1982) 'Involuntary migration and resettlement: causes and contexts', in A. Oliver-Smith and A. Hansen (eds) *Involuntary migration and resettlement: the problems and responses of dislocated persons*, Boulder, Colorado: Westview Press, pp. 1–9.

O'Loughlin, J. (1980) 'Distribution and migration of foreigners in German cities', *Geographical Review* 70: 253–75.

O'Loughlin, J. and Glebe, G. (1984) 'Residential segregation of foreigners in German cities', *Tijdschrift voor Economische en Sociale Geografie* 75: 273–84.

Olsson, G. (1965) 'Distance and human interaction: a migration study', *Geografiska Annaler* 47.

Omvedt, G. (1980) 'Migration in colonial India: the articulation of feudalism and capitalism by the colonial state', *Journal of Peasant Studies* 7: 185–212.

Openshaw, S. (1977) 'Algorithm 3: a procedure to generate pseudo-random aggregations of N zones into M zones, where M is less than N', *Environment and Planning A* 9: 1423–8.

Osmond, J. (1987) 'A million on the move', *Planet* 62: 114–18.

Osmond, J. (1988) *The divided kingdom*, London: Constable.

O'Sullivan, P. (ed) (1992) *Patterns of migration*, Leicester: Leicester University Press.

Otomo, A. (1992) 'Elderly migration and population redistribution in Japan', in A. Rogers (ed.) *Elderly migration and population redistribution*, London: Belhaven, pp. 185–203.

Owen, D. and Green, A. (1992) 'Migration patterns and trends', in A. Champion and A. Fielding (eds) *Migration processes and patterns. Volume 2: Research progress and prospects*, London: Belhaven Press, pp. 17–38.

Paasi, A. and Vartiainen, P. (1981) 'Keskuksen ja jmpariston valinen muuttoliike: tapaustutkimus Joensuusta', *Terra* 93: 57–72.

Pahl, J. and Pahl, R. (1971) *Managers and their wives*, London: Allen Lane.

Parnell, S. (1988) 'Racial segregation in Johannesburg: the Slums Act 1934–39', *South African Geographical Journal* 70: 112–26.

Parnwell, M. (1993) *Population movements and the Third World*, London: Routledge.

Payne, G. and Abbott, P. (1990) *The social mobility of women: beyond male mobility models*, London: Falmer Press.

Peach, C., Robinson, V., Maxted, J. and Chance, J. (1988) 'Immigration and ethnicity', in A. Halsey (ed.) *British Social Trends Since 1900*, Basingstoke: Macmillan, pp. 561–616.

Peck, J. (1989) 'Reconceptualising the local labour market: space, segmentation and the state', *Progress in Human Geography* 13: 42–61.

Peck, J. (1992) ' "Invisible threads": homeworking, labour–market relations, and industrial restructuring in the Australian clothing trade', *Environment and Planning D* 10: 671–89.

Perry, R., Dean, K. and Brown, B. (1986) *Counterurbanisation*, Norwich: Geo Books.

Pessar, P. (1994) 'Sweatshop workers and domestic ideologies: Dominican women in New York's apparel industry', *International Journal of Urban and Regional Research* 18: 127–42.

Petersen, W. (1958) 'A general typology of migration', *American Sociological Review* 23: 256–66.

Phillips, D. and Williams, A. (1984) *Rural Britain: a social geography*, Oxford: Blackwell.

Phipps, A. and Carter, J. (1984) 'An individual-level analysis of the stress-resistance model of household mobility', *Geographical Analysis* 16: 176–89.

Pickles, A. and Davies, R. (1991) 'The empirical analysis of housing careers: a review and a general statistical modelling framework', *Environment and Planning A* 23: 465–84.

Pickles, A., Davies, R. and Crouchley, R. (1982)

'Heterogeneity, nonstationarity, and duration-of-stay effects in migration', *Environment and Planning A* 14: 615–22.

Piore, M. (1979) *Birds of passage: migrant labor in industrial societies*, Cambridge: Cambridge University Press.

Pissarides, C. and Wadsworth, J. (1989) 'Unemployment and the inter-regional mobility of labour', *Economic Journal* 99: 739–55.

Pittin, R. (1984) 'Migration of women in Nigeria: the Hausa case', *International Migration Review* 18: 1293–1315.

Plane, D. (1984) 'Migration space: doubly constrained gravity model mapping of relative interstate separation', *Annals of the Association of American Geographers* 74: 244–56.

Plane, D. (1992) 'Age-composition change and the geographical dynamics of interregional migration in the US', *Annals of the Association of American Geographers* 82: 64–85.

Plane, D. (1993) 'Demographic influences on migration', *Regional Studies* 27: 375–83.

Plane, D. (1994) 'The wax and wane of migration patterns in the US in the 1980s: a demographic effectiveness field perspective', *Environment and Planning A* 26: 1545–61.

Plane, D. and Rogerson, P. (1994) *The geographical analysis of population with applications to planning and business*, New York: John Wiley & Sons.

Plane, D. and Rogerson, P. (1991) 'Tracking the baby boom, the baby bust, and the echo generations: how age composition regulates US migration', *Professional Geographer* 43: 416–30.

Plaut, W. (1996) 'Jewish ethics and international migrations', *International Migration Review* 30: 18–26.

Pocock, D. and Hudson, R. (1978) *Images of the urban environment*, London: Macmillan.

Polèse, M. (1981) 'Regional disparity, migration and economic adjustment: a reappraisal', *Canadian Public Policy* 7: 519–25.

Pomeroy, W. (1986) *Apartheid, imperialism and African freedom*, New York: International Publishers.

Pooley, C. and Whyte, I. (1991a) 'Introduction: approaches to the study of migration and social change', in C. Pooley and I. Whyte (eds) *Migrants, emigrants and immigrants*, London: Routledge, pp. 1–15.

Pooley, C. and Whyte, I. (eds) (1991b) *Migrants, emigrants and immigrants*, London: Routledge.

Poot, J. (1986) 'A system approach to modelling the

inter-urban exchange of workers in New Zealand', *Scottish Journal of Political Economy* 33: 249–74.

Porell, F. (1982) 'Intermetropolitan migration and quality of life', *Journal of Regional Science* 22: 137–58.

Portes, A. (1978a) 'Migration and underdevelopment', *Politics and Society* 8: 1–48.

Portes, A. (1978b) 'Toward a structural analysis of illegal (undocumented) immigration', *International Migration Review* 12: 469–84.

Portes, A. (1981) 'Unequal exchange and the urban informal sector', in A. Portes and J. Walton (eds) *Labor, class and the international system*, New York: Academic Press, pp. 67–103.

Portes, A. and Rumbaut, R. (1990) *Immigrant America: a portrait*, Los Angeles: University of California Press.

Pradilla, E. (1990) 'Las Políticas neoliberales y la cuestión territorial', *Revista Interamericana de Planificación* 22: 77–107.

Prothero, R. and Chapman, M. (eds) (1985) *Circulation in Third World countries*, London: Routledge and Kegan Paul.

Pryer, J. (1992) 'Purdah, patriarchy and population movement: perspectives from Bangladesh', in S. Chant (ed.) *Gender and migration in developing countries*, London: Belhaven, pp. 139–54.

Pryor, R. (1975) 'Migration and the process of modernization', in L. Kosínski and R. Mansell Prothero (eds) *People on the move. Studies of internal migration*, London: Methuen, pp. 23–8.

Pryor, R. (1983) 'Integrating international and internal migration theories', in M. Kritz, C. Keely and S. Tomasi (eds) *Global trends in migration: theory and research on international population movements*, New York: Center for Migration Studies, pp. 110–29.

Punch, M. (1986) *The politics and ethics of fieldwork*, Beverly Hills, California: Sage.

Quayle, R. (1990) 'The Isle of Man constitution', in V. Robinson and D. McCarroll (eds) *The Isle of Man: celebrating a sense of place*, Liverpool: Liverpool University Press, pp. 123–33.

Raban, J. (1974) *Soft city*, London: Fontana.

Radcliffe, S. (1986) 'Gender relations, peasant livelihood strategies and migration: a case study from Cuzco, Peru', *Bulletin of Latin American Research* 5: 29–47.

Radcliffe, S. (1990) 'Between hearth and labour market: the recruitment of peasant women in the Andes', *International Migration Review* 24: 229–49.

Radford, E. (1970) *The new villagers*, Birmingham: Frank Cass.

Rao, A. (1975) 'Some mānuš conceptions and attitudes', in F. Rehfisch (ed.) *Gypsies, tinkers and other travellers*, London: Academic Press, pp. 139–67.

Rapoport, R. and Rapoport, R. (1969) 'The dual-career family: a variant pattern and social change', *Human Relations* 22: 3–30.

Redfern, P. (1981) 'Census 1981 – an historical and international perspective', *Population Trends* 23: 3–15.

Rees, G. and Rees, T. (1981) 'Migration, industrial restructuring and class relations: the case of South Wales', Papers in Planning Research 22, Cardiff: UWIST.

Rees, P. (1992) 'Elderly migration and population redistribution in the United Kingdom', in A. Rogers (ed.) *Elderly migration and population redistribution*, London: Belhaven, pp. 203–26.

Rees, P. and Duke-Williams, O. (1995) 'The story of the British Special Migration Statistics', *Scottish Geographical Magazine* 111: 13–20.

Rees, P., Stillwell, J., Convey, A. and Kupiszewski, M. (eds) (1996) *Population migration in the European Union*, Chichester: John Wiley & Sons.

Reintges, C. (1992) 'Urban (mis)management? A case study of the effects of orderly urbanization on Duncan Village', in D. Smith (ed.) *The apartheid city and beyond*, London: Routledge, pp. 99–111.

Reissman, L. (1964) *The urban process*, New York: Free Press.

Relph, E. (1981) *Rational landscapes and humanist geography*, London: Croom Helm.

Richardson, A. (1967) 'A theory and a method for the psychological study of assimilation', *International Migration Review* 2.

Richmond, A. (1968) 'Return migration from Canada to Britain', *Population Studies* 22: 263–71.

Richmond, A. (1988) 'Sociological theories of international migration: the case of refugees', *Current Sociology* 36: 7–25.

Rigg, J. and Stott, P. (1992) 'The rise of the Naga', in G. Chapman and K. Baker (eds) *The changing geography of Asia*, London: Routledge, pp. 74–121.

Rist, R. (1978) *Guestworkers in Germany: the prospects for pluralism*, New York: Praeger.

Ritchey, P. (1976) 'Explanations of migration', *Annual Review of Sociology* 2: 363–404.

Roberts, R. (1986) 'Historical development of labour recruitment agencies in Britain', Paper presented to the workshop of the IBG Working Party on Skilled International Migration, University College, London, November.

Robinson, G. (1986) 'Migration: aspirations and reality amongst school-leavers in a small Canterbury town', *New Zealand Population Review* 12: 218–34.

Robinson, G. (1990) *Conflict and change in the countryside*, London: Belhaven.

Robinson, V. (1980) 'Lieberson's isolation index: a case study evaluation', *Area* 12: 307–13.

Robinson, V. (1986a) *Transients, settlers and refugees: Asians in Britain*, Oxford: Clarendon.

Robinson, V. (1986b) 'Bridging the Gulf: the economic significance of South Asian labour migration to the Middle East', in R. King (ed.) *Return migration and regional economic development*, Beckenham: Croom Helm, pp. 243–73.

Robinson, V. (1989) 'Up the creek without a paddle? Britain's boat people ten years on', *Geography* 74: 332–8.

Robinson, V. (1990a) 'Social demography', in V. Robinson and D. McCarroll (eds) *The Isle of Man: celebrating a sense of place*, Liverpool: Liverpool University Press, pp. 133–59.

Robinson, V. (1990b) 'Into the next millenium, an agenda for Refugee Studies', *Journal of Refugee Studies* 3: 3–15.

Robinson, V. (1992) 'Une minorité invisible: les Chinois au Royaume-Uni', *Revue Européene des Migrations Internationales* 8: 9–31.

Robinson, V. (1993a) 'Making waves? The contribution of ethnic minorities to local demography', in A. Champion (ed.) *Population matters. The local dimension*, London: Paul Chapman, pp. 150–69.

Robinson, V. (1993b) 'North and South: resettling Vietnamese refugees in Australia and the UK', in R. Black and V. Robinson (eds) *Geography and refugees: patterns and processes of change*, London: Belhaven, pp. 134–55.

Robinson, V. (1993c) 'The nature of the crisis and the academic response', in V. Robinson (ed.) *The international refugee crisis: British and Canadian responses*, Basingstoke: Macmillan, pp. 3–17.

Robinson, V. (1993d) 'Marching into the middle classes? The long term resettlement of East African Asians in the UK', *Journal of Refugee Studies* 6: 230–48.

Robinson, V. (ed.) (1993e) *The international refugee crisis: British and Canadian responses*, Basingstoke: Macmillan.

Robinson, V. (1995) 'The migration of East African Asians to the UK', in R. Cohen (ed.) *The Cambridge survey of world migration*, Cambridge: Cambridge University Press, pp. 331–7.

Robinson, V. (1996a) 'The changing nature and European perceptions of Europe's refugee problem', *Geoforum* 26: 411–27.

Robinson, V. (1996b) 'Indians: onward and upward', in C. Peach (ed.) *Ethnicity in the 1991 Census. Volume 2. The ethnic minority populations of Great Britain*, London: HMSO, pp. 95–120.

Robinson, V. (1997, forthcoming) 'The evolution of refugee resettlement policy in post-war Britain', in M. Kenzer (ed.) *The world refugee atlas*, Toronto: Toronto University Press.

Robinson, V. and Hale, S. (1989) *The geography of Vietnamese secondary migration in the UK*, Warwick: ESRC Centre for Research into Ethnic Relations.

Robson, B. (1987a) 'The enduring city: a perspective on decline', in B. Robson (ed.) *Managing the city*, London: Croom Helm, pp. 6–21.

Robson, B. (1987b) 'The policy framework', in B. Robson (ed.) *Managing the city*, London: Croom Helm, pp. 211–15.

Robson, B. (1988) *Those inner cities*, Oxford: Clarendon.

Rodwin, L. and Hollister, R. (1984) *Cities of the mind*, London: Plenum Press.

Rogers, A. (1967) 'A regression analysis of interregional migration in California', *Review of Economics and Statistics* 49: 262–7.

Rogers, A. (1988) 'Age patterns of elderly migration: an international comparison', *Demography* 25: 355–70.

Rogers, A. (1989) 'The elderly mobility transition: growth, concentration, and tempo', *Research on Aging* 11: 3–32.

Rogers, A. (1992a) 'Introduction', in A. Rogers (ed.) *Elderly migration and population redistribution*, London: Belhaven, pp. 1–17.

Rogers, A. (1992b) 'Elderly migration and population redistribution in the United States', in A. Rogers (ed.) *Elderly migration and population redistribution*, London: Belhaven, pp. 226–49.

Rogers, A. and Castro, L. (1981) 'Age patterns of migration: cause-specific profiles', in A. Rogers (ed.) *Advances in multiregional demography*, Laxenburg: International Institute for Applied Systems Analysis, pp. 125–59.

Rogers, A. and Castro, L. (1986) 'Migration', in A. Rogers and F. Willekens (eds) *Migration and settlement*, Dordrecht: Reidel, pp. 157–209.

Rogers, A. and Watkins, J. (1987) 'General versus elderly interstate migration and population redistribution in the United States', *Research on Aging* 9: 483–529.

Rogers, A. and Willekens, F. (1986) 'A short course on multiregional mathematical demography', in A. Rogers and F. Willekens (eds) *Migration and settlement: a multiregional comparative study*, Dordrecht: Reidel North Holland, pp. 355–84.

Rogerson, R., Morris, A., Findlay, A. and Paddison, R. (1989) *Quality of life in Britain's intermediate cities*, Glasgow: Glasgow University, Department of Geography.

Rogg, E. (1971) 'The influence of a strong refugee community on the economic adjustment of its members', *International Migration Review* 5: 474–81.

Rogge, J. (1977) 'A geography of refugees: some illustrations from Africa', *Professional Geographer* 29: 186–93.

Rogge, J. (1978) 'Some comments on definitions and typologies of Africa's refugees', *Zambian Geographical Journal* 33: 49–60.

Rogge, J. (1985) 'The Indo-Chinese diaspora: where have all the refugees gone?', *Canadian Geographer* 29: 65–72.

Rogge, J. (1986) 'Africa's displaced population; dependency or self-sufficiency?', in J. Clarke, M. Khogali and L. Kosínski (eds) *Population and development projects in Africa*, Cambridge: Cambridge University Press, pp. 68–83.

Rogge, J. (1987) 'When is self-sufficiency achieved? Rural settlements in Sudan', in J. Rogge (ed.) *Refugees. A Third World dilemma*, Totowa: Rowman and Littlefield, pp. 86–99.

Rose, D. (1989) 'A feminist perspective of employment restructuring and gentrification: the case of Montréal', in J. Wolch and M. Dear (eds) *The power of geography*, Boston: Unwin Hyman, pp. 118–38.

Rose, D. and Le Bourdais, C. (1986) 'The changing conditions of female single parenthood in Montréal's inner city and suburban neighborhoods', *Urban Resources* 3: 45–52.

Rose, S. (1976) *Black suburbanization*, Cambridge, Massachusetts: Ballinger Publishing Company.

Roseman, C. (1971) 'Migration as a spatial and temporal process', *Annals of the Association of American Geographers* 61: 589–98.

Roseman, C. (1977) *Changing migration patterns within the United States*, Washington DC: Association of American Geographers.

Roseman, C. and Williams, J. (1980) 'Metropolitan to nonmetropolitan migration: a decision-making perspective', *Urban Geography* 1: 283–94.

Rosen, S. (1979) 'Wage-based indexes of quality of life', in P. Mieszkowski and M. Straszheim (eds) *Current issues in urban economics*, Baltimore: John Hopkins University Press, pp. 74–104.

Rossi, P. (1955) *Why families move: a study in the social psychology of residential mobility*, Glencoe, Illinois: Free Press.

Rowley, G. (1990) 'The Jewish colonization of the Nablus Region – perspectives and continuing developments', *GeoJournal* 21: 349–62.

Rowley, G. (1991) 'Suburbanization and community: the West Bank in transition', *Cambria* 16: 102–22.

Rowlingson, B. and Boyle, P. (1992) 'GIS and migration modelling: an environment for fitting intervening opportunities models', *North West Regional Research Laboratory Research Report* 25, Department of Geography, Lancaster University.

Rowntree, B. (1902) *Poverty: a study of town life*, London: Macmillan.

Rudolph, H. (1994) 'Dynamics of immigration in a nonimmigrant country: Germany', in H. Fassman and R. Münz (eds) *European migration in the late twentieth century*, Aldershot: Edward Elgar, pp. 113–26.

Rudzitis, G. (1989) 'Migration, places, and nonmetropolitan development', *Urban Geography* 10: 396–411.

Rudzitis, G. (1991) 'Migration, sense of place, and nonmetropolitan vitality', *Urban Geography* 12: 80–8.

Rudzitis, G. (1993) 'Nonmetropolitan geography: migration, sense of place and the American west', *Urban Geography* 14: 574–85.

Rudzitis, G. and Johansen, H. (1991) 'How important is wilderness? Results from a United States survey', *Environmental Management* 15: 227–33.

Russell, S. (1986) 'Remittances from international migration: a review in perspective', *World Development* 14: 677–96.

Rutinwa, B. (1996) 'Beyond durable solutions: an appraisal of the new proposals for prevention and solution of refugee crises in the Great Lakes area', *Journal of Refugee Studies* 9: 312–26.

Salt, J. (1987) 'Contemporary trends in international migration study', *International Migration* 25: 241–51.

Salt, J. (1988) 'Highly-skilled international migrants, careers and internal labour markets', *Geoforum* 19: 387–99.

Salt, J. (1989) 'A comparative overview of international trends and types, 1950–80', *International Migration Review* 23: 431–56.

Salt, J. (1990) 'Organisational labour migration: theory and practice in the United Kingdom', in J. Johnson and J. Salt (eds) *Labour migration: the internal geographical mobility of labour in the developed world*, London: David Fulton, pp. 53–69.

Salt, J. (1991) 'Labour migration and housing in the UK: an overview', in J. Allen and C. Hamnett (eds) *Housing and labour markets: building the connection*, London: Unwin Hyman.

Salt, J. (1992a) 'The future of international labour migration', *International Migration Review* 26: 1077–111.

Salt, J. (1992b) 'Migration processes among the highly skilled in Europe', *International Migration Review* 26: 484–505.

Salt, J. (1995) 'International migration report', *New Community* 21: 443–65.

Salt, J. and Kitching, R. (1992) 'The relationship between international and internal labour migration', in A. Champion and A. Fielding (eds) *Migration processes and patterns. Volume 1. Research progress and prospects*, London: Belhaven Press, pp. 148–62.

Salva Tomas, P. (1992) 'The new migration flows in the Spanish Mediterranean area and the Balearic Islands', Paper presented at the International Geographical Congress Commission on Population Geography, Los Angeles.

Samuel, R. (1989) 'Introduction: exciting to be English', in R. Samuel (ed.) *Patriotism. Volume 3. National fictions*, London: Routledge, pp. xviii–lxvii.

Sandell, S. (1977) 'Women and the economics of family migration', *Review of Economics and Statistics* 59: 406–14.

Sant, M. and Simons, P. (1993) 'Counterurbanization and coastal development in New South Wales', *Geoforum* 24: 291–306.

Sassen Koob, S. (1979) 'Economic growth and immigration in Venezuela', *International Migration Review* 13: 455–74.

Sassen, S. (1988) *The mobility of labour and capital*, Cambridge: Cambridge University Press.

Sassen, S. (1991) *Global Cities*, New York, London, Tokyo, Princeton: Princeton University Press.

Saunders, M. (1985) 'The influence of job-vacancy advertising upon migration: some empirical examples', *Environment and Planning A* 17: 1581–89.

Saunders, M. (1990) 'Migration and job vacancy information', in J. Johnson and J. Salt (eds) *Labour migration: the internal geographical mobility of labour in the developed world*, London: David Fulton, pp. 137–54.

Saunders, M. and Flowerdew, R. (1987) 'Spatial aspects of the provision of job information', in M. Fischer and P. Nijkamp (eds) *Regional labour markets*, Amsterdam: North Holland, pp. 205–28.

Savage, M. (1987) 'Spatial mobility and the professional labour market: a case study of employers in Slough', *Working Paper* 56, Urban and Regional Studies, University of Sussex.

Savage, M. (1988) 'The missing link? The relationship between spatial and social mobility', *British Journal of Sociology* 39: 554–77.

Savage, M., Barlow, J., Dickens, P. and Fielding, A. (1992) *Property, bureaucracy and culture*, London: Routledge.

Savage, M., Dickens, P. and Fielding, A. (1988) 'Some social and political implications of the contemporary fragmentation of the 'service class' in Britain', *International Journal of Urban and Regional Research* 12: 455–76.

Savoie, D. (1986) 'Courchene and regional development: beyond the neoclassical approach' *Canadian Journal of Regional Science* 9: 69–77.

Sayer, A. (1984) *Method in social science*, London: Hutchinson.

Schachter, J. and Althaus, P. (1989) 'An equilibrium model of gross migration', *Journal of Regional Science* 29: 143–59.

Schaeffer, P. (1991) 'Guests who stay', *Geographical Analysis* 23: 247–60.

Schaffer, F. (1972) *The New Town story*, London: Paladin.

Scheinman, R. (1983) 'Refugees: goodbye to the Good Old Days', *Annals of the American Academy of Political and Social Sciences* 467: 78–88.

Schnell, I. and Graicer, I. (1993) 'Causes of in-migration to Tel-Aviv inner city', *Urban Studies* 30: 1187–207.

Schnell, I. and Graicer, I. (1994) 'Rejuvenation of population in Tel-Aviv inner city', *Geographical Journal* 160: 185–97.

Scholten, H. and Van der Velde, R. (1989) 'Internal migration: the Netherlands', in J. Stillwell and H. Scholten (eds) *Contemporary research in population geography*, Dordrecht: Kluwer, pp. 75–86.

Scott, A. (1992) 'Low-wage workers in a high-technology manufacturing complex: the southern

Californian electronics assembly industry', *Urban Studies* 29: 1231–46.

Scudder, T. (1981) 'What it means to be dammed', *Engineering and Science* 54: 9–15.

Scudder, T. and Colson, E. (1982) 'From welfare to development: a conceptual framework for the analysis of dislocated people', in A. Oliver-Smith and A. Hansen (eds) *Involuntary migration and resettlement: the problems and responses of dislocated persons*, Boulder, Colorado: Westview Press, pp. 267–87.

Sell, R. (1990) 'Migration and job transfers in the United States', in J. Johnson and J. Salt (eds) *Labour migration: the internal geographical mobility of labour in the developed world*, London: David Fulton, pp. 17–31.

Sellek, Y. (1994) 'Illegal foreign migrant workers in Japan: change and challenge in Japanese society', in J. Brown and R. Foot (eds) *Migration: the Asian experience*, Basingstoke: Macmillan, pp. 169–202.

Senior, M. (1979) 'From gravity modelling to entropy maximising: a pedagogic guide', *Progress in Human Geography* 3: 175–210.

Sennett, R. and Cobb, J. (1973) *The hidden injuries of class*, New York: Vintage.

Serageldin, I., Socknat, S., Birks, S., Li, S. and Sinclair, C. (1981) *Manpower and international labour migration in the Middle East and North Africa*, final report, research project on international labour migration and manpower in the Middle East and North Africa, Washington DC: The World Bank.

Serow, W. (1991) 'Recent trends and future prospects for urban–rural migration in Europe', *Sociologia Ruralis* 31: 269–80.

Seyfrit, C. and Hamilton, L. (1992) 'Who will leave? Oil, migration and Scottish island youth', *Society and Natural Resources* 5: 263–76.

Shah, N. (1983) 'Pakistani workers in the Middle East: volume, trends and consequences', *International Migration Review* 17: 410–24.

Shankman, P. (1976) *Migration and underdevelopment: the case of Western Samoa*, Boulder, Colorado: Westview Press.

Shapiro, D. (1981) 'Migrant workers – theory and practice', in R. Moore (ed.) *Labour migration and oil*, London: Social Science Research Council, pp. 34–53.

Shaw, R. (1974) 'A note on cost-relative calculations and decisions to migrate', *Population Studies* 28: 167–9.

Shen, J. (1994) *Dynamic analysis of spatial population systems*, Norwich: School of Environmental Science, CATMOG 57.

Shen, J. (1995) 'Rural development and rural to urban migration in China 1978–1990', *Geoforum* 26: 395–409.

Shen, J. (1996) 'China's economic reforms and their impacts on the migration process', *Migration Unit Research Paper* 10, Department of Geography, University of Wales Swansea.

Shen, J., Boyle, P. and Flowerdew, R. (1994) Comparing 1991 Census, NHSCR and FHSA migration data sets, *Migration Unit Research Paper* 8, Department of Geography, University of Wales Swansea.

Sherwood, K. (1984) 'Population turnover, migration and social change in the rural environment: a geographical study in south Northants', PhD thesis, Department of Geography, University of Leicester.

Shields, R. (1991) *Places on the margin*, London: Routledge.

Shilhav, Y. (1993) 'Ethnicity and geography in Jewish perspectives', *GeoJournal* 30: 272–77.

Short, J. (1989) 'Yuppies, yuffies and the new urban order', *Transactions of the Institute of British Geographers* 14: 173–88.

Short, J. (1991) *Imagined country*, London: Routledge.

Short, J., Fleming, S. and Witt, S. (1986) *Housebuilding, planning and community action*, London: Routledge & Kegan Paul.

Short, J., Witt, S. and Fleming, S. (1987) 'Conflict and compromise in the built environment: housebuilding in central Berkshire', *Transactions of the Institute of British Geographers* NS 12: 29–42.

Shrestha, N. (1988) 'A structural perspective on labour migration in underdeveloped countries', *Progress in Human Geography* 12: 179–207.

Shuttleworth, I. (1991) 'Graduate emigration from Ireland: a symptom of peripherality?', in R. King (ed.) *Contemporary Irish Migration*, Dublin: Geographical Society of Ireland, pp. 83–95.

Sibley, D. (1981) *Outsiders in urban society*, Oxford: Blackwell.

Sibley, D. (1995) *Geographies of exclusion*, London: Routledge.

Silvers, A. (1977) 'Probabilistic income-maximising behaviour in regional migration', *International Regional Science Review* 2: 29–40.

Simkins, P. (1970) 'Migration as a response to population pressure: the case of the Philippines', in W. Zelinsky, L. Kosínski and R. Mansell Prothero (eds) *Geography and a crowding world*, New York: Oxford University Press, pp. 259–68.

Simmons, A. and Guengant, J. (1992) 'Caribbean exodus and the world system', in M. Kritz, L. Lim and H. Zlotnik (eds) *International migration systems. A global approach*, Oxford: Clarendon Press, pp. 94–114.

Simmons, A., Diaz-Briquets, S. and Laquian, A. (1977) *Social change and internal migration: a review of research findings from Africa, Asia and Latin America*, Ottawa: International Development Research Centre.

Simon, D. (1989) 'Rural–urban interaction and development in Southern Africa: the implications of reduced labour migration', in R. Potter and T. Unwin (eds) *The geography of urban–rural interaction in developing countries: essays for Alan B. Mountjoy*, London: Routledge, pp. 141–68.

Simon, D. and Preston, R. (1993) 'Return to the promised land: repatriation and resettlement of Namibian refugees, 1989–90', in R. Black and V. Robinson (eds) *Geography and refugees: patterns and processes of change*, London: Belhaven, pp. 47–64.

Simon, H. (1957) *Models of man*, New York: Wiley.

Simon, J. (1989) *The economic consequences of immigration*, Oxford: Basil Blackwell.

Simpson, E. (1987) *The developing world: an introduction*, Harlow: Longman.

Simpson, S. and Dorling, D. (1994) 'Those missing millions – implications for social statistics of non-response to the 1991 census', *Journal of Social Policy* 23: 543–67.

Sjaastad, L. (1962) 'The costs and returns of migration', *The Journal of Political Economy* 70: 80–93.

Skeldon, R. (1986) 'On migration patterns in India during the 1970s', *Population and Development Review* 12: 759–79.

Skeldon, R. (1990) *Population mobility in developing countries: a reinterpretation*, London: Belhaven Press.

Skeldon, R. (1994) 'East Asian migration and the changing world order', in W. Gould and A. Findlay (eds) *Population migration and the changing world order*, London: Belhaven, pp. 173–203.

Skeldon, R. (1995) 'The challenge facing migration research: the case for greater awareness', *Progress in Human Geography* 19: 91–6.

Skilling, H. (1964) 'An operational view', *American Scientist* 52: 388–96.

Smailes, P. and Hugo, G. (1985) 'A process view of the population turnaround: an Australian rural case study', *Journal of Rural Studies* 1: 31–43.

Smith, D. (1982) 'Urbanization and social change under apartheid', in D. Smith (ed.) *Living under apartheid: aspects of urbanization and social change*, London: Allen & Unwin, pp. 24–47.

Smith, D. (1991) *National identity*, Harmondsworth: Penguin.

Smith, N. (1979) 'Towards a theory of gentrification: a back to the city movement by capital not people', *American Planning Association Journal* 45: 538–48.

Smith, N. (1982) 'Gentrification and uneven development', *Economic Geography* 58: 139–55.

Smith, N. (1987) 'Of yuppies and housing: gentrification, social restructuring, and the urban dream', *Environment and Planning D. Society and Space* 5: 151–72.

Smith, N. and Williams, P. (eds) (1986) *Gentrification of the city*, Boston: Allen & Unwin.

Smith, S. (1993) *Electoral registration in 1991*, London: HMSO.

Snaith, J. (1990) 'Migration and dual career households', in J. Johnson and J. Salt (eds) *Labour migration: the internal geographical mobility of labour in the developed world*, London: David Fulton, pp. 155–71.

Solomos, J. (1993) *Race and racism in Britain*, second edition, Basingstoke: Macmillan.

Solomos, J. and Back, L. (1996) *Racism and society*, Basingstoke: Macmillan.

Soni, D. and Maharaj, B. (1991) 'Emerging urban forms in rural South Africa', *Antipode* 23: 47–67.

Sorokin, P., Zimmerman, C. and Galpin, C. (1930) *A systematic sourcebook in rural sociology*, Minneapolis: University of Minnesota Press.

Spanier, G., Sauer, W. and Larzelere, R. (1979) 'An empirical evaluation of the family life-cycle', *Journal of Marriage and the Family* 41: 27–38.

Speare, A. (1971) 'A cost–benefit model of rural to urban migration in Taiwan', *Population Studies* 25: 117–30.

Speare, A., Goldstein, S. and Frey, W. (1975) *Residential mobility, migration, and metropolitan change*, Cambridge, Massachusetts: Ballinger.

Spencer, S. (1994a) *Strangers and citizens: a positive approach to migrants and refugees*, London: Rivers Oram Press.

Spencer, S. (ed.) (1994b) *Immigration as an economic asset: the German experience*, Stoke-on-Trent: Trentham Books.

Spradley, J. (1970) *You owe yourself a drunk: an*

ethnography of urban nomads, Boston: Little, Brown.

Stanfield, J. (1993) 'Ethnic modelling in qualitative research', in N. Denzin and Y. Lincoln (eds) *Handbook of qualitative research*, Thousand Oaks, California: Sage, pp. 175–88.

Stanley, L. and Wise, S. (1993) *Breaking out again: feminist ontology and epistemology*, London: Routledge.

Stanley, W. (1987) 'Economic migrants or refugees from violence? A time-series analysis of Salvadoran migration to the US', *Latin American Research Bulletin* 22: 132–54.

Stark, O. (1991) *The migration of labor*, Oxford: Blackwell.

Stark, O. (1992) 'Migration in LDCs: risk, remittances and the family', *Finance and Development* 28: 39–41.

Stark, O. and Bloom, D. (1985) 'The new economics of labor migration', *American Economic Review* 75: 173–8.

Stark, O. and Levhari, D. (1982) 'On migration and risk in LDCs', *Economic Development and Cultural Change* 31: 191–6.

Stave, B. (1989) 'The future of suburbia', in B. Kelly (ed.) *Suburbia re-examined*, New York: Greenwood Press, pp. 35–8.

Stegner, W. (1992) 'A geography of hope', in G. Holthaus, P. Limerick, C. Wilkinson and E. Munson (eds) *A society to match the scenery: personal visions of the future of the American West*, Boulder: University of Colorado Press, pp. 218–29.

Stein, B. (1981) 'The refugee experience; defining the parameters of a field of study', *International Migration Review* 15: 320–30.

Stein, B. (1983) 'The commitment to refugee resettlement', *Annals of the American Academy of Political and Social Science* 467: 187–201.

Steinbeck, J. (1939) *The grapes of wrath*, New York: Viking Press.

Sternlieb, G. and Hughes, J. (1977) 'New regional and metropolitan realities of America', *Journal, American Institute of Planners* July: 227–41.

Stevens, J. (1980) 'The demand for public goods as a factor in the nonmetropolitan migration turn-around', in D. Brown and J. Wardwell (eds) *New directions in urban–rural migration*, New York: Academic Press, pp. 115–35.

Stewart, J. (1948) 'Concerning social physics', *Scientific American* 178(5): 20–3.

Stillwell, J. (1991) 'Spatial interaction models and the propensity to migrate over distance', in J. Stillwell and P. Congdon (eds) *Migration models: macro and micro approaches*, London: Belhaven Press, pp. 34–56.

Stillwell, J., Rees, P. and Boden, P. (1992a) 'Internal migration trends: an overview', in J. Stillwell, P. Rees and P. Boden (eds) *Migration processes and patterns. Volume 2. Population redistribution in the United Kingdom*, London: Belhaven Press, pp. 28–55.

Stillwell, J., Rees, P. and Boden, P. (eds) (1992b) *Migration processes and patterns. Volume 2. Population redistribution in the United Kingdom*, London: Belhaven Press.

Stock, R. (1995) *Africa south of the Sahara. A geographical interpretation*, New York: Guilford Press.

Stouffer, S. (1940) 'Intervening opportunities: a theory relating mobility and distance', *American Sociological Review* 5: 845–67.

Stouffer, S. (1960) 'Intervening opportunities and competing migrants', *Journal of Regional Science* 2: 1–26.

Stubbs, J. (1984) 'Some thoughts on the life story method in labour history and research on rural women', *Institute of Development Studies Bulletin* 15: 34–7.

Sullivan, O. (1986) 'Housing movements of the divorced and separated', *Housing Studies* 1: 35–48.

Sutherland, A. (1975a) *Gypsies. The hidden Americans*, Prospect Heights, Illinois: Waveland Press.

Sutherland, A. (1975b) 'The American Rom: a case of economic adaption', in F. Rehfisch (ed.) *Gypsies, tinkers and other travellers*, London: Academic Press, pp. 1–39.

Svart, L. (1976) 'Environmental preference migration: a review', *Geographical Review* 66: 314–30.

Svenson, O. (1979) 'Process descriptions of decision making', *Organizational Behavior and Human Performance* 23: 86–112.

Swinerton, N., Kuepper, W. and Lackey, L. (1975) *Ugandan Asians in Great Britain*, London: Croom Helm.

Sword, K., Davies, N. and Ciechanowski, J. (1989) *The formation of the Polish community in Great Britain, 1939–50*, London: University of London.

Taylor, G. (ed.) (1969) 'Philadelphia in slices: the diary of George G. Foster', *The Pennsylvania Magazine of History and Biography* XCIII: 34, 39, 41.

Taylor, J. (1986) 'Measuring circulation in Botswana', *Area* 18: 203–8.

Taylor, J. and Bell, M. (1994) 'The relative mobility status of indigenous Australians: setting the research agenda', *Discussion Paper* 77, Centre for Aboriginal Economic Policy Research, Australian National University, Canberra.

Taylor, J. and Bell, M. (1996) 'Population mobility and indigenous peoples: the view from Australia', *International Journal of Population Geography* 2: 153–69.

Taylor, N. (1990) 'Glasgow: a personal view of the city', *Catalyst* 3: 2–4.

Taylor, P. (1989) *Political geography*, third edition, Harlow: Longman.

Thomas, B. (1954) *Migration and economic growth*, Cambridge: Cambridge University Press.

Thomas, B. (1973) *Migration and economic growth*, second edition, Cambridge: Cambridge University Press.

Thomas, D. (1941) *Social and economic aspects of Swedish population movements: 1750–1933*, New York: Macmillan.

Thomas, G. (1991) 'The gentrification of paradise: St John's, Antigua', *Urban Geography* 12: 469–87.

Thränhardt, D. (1995) 'Germany: an undeclared immigrant country', *New Community* 21: 19–36.

Thrift, N. (1985a) 'Flies and germs: a geography of knowledge', in D. Gregory and J. Urry (eds) *Social relations and spatial structures*, Basingstoke: Macmillan, pp. 366–403.

Thrift, N. (1985b) 'Bear and mouse or bear and tree? Anthony Giddens's reconstitution of social theory', *Sociology* 19: 609–623.

Thrift, N. (1986) 'Little games and big stories: accounting for the practice of personality and politics in the 1945 general election', in K. Hoggart and E. Kofman (eds) *Politics, geography and social stratification*, London: Croom Helm, pp. 86–143.

Thrift, N. (1987) 'Introduction: the geography of late twentieth century class formation', in N. Thrift and P. Williams (eds) *Class and space*, London: Routledge & Kegan Paul, pp. 207–53.

Tinker, H. (1977) *The Banyan tree: overseas emigrants from India, Pakistan and Bangladesh*, Oxford: Oxford University Press.

Tisdell, C. and Fairburn, T. (1984) 'Subsistence economies and unsustainable development and trade: some simple theory', *Journal of Development Studies* 20: 227–41.

Tittle, C. and Stafford, M. (1992) 'Urban theory, urbanism, and suburban residence', *Social Forces* 70: 725–44.

Todaro, M. (1969) 'A model of labor migration and urban unemployment in less developed countries', *American Economic Review* 59: 138–48.

Todaro, M. (1976) *Internal migration in developing countries*, Geneva: International Labour Office.

Tolstoy, N. (1981) *Stalin's secret war*, London: Jonathan Cape.

Townsend, J. and Momsen, J. (1987) 'Towards a geography of gender in the third world', in J. Momsen and J. Townsend (eds) *Geography of gender in the Third World*, London: Hutchinson, pp. 27–81.

Trager, L. (1984) 'Family strategy and the migration of women: migration to Dagupa City, Philippines', *International Migration Review* 18: 1264–78.

Trager, L. (1988) *The city connection. Migration and family interdependence in the Philippines*, Ann Arbor: University of Michigan Press.

Trost, J. (1977) 'The family life cycle; a problematic approach', in J. Cuisenier (ed.) *Le cycle de vie familiale*, Paris: Mouton.

Tsuya, N. and Kuroda, T. (1989) 'Japan: the slowing of urbanization and metropolitan concentration', in A. Champion (ed.) *Counterurbanization*, London: Edward Arnold, pp. 207–29.

Tucker, R. (ed.) (1972) *The Marx–Engels reader*, New York: W.W. Norton and Co.

Turner, F. (1894) 'The significance of the frontier in American history', *Annual Report of the American Historical Association for 1893*, Washington DC: United States Government Printing Office.

Twaddle, M. (1995) 'The settlement of South Asians in East Africa', in R. Cohen (ed.) *The Cambridge survey of migration*, Cambridge: Cambridge University Press, pp. 74–7.

Uhlenberg, P. (1969) 'A study of cohort life cycles: cohorts of native born Massachusetts women, 1830–1920', *Population Studies* 23: 407–20.

Uhlenberg, P. (1974) 'Cohort variations in family life cycle experiences of US females', *Journal of Marriage and the Family* 36: 284–92.

UNHCR (1993) *The state of the world's refugees: the challenge of protection*, Harmondsworth: Penguin.

UNHCR (1995) *The state of the world's refugees: in search of solutions*, Oxford: Oxford University Press.

UNICEF (1994) *I dream of peace*, London: HarperCollins.

United States Bureau of the Census (1908) *Heads of families at the first census of the United States taken in the year 1790. Pennsylvania*, Washington: Government Printing Office.

United States Committee for Refugees (1994) *World refugee survey*, New York: USCR.

Valeny, R. (1996) 'From pariah to paragon? The long term resettlement of Ugandan Asian refugees in Britain', Paper presented at the annual conference of the Institute of British Geographers, Strathclyde University.

Van Binsbergen, W. and Meilink, H. (1978) 'Migration and the transformation of modern African society', *African Perspectives* 1.

Van der Laan, L. (1992) 'Structural determinants of spatial labour markets: a case study of the Netherlands', *Regional Studies* 26: 485–98.

Van de Walle, E. (1979) 'France', in W. Lee (ed.) *European demography and economic growth*, London: Croom Helm, pp. 123–43.

Van Dijk, J., Folmer, H., Herzog, H. and Schlottman, A. (1989) 'Labor market institutions and the efficiency of interregional migration: a cross nation comparison', in J. van Dijk, H. Folmer, H. Herzog and A. Schlottman (eds) *Migration and labor market adjustment*, Dordrecht: Kluwer, pp. 61–84.

Vandsemb, B. (1995) 'The place of narrative in the study of Third World migration: the case of spontaneous rural migration in Sri Lanka', *Professional Geographer* 47: 411–25.

Van Hear, N (1993) 'Mass flight in the Middle East: involuntary migration and the Gulf conflict, 1990–91', in R. Black and V. Robinson (eds) *Geography and refugees: patterns and processes of change*, London: Belhaven, pp. 64–85.

Van Hear, N. (1995) 'The impact of the involuntary mass return to Jordan in the wake of the Gulf crisis', *International Migration Review* 29: 352–74.

Van Weesep, J. (1994) 'Gentrification as a research frontier', *Progress in Human Geography* 18: 74–83.

Van Weesep, J. and Musterd, S. (eds) (1991) *Urban housing for the better-off: gentrification in Europe*, Utrecht: Stedelijke Netwerken.

Van Westen, A. and Klute, M. (1986) 'From Bamako, with love: a case study of migrants and their remittances', *Tijdschrift voor Economische en Sociale Geografie* 77: 42–49.

Vartiainen, P. (1989a) 'Counterurbanisation: a challenge for socio-theoretical geography', *Journal of Rural Studies* 5: 217–25.

Vartiainen, P. (1989b) 'The end of drastic depopulation in rural Finland: evidence of counterurbanisation?', *Journal of Rural Studies* 5: 123–36.

Vergoossen, W. and Warnes, A. (1989) 'Mobility of the elderly in the Netherlands and Great Britain', in J. Stillwell and H. Scholten (eds) *Contemporary research in population geography*, Dordrecht: Kluwer, pp. 129–46.

Vining, D. and Kontuly, T. (1978) 'Population dispersal from major metropolitan regions: an international comparison', *International Regional Science Review* 3: 49–73.

Vining, D. and Pallone, R. (1982) 'Migration between core and peripheral regions: a description and tentative explanation of the patterns in 20 countries' *Geoforum* 13: 339–410.

Vital, D. (1990) *The future of the Jews*, Cambridge, Massachusetts: Harvard University Press.

Walby, S. (1986) *Patriarchy at work*, Cambridge: Polity Press.

Walby, S. (1989) 'Theorising patriarchy', *Sociology* 23: 213–34.

Walby, S. (1990) *Theorizing patriarchy*, Oxford: Blackwell.

Walcott, D. (1990) *Omeros*, London: Faber & Faber.

Waldinger, R. (1985) 'Immigrant enterprise and the structure of the labour market', in B. Roberts, R. Finnegan and D. Gallie (eds) *New approaches in economic life*, Manchester: Manchester University Press, pp. 213–28.

Walker, R. (1981) 'A theory of suburbanization: capitalism and the construction of urban space in the United States', in M. Dear and A. Scott (eds) *Urbanization and urban planning in capitalist society*, London: Methuen, pp. 383–429.

Walker, R., Ellis, M. and Barff, R. (1992) 'Linked migration systems: immigration and internal labor flows in the United States', *Economic Geography* 68: 234–48.

Wallerstein, I. (1974) *The modern world-system*, New York: Academic Press.

Wallerstein, I. (1979) *The capitalist world economy*, Cambridge: Cambridge University Press.

Wallerstein, I. (1983) *Historical capitalism*, London: Verso.

Walsh, J. (1991) 'Population change in the Republic of Ireland in the 1980s', *Geographical Viewpoint* 19: 89–98.

Walter, B. (1991) 'Gender and recent migration Irish migration to Britain', in R. King (ed.) *Contemporary Irish Migration*, Dublin: Geographical Society of Ireland, pp. 11–20.

Ward, S. (1994) *Planning and urban change*, London: Paul Chapman.

Warde, A. (1991) 'Gentrification as consumption:

issues of class and gender', *Environment and Planning D. Society and Space* 9: 223–32.

Warnes, A. (1983a) 'Migration in late working age and early retirement', *Socio-Economic Planning Sciences* 17: 291–302.

Warnes, A. (1983b) 'Variations in the propensity among older persons to migrate', *Journal of Applied Gerontology* 2: 20–7.

Warnes, A. (1991) 'Migration to and seasonal residence in Spain of North European elderly people', *European Journal of Gerontology and Geriatrics* 1: 53–60.

Warnes, A. (1992a) 'Migration and the life course', in A. Champion and A. Fielding (eds) *Migration processes and patterns. Volume 1. Research progress and prospects*, London: Belhaven, pp. 175–87.

Warnes, A. (1992b) 'Age-related variation and temporal change in elderly migration', in A. Rogers (ed.) *Elderly migration and population redistribution*, London: Belhaven, pp. 35–57.

Warnes, A. and Law, C. (1984) 'The elderly population of Great Britain: locational trends and policy implications', *Transactions of the Institute of British Geographers* 9: 37–59.

Waterman, S. and Kosmin, B. (1987) 'Ethnic identity, residential concentration and social welfare: the Jews in London', in P. Jackson (ed.) *Race and racism*, London: Allen & Unwin, pp. 254–71.

Waterman, S. and Schmool, M. (1995) 'Literary perspectives on Jews in Britain in the early twentieth century', in R. King, J. Connell and P. White (eds) *Writing across worlds. Literature and migration*, London: Routledge, pp. 180–97.

Watson, W. (1964) 'Social mobility and social class in industrial societies', in M. Glucksmann and E. Devons (eds) *Closed systems and open minds*, Edinburgh: Oliver and Boyd, pp. 129–57.

Webber, F. (1991) 'From ethno-centrism to Euro-racism', *Ethnic and Racial Studies* 32: 11–17.

Weber, E. (1977) *Peasants into Frenchmen: the modernization of rural France 1870–1914*, London: Chatto & Windus.

Weekley, I. (1988) 'Rural depopulation and counterurbanisation: a paradox', *Area* 20: 127–34.

Weist, K. (1995) 'Development refugees; Africans, Indians and the big dams', *Journal of Refugee Studies* 8: 163–85.

Wells, W. and Gubar, G. (1966) 'Life-cycle concepts in marketing research', *Journal of Marketing Research* 3: 355–63.

Wendt, A. (1973) *Sons for the return home*, Auckland: Longman.

Western, J. (1981) *Outcast Cape Town*, London: Allen & Unwin.

Wheatley, P. (1971) *The pivot of the four quarters*, Chicago: Aldine.

White, H. (1970) *Chains of opportunity: system models of mobility in organizations*, Cambridge, Massachusetts: Harvard University Press.

White, M. (1986) 'Segregation and diversity measures in population distribution', *Population Index* 52: 198–221.

White, M. and White, L. (1962) *The intellectual versus the city*, Cambridge, Massachusetts: Harvard University and MIT Press.

White, P. (1984) *The West European city*, Harlow: Longman.

White, P. (1989) 'Internal migration in the nineteenth and twentieth centuries', in P. Ogden and P. White (eds) *Migrants in modern France: population mobility in the later nineteenth and twentieth centuries*, London: Unwin Hyman, pp. 13–33.

White, P. (1995) 'Geography, literature and migration', in R. King, J. Connell and P. White (eds) *Writing across worlds. Literature and migration*, London: Routledge, pp. 1–19.

White, P. and Jackson, P. (1995) '(Re)theorising population geography', *International Journal of Population Geography* 1: 111–23.

White, S. (1980) 'A philosophical dichotomy in migration research', *Professional Geographer* 32: 6–13.

Whyte, W. (1957) *The organisation man*, New York: Touchstone.

Wiener, M. (1981) *English culture and the decline of the industrial spirit*, Cambridge: Cambridge University Press.

Wilkinson, R. (1983) 'Migration in Lesotho: some comparative aspects with particular reference to the role of women', *Geography* 68: 208–24.

Willekens, F. and Rogers, A. (1978) 'Spatial population analysis: methods and computer programs', *Report* 4–24, International Union for the Scientific Study of Population, Laxenburg.

Willems, W. and Lucassen, L. (1992) 'A silent war: foreign Gypsies and Dutch government policy, 1969–89', *Immigrants and Minorities* 11: 81–101.

Williams, A. and Jobes, P. (1990) 'Economic and quality-of-life considerations in urban–rural migration', *Journal of Rural Studies* 6: 187–94.

Williams, G. (1985) 'Recent social changes in Mid Wales', *Cambria* 12(2): 117–38.

Williams, J. and McMillen, D. (1980) 'Migration decision making among nonmetropolitan bound migrants', in D. Brown and J. Wardwell (eds) *New directions in urban-rural migration*, New York: Academic Press, pp. 189–211.

Williams, J. and McMillen, D. (1983) 'Location-specific capital and destination selection among migrants to non-metropolitan areas', *Rural Sociology* 48: 447–57.

Williams, J. and Sofranko, A. (1979) 'Motivations for the inmigration component of population turn-around in nonmetropolitan areas', *Demography* 16: 39–55.

Williams, N., Sewel, J. and Twine, F. (1986) 'Council house sales and residualisation', *Journal of Social Policy* 15: 273–92.

Williams, P. (1986) 'Class constitution through spatial construction? A re-evaluation of gentrification in Australia, Britain, and the United States', in N. Smith and P. Williams (eds) *Gentrification of the city*, Boston: Allen & Unwin, pp. 56–77.

Williams, R. (1973) *The country and the city*, London: Chatto & Windus.

Wilpert, C. (1992) 'The use of social networks in Turkish migration to Germany', in M. Kritz, L. Lim and H. Zlotnik (eds) *International migration systems*, Oxford: Clarendon Press, pp. 177–89.

Wilson, A. (1967) 'A statistical theory of spatial distribution models', *Transportation Research* 1: 253–69.

Wilson, A. (1970) *Entropy in urban and regional modelling*, London: Pion.

Wilson, K. (1985) 'Impact of refugees in Yei River District of Southern Sudan', *Horn of Africa* 8: 73–8.

Wilson, K. (1995) 'Refugees, displaced people and returnees in southern Africa', in R. Cohen (ed.) *The Cambridge survey of world migration*, Cambridge: Cambridge University Press, pp. 434–41.

Wilson, W. (1990) 'Residential relocation and settlement adjustment of Vietnamese refugees in Sydney', *Australian Geographical Studies* 28: 155–77.

Wiltshire, R. (1990) 'Employee movement in large Japanese organisations', in J. Johnson and J. Salt (eds) *Labour migration: the internal geographical mobility of labour in the developed world*, London: David Fulton, pp. 32–52.

Winchester, H. (1989) 'The structure and impact of the postwar rural revival: Isère', in P. Ogden and P. White (eds) *Migrants in modern France*, London: Unwin Hyman, pp. 142–59.

Winchester, H. and Ogden, P. (1989) 'France: decentralization and deconcentration in the wake of late urbanization', in A. Champion (ed.) *Counterurbanization*, London: Edward Arnold, pp. 162–86.

Wolfe, T. (1989) [1968] *The electric kool-aid acid test*, London: Black Swan.

Wolpert, J. (1964) 'The decision process in a spatial context', *Annals of the Association of American Geographers* 54: 537–58.

Wolpert, J. (1965) 'Behavioural aspects of the decision to migrate', *Papers of the Regional Science Association* 15: 159–69.

Wolpert, J. (1966) 'Migration as an adjustment to environmental stress', *Journal of Social Issues* 22: 92–102.

Wood, C. (1981) 'Structural changes and household strategies: a conceptual framework for the study of rural migration', *Human Organization* 40: 338–44.

Wood, C. (1982) 'Equilibrium and historical–structural perspectives on migration', *International Migration Review* 16: 298–319.

Wood, W. (1989) 'Long time coming: the repatriation of Afghan refugees', *Annals of the Association of American Geographers* 79: 345–69.

Wood, W. (1995) 'Hazardous journeys: ecomigrants in the 1990s', in D. Conway and J. White (eds) *Global change. How vulnerable are North and South communities?*, Environment and Development Monograph 27, Indiana University Centre on Global Change and World Peace, pp. 33–62.

Woods, R. (1982) *Theoretical population geography*, London: Longman.

Woods, R. (1993) 'Commentary on Zelinsky's model', *Progress in Human Geography* 17: 213–15.

World Bank (1992) *World Development Report 1992: development and the environment*, Washington DC: World Bank.

Wright, P. (1985) *On living in an old country*, London: Verso.

Yang, X. and Goldstein, S. (1990) 'Population movement in Zheijiang Province, China: the impact of government policies', *International Migration Review* 24: 509–33.

Young, C. (1977) *The family life cycle*, Australian Family Formation Project Monograph 6, Canberra: Australian National University Press.

Young, E. and Doohan, K. (1989) *Mobility for survival: a process analysis of Aboriginal population movement in central Australia*, Darwin: Australian

National University, North Australia Research Unit.

Zeigler, D. and Johnson, J. (1984) 'Evacuation behaviour in response to nuclear power plant accidents', *Professional Geographer* 36: 207–15.

Zeigler, D., Brunn, S. and Johnson, J. (1981) 'Evacuation from a nuclear technological disaster', *Geographical Review* 71: 1–17.

Zelinsky, W. (1971) 'The hypothesis of the mobility transition', *Geographical Review* 61: 219–49.

Zelinsky, W. (1983) 'The impasse in migration theory: a sketch map for potential escapees', in P. Morrison (ed.) *Population movements: their forms and functions in urbanization and development*, Liège: Ordina Editions, pp. 19–46.

Zelinsky, W. (1993) 'Reply to commentary on Zelinsky's model', *Progress in Human Geography* 17: 217–19.

Zelinsky, W. and Kosinski, L. (1991) *The emergency evacuation of cities*, Savage: Rowman and Littlefield.

Zetter, R. (1991) 'Labelling refugees: forming and transforming a bureaucratic identity', *Journal of Refugee Studies* 4: 39–62.

Zipf, G. (1946) 'The P1 P2 / D hypothesis: on the intercity movement of persons', *American Sociological Review* 11: 677–86.

Zipf, G. (1949) *Human behavior and the principle of least effort*, New York: Hafner.

Zlotnik, H. (1996) 'Migration to and from developing regions: a review of past trends', in W. Lutz (ed.) *The future population of the world*, London: Earthscan, pp. 299–35.

Zolberg, A. (1983) 'The formation of new states as a refugee-generating process', *Annals of the American Academy of Political and Social Sciences* 467: 24–38.

Zolberg, A. (1989) 'The next waves: migration theory for a changing world', *International Migration Review* 23: 403–30.

Zucker, N. and Zucker, N. (1994) 'United States admission policies toward Cuban and Haitian migrants', Paper presented at the International Refugee Advisory Panel Conference, Oxford.

Zuiches, J. (1980) 'Residential preferences in migration theory', in D. Brown and J. Wardwell (eds) *New directions in urban–rural migration*, New York: Academic Press, pp. 163–88.

Zukin, S. (1987) 'Gentrification: culture and capital in the urban core', *Annual Review of Sociology* 13: 129–47.

Zukin, S. (1989) *Loft living*, second edition, New Brunswick, New Jersey: Rutgers University Press.

GENERAL INDEX

INDEX OF COUNTRIES, GROUPINGS OF COUNTRIES AND NAMED ETHNIC GROUPS